The Master Switch

The Master Switch

The Rise and Fall of Information Empires

Tim Wu

 ALFRED A. KNOPF · NEW YORK · 2010

THIS IS A BORZOI BOOK
PUBLISHED BY ALFRED A. KNOPF

Copyright © 2010 by Tim Wu

Library of Congress Cataloging-in-Publication Data
Wu, Tim.
The master switch : the rise and fall of information empires / Tim Wu.
p.   cm.
Includes bibliographical references and index.
ISBN 978-0-307-26993-5
1. Telecommunication—History.   2. Information technology—
History.   I. Title.
HE7631.W8 2010
384'.041—dc22   2010004137

Manufactured in the United States of America

First Edition

For Kate

At stake is not the First Amendment or the right of
free speech, but exclusive custody of the master switch.

—FRED FRIENDLY

———

Every age thinks it's the modern age, but this one really is.

—TOM STOPPARD, *The Invention of Love*

Contents

Introduction 3

PART I The Rise 15

    1 · The Disruptive Founder 17

    2 · Radio Dreams 33

    3 · Mr. Vail Is a Big Man 45

    4 · The Time Is Not Ripe for Feature Films 61

    5 . Centralize All Radio Activities 74

    6 · The Paramount Ideal 86

PART II Beneath the All-Seeing Eye 99

    7 · The Foreign Attachment 101

    8 · The Legion of Decency 115

    9 · FM Radio 125

    10 · We Now Add Sight to Sound 136

PART III The Rebels, the Challengers, and the Fall 157

    11 · The Right Kind of Breakup 159

    12 · The Radicalism of the Internet Revolution 168

    13 · Nixon's Cable 176

    14 · Broken Bell 187

    15 · Esperanto for Machines 196

PART IV   Reborn Without a Soul 205⁻

       16 · Turner Does Television 207

       17 · Mass Production of the Spirit 217

       18 · The Return of AT&T 238

PART V   The Internet Against Everyone 255

       19 · A Surprising Wreck 257

       20 · Father and Son 269

       21 · The Separations Principle 299

       Acknowledgments 321

       Notes 323

       Index 355

The Master Switch

Introduction

On March 7, 1916, Theodore Vail arrived at the New Willard Hotel in Washington, D.C., to attend a banquet honoring the achievements of the Bell system.[1] Hosted by the National Geographic Society, the festivities were of a scale and grandeur to match American Telephone and Telegraph's vision of the nation's future.

The Willard's dining room was a veritable cavern of splendor, sixty feet wide and a city block long. At one end of the room was a giant electrified map showing the extent of AT&T's "long lines," and before it sat more than eight hundred men in stiff dinner clothes at tables individually wired with telephones. Private power mingled with public: there were navy admirals, senators, the founders of Bell, and all of its executives, as well as much of Woodrow Wilson's cabinet. "From the four corners of the country had come a country's elite" wrote the Society's magazine, "to crown with the laurels of their affection and admiration the brilliant men whose achievements had made possible the miracles of science that were to be witnessed."

Then seventy-one years old, his hair and mustache white, Vail was the incarnation of Bell, the Jack Welch of his time, who had twice rescued his colossal company from collapse. As Alan Stone, Bell's chronicler, writes, "Few large institutions have ever borne the imprint of one person as thoroughly as Vail's on AT&T." In an age when many industrial titans were feared or hated, Vail was widely respected. He styled himself a private sector Theodore Roosevelt, infusing his imperial instincts with a sense of civic duty. "We recognize a 'responsibility' and 'accountability' to the public on our part," wrote Vail, as the voice of AT&T, "which is something different from and something more than the obligation of other public service companies not so closely interwoven with the daily life of the whole community." Serving whatever good, his taste for grandeur was unmistakable. "He could do nothing in a small way," writes his biographer, Albert Paine. "He might start to build a squirrel cage, but it would end by becoming a menagerie." Thomas Edison said of him, simply, "Mr. Vail is a big man."[2]

"Voice voyages" was the theme of the Bell banquet. It would be a riveting demonstration of how AT&T planned to wire America and the world as never before, using a technological marvel we now take for granted: long distance telephone calls.

After dinner, the guests were invited to pick up their receivers from the phones resting on the table. They would travel over the phone line to El Paso, on the Mexican border, to find General John Pershing, later to command the American forces in World War I.

"Hello, General Pershing!"

"Hello, Mr. Carty!"

"How's everything on the border?"

"All's quiet on the border."

"Did you realize you were talking with eight hundred people?"

"No, I did not," answered General Pershing. "If I had known it, I might have thought of something worthwhile to say."

The audience was visibly stunned. "It was a latter-day miracle," reported the magazine. "The human voice was speeding from ocean to ocean, stirring the electric waves from one end of the country to the other."

The grand finale was a demonstration of Bell's newest and perhaps most astonishing invention yet: a "wireless telephone," the ancestor of our mobile phone, of which, by 1916, Bell already had a working prototype. To show it off, Bell mounted what might be called one of history's first multimedia presentations, combining radio, the phonograph, the telephone, and the motion picture projector—the most dazzling inventions of the early twentieth century.

Miles away, in a radio station in Arlington, a record player began

"The Star-Spangled Banner." The sound came wirelessly to the Willard banquet hall over the eight-hundred receivers, while a motion picture projector beamed a waving Old Glory onto a screen. The combination of sight and sound "brought the guests to their feet with hearts beating fast, souls aflame with patriotism, and minds staggered." AT&T, it seemed, had powers to rival the gods: "Perhaps never before in the history of civilization," opined *National Geographic,* had "there been such an impressive illustration of the development and power of the human mind over mundane matter."

It may seem a bit incongruous to begin a book whose ultimate concern future of information with a portrait of Theodore Vail, the greatest monopolist in the history of the information industries, basking in the glories of the nation's most vital communications network under his absolute control. After all, these are far different times: our own most important network, the Internet, would seem to be the antithesis of Vail's Bell system: diffusely organized—even chaotic—where his was centrally controlled; open to all users and content (voice, data, video, and so on.) The Internet is the property of no one where the Bell system belonged to a private corporation.

Indeed, thanks mainly to this open character of the Internet, it has become a commonplace of the early twenty-first century that, in matters of culture and communications, ours is a time without precedent, outside history. Today information zips around the nation and around the globe at the speed of light, more or less at the will of anyone who would send it. How could anything be the same after the Internet Revolution? In such a time, an information despot like Vail might well seem antediluvian.

Yet when we look carefully at the twentieth century, we soon find that the Internet wasn't the first information technology supposed to have changed everything forever. We see in fact a succession of optimistic and open media, each of which, in time, became a closed and controlled industry like Vail's. Again and again in the past hundred years, the radical change promised by new ways to receive information has seemed, if anything, more dramatic than it does today. Thanks to radio, predicted Nikola Tesla, one of the fathers of commercial electricity, in 1904, "the entire earth will be converted into a huge brain, as it were, capable of response in every one of its parts." The invention of film, wrote D. W. Griffith in the 1920s, meant that "children in the public schools will be taught practically everything by moving pictures. Certainly they will never be obliged to read history again." In 1970, a Sloan Foundation report compared the advent of cable television to that of movable type: "the revolution now in sight may be nothing less . . . it may conceivably be more." As a character in Tom Stoppard's *The Invention of Love,* set in 1876, remarks, "Every age thinks it's the modern age, but this one really is."[3]

Each of these inventions to end all inventions, in time, passed through a phase of revolutionary novelty and youthful utopianism; each would change our lives, to be sure, but not the nature of our existence. For whatever social transformation any of them might have effected, in the end, each would take its place to uphold the social structure that has been with us since the Industrial Revolution. Each became, that is, a highly centralized and integrated new industry. Without exception, the brave new technologies of the twentieth century—free use of which was originally encouraged, for the sake of further invention and individual expression—eventually evolved into privately controlled industrial behemoths, the "old media" giants of the twenty-first, through which the flow and nature of content would be strictly controlled for reasons of commerce.

History shows a typical progression of information technologies: from somebody's hobby to somebody's industry; from jury-rigged contraption to slick production marvel; from a freely accessible channel to one strictly controlled by a single corporation or cartel—from open to closed system. It is a progression so common as to seem inevitable, though it would hardly have seemed so at the dawn of any of the past century's transformative technologies, whether telephony, radio, television, or film. History also shows that whatever has been closed too long is ripe for ingenuity's assault: in time a closed industry can be opened anew, giving way to all sorts of technical possibilities and expressive uses for the medium before the effort to close the system likewise begins again.

This oscillation of information industries between open and closed is so typical a phenomenon that I have given it a name: "the Cycle." And to understand why it occurs, we must discover how industries that traffic in information are naturally and historically different from those based on other commodities.

Such understanding, I submit, is far from an academic concern. For if the Cycle is not merely a pattern but an inevitability, the fact that the Internet, more than any technological wonder before it, has truly become the fabric of our lives means we are sooner or later in for a very jarring turn of history's wheel. Though it's a cliché to say so, we do have an information-based economy and society. Our past is one of far less reliance on information than we experience today, and that lesser reliance was served by several information industries at once. Our future, however, is almost certain to be an intensification of our present reality: greater and greater information dependence in every matter of life and work, and all that needed information increasingly traveling a single network we call the Internet. If the Internet, whose present openness has become a way of life, should prove as much subject to the Cycle as every other information network before it, the practical consequences will be staggering. And already there are signs that the good old days of a completely open network are ending.

To understand the forces threatening the Internet as we know it, we must understand how information technologies give rise to industries, and industries to empires. In other words, we must understand the nature of the Cycle, its dynamics, what makes it go, and what can arrest it. As with any economic theory, there are no laboratories but past experience.

Illuminating the past to anticipate the future is the raison d'être of this book. Toward that end, the story rightly begins with Theodore Vail. For in the Bell system, Vail founded the Ur–information network, the one whose working assumptions and ideology have influenced every information industry to follow it.

Vail was but one of many speakers that evening at the Willard, along with Alexander Graham Bell and Josephus Daniels, secretary of the navy. But among these important men, Vail was in a class by himself. For it was his idea of enlightened monopoly in communications that would dominate the twentieth century, and it is an idea whose attraction has never really waned, even if few will admit to their enduring fondness for it. Vail believed it was possible to build a perfect system and devoted his life to that task. His efforts and the history of AT&T itself are a testament to both the possibilities and the dangers of an information empire. As we shall see, it is the enigma posed by figures like Vail—the greatest, to be sure, but only the first of a long line of individuals who sought to control communications for the greater good—that is the preoccupation of this book.

Vail's ideas, while new to communications, were of his times. He came to power in an era that worshipped size and speed (the *Titanic* being among the less successful exemplars of this ideal), and in which there prevailed a strong belief in both human perfectibility and the unique optimal design of any system. It was the last decades of Utopia Victoriana, an era of faith in technological planning, applied science, and social conditioning that had seen the rise of eugenics, Frederick Taylor's "scientific management," socialism, and Darwinism, to name but a few disparate systematizing strains of thought. In those times, to believe in man's ability to perfect communications was far from a fantastical notion. In a sense, Vail's extension of social thinking to industry was of a piece with Henry Ford's assembly lines, his vision of a communications empire of a piece, too, with the British Empire, on which the sun never set.[4]

Vail's dream of a perfected, centralized industry was predicated on another contemporary notion as well. It may sound strange to our ears, but Vail, a full-throated capitalist, rejected the whole idea of "competition." He had professional experience of both monopoly and competition at different times, and he judged monopoly, when held in the right hands, to be the superior arrangement. "Competition," Vail had written, "means *strife*, industrial warfare; it means contention; it oftentime means taking advantage of or resorting to any means that the conscience of the contestants . . . will permit." His reasoning was moralistic: competition was giving American business a bad name. "The vicious acts associated with aggressive competition are responsible for much, if not all, of the present antagonism in the public mind to business, particularly to large business."[5]

Adam Smith, whose vision of capitalism is sacrosanct in the United States, believed that individual selfish motives could produce collective goods for humanity, by the operation of the "invisible hand." But Vail didn't buy it. "In the long run . . . the public as a whole has never benefited by destructive competition." Smith's key to efficient markets was Vail's cause of waste. "All costs of aggressive, uncontrolled competition are eventually borne, directly or indirectly, by the public." In his heterodox vision of capitalism, shared by men like John D. Rockfeller, the right corporate titans, monopolists in each industry, could, and should, be trusted to do what was best for the nation.[6]

But Vail also ascribed to monopoly a value beyond mere efficiency and this was born of a high-mindedness that was his own. With the security of monopoly, Vail believed, the dark side of human nature would shrink, and natural virtue might emerge. He saw a future free of capitalism's form of Darwinian struggle, in which scientifically organized corporations, run by good men in close cooperation with the government, would serve the public best.

Henry Ford wrote in *My Life and Work* that his cars were "concrete evidence of the working out of a theory of business"; and so was the Bell system the incarnation of Vail's ideas about communications. AT&T was building a privately held monopoly yet one that pledged commitment to the public good. It was building the world's mightiest network, yet it promised to reach even the humblest American with a telephone line. Vail called for "*a universal wire system* for the *electrical transmission of intelligence* (*written or personal communication*), from every one in every place to every one in every other place, a system as universal and as extensive as the highway system of the country which extends from every man's door to every other man's door." As he correctly foretold at that dinner, one day "we will be able to telephone to every part of the world."[7]

As he spoke at the National Geographic banquet, Vail was just four years from death. But he had already realized an ideology—the Bell ideology—and built a system of communications that would profoundly influence not just how people spoke over distances, but the shape of the television, radio, and film industries as well: in other words, all of the new media of the twentieth century.

To see specifically how Vail's ideology shaped the course of telephony and all subsequent information industries—serving as, so to speak, the spiritual source of the Cycle—it will be necessary to tell some stories, about Vail's own firm and others. There are, of course, enough to fill a book about each, and there have been no few such volumes. But this book will focus on chronicling the turning points of the twentieth century's information landscape: those particular, decisive moments when a medium opens or closes. The pattern is distinctive. Every few decades, a new communications technology appears, bright with promise and possibility. It inspires a generation to dream of a better society, new forms of expression, alternative types of journalism. Yet each new technology eventually reveals its flaws, kinks, and limitations. For consumers, the technical novelty can wear thin, giving way to various kinds of dissatisfaction with the quality of content (which may tend toward the chaotic and the vulgar) and the reliability or security of service. From industry's perspective, the invention may inspire other dissatisfactions: a threat to the revenues of existing information channels that the new technology makes less essential, if not obsolete; a difficulty commoditizing (i.e., making a salable product out of) the technology's potential; or too much variation in standards or protocols of use to allow one to market a high quality product that will answer the consumers' dissatisfactions.

When these problems reach a critical mass, and a lost potential for substantial gain is evident, the market's invisible hand waves in some great mogul like Vail or band of them who promise a more orderly and efficient regime for the betterment of all users. Usually enlisting the federal government, this kind of mogul is special, for he defines a new type of industry, integrated and centralized. Delivering a better or more secure product, the mogul heralds a golden age in the life of the new technology. At its heart lies some perfected engine for providing a steady return on capital. In exchange for making the trains run on time (to hazard an extreme comparison), he gains a certain measure of control over the medium's potential for enabling individual expression and technical innovation—control such as the inventors never dreamed of, and necessary to perpetuate itself, as well as the attendant profits of centralization. This, too, is the Cycle.

Since the stories of these individual industries take place concurrently and our main purpose in recounting them is to observe the operations of the Cycle, the narrative is arranged in the following way:

Part I traces the genesis of cultural and communications empires, the first phase of the Cycle, and shows how each of the early twentieth century's new information industries—telephony, radio broadcast, and film—evolved from a novel invention.

By the 1940s, every one of the twentieth century's new information industries, in the United States and elsewhere, would reach an established, stable, and seemingly permanent form, excluding all potential entrants. Communications by wire became the sole domain of the Bell system. The great networks, NBC and CBS, ruled radio broadcasting, as they prepared, with the help of the Federal Communications Commission, to launch in their own image a new medium called television. The Hollywood studios, meanwhile, closed a vise grip on every part of the film business, from talent to exhibition. And so in Part II, we will focus on the consolidation of information empire, often with state support, and the consequences, particularly for the vitality of free expression and technical innovation. For while we may rightly feel a certain awe for what the information industries manage to accomplish thanks to the colossal centralized structures created through the 1930s, we will also see how the same period was one of the most repressive in American history vis-à-vis new ideas and forms.

But as we have said, that which is centralized also eventually becomes a target for assault, triggering the next phase of the Cycle. Sometimes this takes the form of a technological innovation that breaks through the defenses and becomes the basis of an insurgent industry. The advent of personal computing and the Internet revolution it will eventually beget are both instances of such game-changing developments. And though less endowed with the romantic lore of invention, so too is the rise of cable television. But sometimes it is not invention—or invention alone—that drives the Cycle, but rather the federal government suddenly playing the role of giant-slayer of information cartels and monopolies that it had long tolerated. In Part III, we explore the ways in which the stranglehold of information monopoly is broken after decades.

Through the 1970s each of the great information empires of the twentieth century was fundamentally challenged or broken into pieces, if not blown up altogether, leading to a new period of openness. And a new run of the Cycle. The results were unmistakably invigorating for both commerce and culture. But like the T-1000 killer robot of *Terminator 2* the shattered powers would reconstitute themselves, either in uncannily similar form (as with AT&T) or in the guise of a new corporate species called the conglomerate (as with the revenge of the broadcasters and of Hollywood). In Part IV we will see how the perennial lure of size and scale that led to the original information leviathans in the first half of the century spawned a new generation in the latter part.

By the dawn of the twenty-first century, the second great closing will be complete. The one exception to the hegemony of the latter-day information monopolists will be a new network to end all networks. While all else was being consolidated, the 1990s would also see the so-called Internet revolution, though amid its explosive growth no one could see where the wildly open new medium would lead. Would the Internet usher in a reign of industrial openness without end, abolishing the Cycle? Or would it, despite its radically decentralized design, become in time simply the next logical target for the insuperable forces of information empire, the object of the most consequential centralization yet? Part V will lead us to that ultimate question, the answer to which is as yet a matter of conjecture, for which, I argue, our best basis is history.

Reading all this, you may yet be wondering, "Why should I care?" After all, the flow of information is invisible, and its history lacks the emotional immediacy of, say, the Second World War or the civil rights movement. The fortunes of information empires notwithstanding, life goes on. It hardly occurred to anyone as a national problem when, in the 1950s, a special episode of *I Love Lucy* could attract more than 70 percent of households. And yet, almost like the weather, the flow of information defines the basic tenor of our times, the ambience in which things happen, and, ultimately, the character of a society.

Sometimes it takes an outsider to make this clear. Steaming from Malaysia to the United States in 1926, a young English writer named Aldous Huxley came across something interesting in the ship's library, a volume entitled *My Life and Work,* by Henry Ford.[8] Here was the vivid story of Ford's design of mass production techniques and giant centralized factories of unexampled efficiency. Here, too, were Ford's ideas on things like human equality: "There can be no greater absurdity and no greater disservice to humanity in general than to insist that all men are equal."[9] But what really interested Huxley, the future author of *Brave New World,* was Ford's belief that his systems might be useful not just for manufacturing cars, but for all forms of social ordering. As Ford wrote, "the ideas we have put into practice are capable of the largest application—that they have nothing peculiarly to do with motor cars or tractors but form something in the nature of a universal code. I am quite certain that it is the natural code . . ."

When Huxley arrived in the States, Ford's ideas fresh in mind, he realized something both intriguing and terrifying: Ford's future was already becoming a reality. The methods of the steel factory and car assembly plant had been imported to the cultural and communications industries. Huxley witnessed in the America of 1926 the prototypes of structures that had not yet reached the rest of the world: the first commercial radio networks, rising studios for film production, and a powerful private communications monopoly called AT&T.

When he returned to England, Huxley declared in an essay for *Harper's Magazine* called "The Outlook for American Culture" that "the future of America is the future of the World." He had seen that future and been more than a little dismayed by it. "Mass production," he wrote, "is an admirable thing when applied to material objects; but when applied to the things of the spirit it is not so good."[10]

Seven years later, the question of the spirit would occur to another student of culture and theorist of information. "The radio is the most influential and important intermediary between a spiritual movement and the nation," wrote Joseph Goebbels, quite astutely, in 1933. "Above all," he said, "it is necessary to clearly centralize all radio activities."[11]

It is an underacknowledged truism that, just as you are what you eat, how and what you think depends on what information you are exposed to. How do you hear the voice of political leaders? Whose pain do you feel? And where do your aspirations, your dreams of good living, come from? All of these are products of the information environment.

My effort to consider this process is also an effort to understand the practical realities of free speech, as opposed to its theoretical life. We can sometimes think that the study of the First Amendment is the same as the study of free speech, but in fact it forms just a tiny part of the picture. Americans idealize what Justice Oliver Wendell Holmes called the "marketplace of ideas," a space where every member of society is, by right, free to peddle his creed. Yet the shape or even existence of any such marketplace depends far less on our abstract values than on the structure of the communications and culture industries. We sometimes treat the information industries as if they were like any other enterprise, but they are not, for their structure determines who gets heard. It is in this context that Fred Friendly, onetime CBS News president, made it clear that before any question of free speech comes the question of "who controls the master switch."

The immediate inspiration for this book is my experience of the long wave of easy optimism created by the rise of information technologies in the late twentieth and early twenty-first centuries, a feeling of almost utopian possibility and idealism. I shared in that excitement, both working in Silicon Valley and writing about it. Yet I have always been struck by what I feel is too strong an insistence that we are living in unprecedented times. In fact, the place we find ourselves now is a place we have been before, albeit in different guise. And so understanding how the fate of the technologies of the twentieth century developed is important in making the twenty-first century better.

PART I

The Rise

CHAPTER 1

The Disruptive Founder

Exactly forty years before Bell's National Geographic banquet, Alexander Bell was in his laboratory in the attic of a machine shop in Boston, trying once more to coax a voice out of a wire. His efforts had proved mostly futile, and the Bell Company was little more than a typically hopeless start-up.*

Bell was a professor and an amateur inventor, with little taste for business: his expertise and his day job was teaching the deaf. His main investor and the president of the Bell Company was Gardiner Green Hubbard, a patent attorney and prominent critic of the telegraph monopoly Western Union. It is Hubbard who was responsible for Bell's most valuable asset: its telephone patent, filed even before Bell had a working prototype. Besides Hubbard, the company had one employee, Bell's assistant, Thomas Watson. That was it.[1]

If the banquet revealed Bell on the cusp of monopoly, here is the opposite extreme from which it began: a stirring image of Bell and

---

* I use "the Bell Company," "Bell," and "AT&T" interchangeably in this book. The Bell Company was the name of the company founded by Alexander Bell and his financiers in 1877. The American Telephone and Telegraph Company (AT&T) was created in 1884, as a subsidiary of Bell to provide long distance services. In 1903, after a reorganization, AT&T became a holding company for what were by then dozens of "Bell Companies," with names like Northeastern Bell and Atlantic Bell, that offered local service. That basic structure lasted until the breakup of 1984.

Watson toiling in their small attic laboratory. It is here that the Cycle begins: in a lonely room where one or two men are trying to solve a concrete problem. So many revolutionary innovations start small, with outsiders, amateurs, and idealists in attics or garages. This motif of Bell and Watson alone will reappear throughout this account, at the origins of radio, television, the personal computer, cable, and companies like Google and Apple. The importance of these moments makes it critical to understand the stories of lone inventors.

Over the twentieth century, most innovation theorists and historians became somewhat skeptical of the importance of creation stories like Bell's. These thinkers came to believe the archetype of the heroic inventor had been over-credited in the search for a compelling narrative. As William Fisher puts it, "Like the romantic ideal of authorship, the image of the inventor has proved distressingly durable."[2] These critics undeniably have a point: even the most startling inventions are usually arrived at, simultaneously, by two or more people. If that's true, how singular could the genius of the inventor really be?

There could not be a better example than the story of the telephone itself. On the very day that Alexander Bell was registering his invention, another man, Elisha Gray, was also at the patent office filing for the very same breakthrough.* The coincidence takes some of the luster off Bell's "eureka." And the more you examine the history, the worse it looks. In 1861, sixteen years before Bell, a German man named Johann Philip Reis presented a primitive telephone to the Physical Society of Frankfurt, claiming that "with the help of the galvanic current, [the inventor] is able to reproduce at a distance the tones of instruments and even, to a certain degree, the human voice." Germany has long considered Reis the telephone's inventor. Another man, a small-town Pennsylvania electrician named Daniel Drawbaugh, later claimed that by 1869 he had a working telephone in his house. He produced prototypes and seventy witnesses who testified that they had seen or heard his invention

---

* Consequently, many books have been dedicated to the question of who actually invented the telephone. and the majority seem to side against Bell, though of course to do so furnishes a revisionist the more interesting conclusion. Most damning to Bell is the fact that his telephone, in its specifications, is almost identical to the one described in Gray's patent. On the other hand, Bell was demonstrably first to have constructed a phone that was functional, if not yet presentable enough to patent. A final bit of evidence against Bell: the testimony of a patent examiner, Zenas F. Wilbur, who admitted to accepting a $100 bribe to show Gray's design to one of Alexander Bell's lawyers. (*New York Times,* May 22, 1886.)

at that time. In litigation before the Supreme Court in 1888, three Justices concluded that "overwhelming evidence" proved that "Drawbaugh produced and exhibited in his shop, as early as 1869, an electrical instrument by which he transmitted speech. . . ."*[3]

There was, it is fair to say, no single inventor of the telephone. And this reality suggests that what we call invention, while not easy, is simply what happens once a technology's development reaches the point where the next step becomes available to many people. By Bell's time, others had invented wires and the telegraph, had discovered electricity and the basic principles of acoustics. It lay to Bell to assemble the pieces: no mean feat, but not a superhuman one. In this sense, inventors are often more like craftsmen than miracle workers.

Indeed, the history of science is full of examples of what the writer Malcolm Gladwell terms "simultaneous discovery"—so full that the phenomenon represents the norm rather than the exception. Few today know the name Alfred Russel Wallace, yet he wrote an article proposing the theory of natural selection in 1858, a year before Charles Darwin published *The Origin of Species*. Leibnitz and Newton developed calculus simultaneously. And in 1610 four others made the same lunar observations as Galileo.[4]

Is the loner and outsider inventor, then, merely a figment of so much hype, with no particular significance? No, I would argue his significance is enormous; but not for the reasons usually imagined. The inventors we remember are significant not so much as inventors, but as founders of "disruptive" industries, ones that shake up the technological status quo. Through circumstance or luck, they are exactly at the right distance both to imagine the future and to create an independent industry to exploit it.

Let's focus, first, on the act of invention. The importance of the outsider here owes to his being at the right remove from the prevailing currents of thought about the problem at hand. That distance affords a perspective close enough to understand the problem, yet far enough for

---

* Unfortunately for Drawbaugh, four Justices found his testimony and that of his seventy witnesses not credible and dismissed his case. The dissenting Justices accused the majority of siding with Bell, essentially owing to his fame. "It is perfectly natural for the world to take the part of the man who has already achieved eminence. . . . It is regarded as incredible that so great a discovery should have been made by the plain mechanic, and not by the eminent scientist and inventor."

greater freedom of thought, freedom from, as it were, the cognitive distortion of what is as opposed to what could be. This innovative distance explains why so many of those who turn an industry upside down are outsiders, even outcasts.

To understand this point we need grasp the difference between two types of innovation: "sustaining" and "disruptive," the distinction best described by innovation theorist Clayton Christensen. *Sustaining innovations* are improvements that make the product better, but do not threaten its market. The *disruptive innovation,* conversely, threatens to displace a product altogether. It is the difference between the electric typewriter, which improved on the typewriter, and the word processor, which supplanted it.[5]

Another advantage of the outside inventor is less a matter of the imagination than of his being a disinterested party. Distance creates a freedom to develop inventions that might challenge or even destroy the business model of the dominant industry. The outsider is often the only one who can afford to scuttle a perfectly sound ship, to propose an industry that might challenge the business establishment or suggest a whole new business model. Those closer to—often at the trough of—existing industries face a remarkably constant pressure not to invent things that will ruin their employer. The outsider has nothing to lose.

But to be clear, it is not mere distance, but the right distance that matters; there is such a thing as being too far away. It may be that Daniel Drawbaugh actually did invent the telephone seven years before Bell. We may never know; but even if he did, it doesn't really matter, because he didn't do anything with it. He was doomed to remain an inventor, not a founder, for he was just too far away from the action to found a disruptive industry. In this sense, Bell's alliance with Hubbard, a sworn enemy of Western Union, the dominant monopolist, was all-important. For it was Hubbard who made Bell's invention into an effort to unseat Western Union.

I am not saying, by any means, that invention is solely the province of loners and that everyone else's inspiration is suppressed. But this isn't a book about better mousetraps. The Cycle is powered by disruptive innovations that upend once thriving industries, bankrupt the dominant powers, and change the world. Such innovations are exceedingly rare, but they are what makes the Cycle go.

Let's return to Bell in his Boston laboratory. Doubtless he had some critical assets, including a knowledge of acoustics. His laboratory note-book, which can be read online, suggests a certain diligence. But his greatest advantage was neither of these. It was that everyone else was obsessed with trying to improve the telegraph. By the 1870s inventors and investors understood that there could be such a thing as a tele-phone, but it seemed a far-off, impractical thing. Serious men knew that what really mattered was better telegraph technology. Inventors were racing to build the "musical telegraph," a device that could send multiple messages over a single line at the same time. The other holy grail was a device for printing telegrams at home.*

Bell was not immune to the seduction of these goals. One must start somewhere, and he, too, began his experiments in search of a better telegraph; certainly that's what his backers thought they were paying for. Gardiner Hubbard, his primary investor, was initially skeptical of Bell's work on the telephone. It "could never be more than a scientific toy," Hubbard told him. "You had better throw that idea out of your mind and go ahead with your musical telegraph, which if it is successful will make you a millionaire."[6]

But when the time came, Hubbard saw the potential in the telephone to destroy his personal enemy, the telegraph company. In contrast, Eli-sha Gray, Bell's rival, was forced to keep his telephone research secret from his principal funder, Samuel S. White. In fact, without White's opposition, there is good reason to think that Gray would have both created a working telephone and patented it long before Bell.[7]

The initial inability of Hubbard, White, and everyone else to recog-nize the promise of the telephone represents a pattern that recurs with a frequency embarrassing to the human race. "All knowledge and habit once acquired," wrote Joseph Schumpeter, the great innovation theo-rist, "becomes as firmly rooted in ourselves as a railway embankment in the earth." Schumpeter believed that our minds were, essentially, too lazy to seek out new lines of thought when old ones could serve. "The very nature of fixed habits of thinking, their energy-saving function, is founded upon the fact that they have become subconscious, that they yield their results automatically and are proof against criticism and even against contradiction by individual facts."[8]

---

* In this yearning for "home telegraphs" was the first intimation of what would one day flower as email and text messages.

The men dreaming of a better telegraph were, one might say, mentally warped by the tangible demand for a better telegraph. The demand for a telephone, meanwhile, was purely notional. Nothing, save the hangman's noose, concentrates the mind like piles of cash, and the obvious rewards awaiting any telegraph improver were a distraction for anyone even inclined to think about telephony, a fact that actually helped Bell. For him the thrill of the new was unbeatably compelling, and Bell knew that in his lab he was closing in on something miraculous. He, nearly alone in the world, was playing with magical powers never seen before.

On March 10, 1876, Bell, for the first time, managed to transmit speech over some distance. Having spilled acid on himself, he cried out into his telephone device, "Watson, come here, I want you." When he realized it had worked, he screamed in delight, did an Indian war dance, and shouted, again over the telephone, "God save the Queen!"*[9]

The Plot to Destroy Bell

Eight months on, late on the night of the 1876 presidential election, a man named John Reid was racing from the *New York Times* offices to the Republican campaign headquarters on Fifth Avenue. In his hand he held a Western Union telegram with the potential to decide who would be the next president of the United States.

While Bell was trying to work the bugs out of his telephone, Western Union, telephony's first and most dangerous (though for the moment unwitting) rival, had, they reckoned, a much bigger fish to fry: making their man president of the United States. Here we introduce the nation's first great communications monopolist, whose reign provides history's first lesson in the power and peril of concentrated control over the flow of information. Western Union's man was one Rutherford B. Hayes, an obscure Ohio politician described by a contemporary journalist as "a third rate nonentity." But the firm and its partner newswire, the Associated Press, wanted Hayes in office, for several reasons. Hayes was a close friend of William Henry Smith, a former politician who was now the key political operator at the Associated Press. More generally, since the Civil War, the Republican Party and the telegraph industry had enjoyed

---

* This second statement has been omitted from most American histories of the telephone.

a special relationship, in part because much of what were eventually Western Union's lines were built by the Union army.

So making Hayes president was the goal, but how was the telegram in Reid's hand key to achieving it?

The media and communications industries are regularly accused of trying to influence politics, but what went on in the 1870s was of a wholly different order from anything we could imagine today. At the time, Western Union was the exclusive owner of the only nationwide telegraph network, and the sizable Associated Press was the unique source for "instant" national or European news. (Its later competitor, the United Press, which would be founded on the U.S. Post Office's new telegraph lines, did not yet exist.) The Associated Press took advantage of its economies of scale to produce millions of lines of copy a year and, apart from local news, its product was the mainstay of many American newspapers.

With the common law notion of "common carriage" deemed inapplicable, and the latter-day concept of "net neutrality" not yet imagined, Western Union carried Associated Press reports exclusively.[10] Working closely with the Republican Party and avowedly Republican papers like *The New York Times* (the ideal of an unbiased press would not be established for some time, and the minting of the *Times*'s liberal bona fides would take longer still), they did what they could to throw the election to Hayes. It was easy: the AP ran story after story about what an honest man Hayes was, what a good governor he had been, or just whatever he happened to be doing that day. It omitted any scandals related to Hayes, and it declined to run positive stories about his rivals (James Blaine in the primary, Samuel Tilden in the general). But beyond routine favoritism, late that Election Day Western Union offered the Hayes campaign a secret weapon that would come to light only much later.

Hayes, far from being the front-runner, had gained the Republican nomination only on the seventh ballot. But as the polls closed his persistence appeared a waste of time, for Tilden, the Democrat, held a clear advantage in the popular vote (by a margin of over 250,000) and seemed headed for victory according to most early returns; by some accounts Hayes privately conceded defeat. But late that night, Reid, the *New York Times* editor, alerted the Republican Party that the Democrats, despite extensive intimidation of Republican supporters, remained unsure of their victory in the South. The GOP sent some telegrams of its own to the Republican governors in the South with special instructions for manipulating state electoral commissions. As a result the Hayes campaign abruptly claimed victory, resulting in an electoral dispute that would make *Bush v. Gore* seem a garden party. After a few brutal months, the Democrats relented, allowing Hayes the presidency—in exchange, most historians believe, for the removal of federal troops from the South, effectively ending Reconstruction.

The full history of the 1876 election is complex, and the power of the Western Union network was just one factor, to be sure. But while mostly studied by historians and political scientists, the dispute should also be taken as a crucial parable for communications policy makers. More than anything, it showed what kind of political advantage a discriminatory network can confer. When the major channels for moving information are loyal to one party, its effects, while often invisible, can be profound.

It also showed how a single communications monopolist can use its power not just for discrimination, but for outright betrayal of trust, revealing for the first time why what we now call "electronic privacy" might matter. Hayes might never have been president but for the fact that Western Union provided secret access to the telegrams sent by his rivals. Western Union's role was a blatant instance of malfeasance: despite its explicit promise that "all messages whatsoever" would be kept "strictly private and confidential," the company regularly betrayed the public trust by turning over private, and strategically actionable, communications to the Hayes campaign.

Today Western Union's name remains familiar, but the company that survives is the shriveled rump of what was in 1876 among the most powerful corporations on earth. But power is never entirely secure in any tyranny. Western Union, despite its size, had come under episodic attack from speculators, putting into question whether it was really a "natural" monopoly. And in two years' time Bell's three-man company, though embryonic, would pose an even more devastating threat to the firm's rule over American communications.

In antiquity, Kronos, the second ruler of the universe according to Greek mythology, had a problem. The Delphic oracle having warned him that one of his children would dethrone him, he was more than troubled to hear his wife was pregnant. He waited for her to give birth, then took the child and ate it. His wife got pregnant again and again, so he had to eat his own more than once.

And so derives the *Kronos Effect:* the efforts undertaken by a dominant company to consume its potential successors in their infancy. Understanding this effect is critical to understanding the Cycle, and for that matter, the history of information technology. It may sometimes seem that invention and technological advance are a natural, orderly process, but this is an illusion. Whatever technological reality we live with is the result of tooth-and-claw industrial combat. And the battles are more decisive than those in which the dominant power attempts to co-opt the technologies that could destroy it, Goliath attempting to seize the slingshot.

Western Union, despite its great size and scale, was vulnerable to the same force as every other business: disruptive innovation. No sooner had the firm realized the potential of the Bell company's technology to overthrow the telegraph monopoly than it went into Kronos mode, attempting to kill or devour Bell. It did not happen instantaneously. At the very beginning, in 1877, the Bell Company probably seemed more a source of comic relief than a threat to Western Union. Bell's very first advertisement for the telephone, in May 1877, betrays a distinct lack of confidence in the product:

> The proprietors of the Telephone . . . are now prepared to furnish Telephones for the transmission of articulate speech through instruments not more than twenty miles apart. Conversation can be easily carried on after slight practice and with the occasional repetition of a word or sentence. On first listening to the Telephone . . . the articulation seems to be indistinct; but after a few trials the ear becomes accustomed to the peculiar sound.[11]

Bell's first telephone simply did not work very well. The Bell Company's most valuable asset would remain, for some time, the principal patent, for actual telephones were more like toys than devices adults could depend on. Finding investors, let alone customers, was such tough going that at one point, according to most accounts, Hubbard, acting as Bell's president, offered Western Union all of Bell's patents for $100,000. William Orton, president of Western Union, refused, in one of history's less prudent exercises of business judgment.[12]

In a year, however, as Bell began to pick up customers, Western Union realized its mistake. In 1878 it reversed course and proceeded full steam into the phone business. Against tiny Bell, Western Union brought overwhelming advantages: capital, an existing nationwide network of wires, and a close relationship with newspapers, hotels, and politicians. "With all the bulk of its great wealth and prestige," as the historian Herbert N. Casson wrote in 1910, "it swept down upon Bell and his little bodyguard." The decision, once taken, was implemented quickly. Ignoring Bell's shoddy equipment, Western Union commissioned a promising young inventor named Thomas Edison to design a better telephone. Edison's version would prove a major advance over Bell's, including a much more sensitive transmitter that didn't require one to shout. For that reason, depending on how you define "invention," there is a strong case to be made for giving Bell and Edison, at a minimum, joint credit.

By the end of 1878 Western Union had deployed 56,000 telephones, rendering Bell a bit player.[13] For a brief moment, the telephone industry came under domination by Western Union's subsidiary, the American Speaking Telephone Company. In an 1880 *Scientific American* article we see a drawing of an AST exchange in New York, staffed by boys with Edison phones. In some alternate universe, AST, rather than Ma Bell, would go on to rule communications by wire.

We can stop here to imagine that future. The telephone could easily have been born as what Harvard professor Jonathan Zittrain calls a *tethered* technology: that is, a technology tied directly to its owner, and limited in what it might do.[14] Western Union's telephone network was designed not to pose any threat to the telegraph business. In an oft-exampled way, a dominant power must disable or neuter its own inventions to avoid cannibalizing its core business. In the 1980s and 1990s, General Motors, famously, was fully equipped to take over the electric car market, but was restrained by disinclination to create a rival to the internal combustion engine, its main business.

Western Union's version of the telephone would have remained a feeder business for the telegraph, and another tool for discrimination. Most likely we would have seen a telephone system that was primarily local, used to call in telegraph messages for nationwide communications, and as such always a complement to the telegraph, not a substitute for it. Alexander Bell would be as obscure as the inventors of cable or broadcast television, to name two other initially suppressed inventions—but let us not get ahead of ourselves. For now it is enough to imagine how the retardation of telephony in an alternative run-through of history might have altered the narrative. It might even have affected the development of American economic supremacy, if other nations better grasped the importance of the telephone.

In 1878 the future so described was likelier than not. For months, Bell suffered under the onslaught of Western Union. As if mourning his company, Alexander Bell became a bedridden invalid, in the grip of such a depression that he checked himself in to Massachusetts General Hospital.[15]

CYCLES OF BIRTH AND DEATH

The struggle between Bell and Western Union over the fate of the telephone was, in retrospect, a match to the death. The victor would go on to prosper, while the loser would wilt away and die. This is how the Cycle turns. No thinker of the twentieth century better understood that such winner-take-all contests were the very soul of the capitalist system than did the economist Joseph Schumpeter, the "prophet of innovation."

Schumpeter's presence in the history of economics seems designed to displease everyone. His prose, his personality, and his ideas were infuriatingly provocative and confounding, and quite deliberately so. He bragged of sexual exploits at faculty meetings, and while living in the United States during World War II, he voiced support for Germany, supposedly out of dislike for Russians.

Nonetheless, Schumpeter is the source of a very simple economic theory that has proved itself particularly virulent. At the most basic level, Schumpeter believed that innovation and economic growth are one and the same. Countries that innovated would grow wealthier; those that did not would stagnate. And in Schumpeter's vision innovation was no benignly gradual process, but a merciless cycle of industrial destruction and birth, as implacable as the way of all flesh. This dynamic was, to Schumpeter, the essence of capitalism.[16]

He described innovation as a perennial state of unrest: a "process of industrial mutation . . . that incessantly revolutionizes the economic structure from within, incessantly destroying the old one, incessantly creating a new one." In the age of carts, what mattered was not a cheaper cart, but the Mack truck that runs the cart over. Bell's telephone was a quintessentially Schumpeterian innovation: it promised not improvement of the telegraph industry, but rather its annihilation.

To understand Schumpeter we need to reckon with his very peculiar idea of "competition." He had no patience for what he deemed Adam Smith's fantasy of price warfare, growth through undercutting your competitor and improving the market's overall efficiency thereby. "In capitalist reality as distinguished from its textbook picture, it is not that kind of competition which counts," argued Schumpeter, but rather, "the competition from the new commodity, the new technology, the new source of supply, the new type of organization." It is a vision to out-Darwin Darwin: "competition which commands a decisive cost or quality advantage and which strikes not at the margins of the profits and the outputs of the existing firms but at their foundations and their very lives." Schumpeter termed this process "creative destruction." As he put it, "Creative Destruction is the essential fact about capitalism. It is what capitalism consists in and what every capitalist concern has got to live in."*

Schumpeter's cycle of industrial life and death is an inspiration for this book. His thesis is that in the natural course of things, the new only rarely supplements the old; it usually destroys it. The old, however, doesn't, as it were, simply give up but rather tries to forestall death or co-opt its usurper—à la Kronos—with important implications. In particular Schumpeter's theory did not account for the power of law or the government to stave off industrial death, and (for our particular purposes) arrest the Cycle. As we shall see in future chapters, allying itself with the state, a dominant industrial force can turn a potentially destructive technology into a tool for perpetuating domination and delaying death.

---

* All this may make Schumpeter sound like a hero to free market libertarians, but he is not so easily domesticated. His most famous work, *Capitalism, Socialism, and Democracy,* published in 1942, reads, in part, as a repudiation of the market and a lauding of socialism. He praises Marx and asks, "Can capitalism survive?" His answer: "No. I do not think it can." It may seem paradoxical that an icon of capitalism should be praising Marx and predicting the success of socialism. As with the end of Shakespeare's *The Taming of the Shrew,* a plain reading of the text has caused Schumpeter's fans much discomfort. Whether Schumpeter's true purpose was to praise or to bury capitalism, or to leave his main point so perversely ambiguous, is an indication of the maddening nature of the man.

But before describing such corporate contortions, let us return to the sorrows of Mr. Bell.

ENTER VAIL

In 1878, Theodore Vail was an ambitious and driven thirty-three-year-old working at the U.S. Post Office. He was very good at his job—he pioneered a more efficient form of railroad mail, and he supervised more than thirty-five hundred men—but he was obviously bored. And so when Gardiner Hubbard, Bell's founding father, legal counsel, and first president, showed him the Bell prototype, Vail spied the chance of a lifetime. He was in precisely the position of anyone who leaves a steady job for the promise held out by some start-up. "I can scarce believe that a man of your sound judgment," wrote his boss, "should throw it up for a damned old Yankee notion called a telephone!" It would have seemed imprudent, in a time when Americans did not change jobs as regularly as they do today, to leave a secure situation and hitch one's wagon to what seemed a novelty item, and a rather buggy one. Yet something in Vail's nature allowed him to see the grand potential of the telephone, and the lure was irresistible to him.[17]

We must try to understand Theodore Vail, for his basic character type recurs in other "Defining Moguls," the men who drive the Cycle and populate this book. Schumpeter theorized that men like Vail were far, a special breed, with unusual talents and ambitions. Their motivation was not money, but rather "the dream and the will to found a private kingdom"; "the will to conquer: the impulse to fight, to prove oneself superior to others"; and finally the "joy of creating." Vail was that type. As his biographer put it, "he always had a taste for conquest . . . here was a new world to subjugate."[18]

When Vail arrived at Bell, Hubbard soon recognized where his potential lay and made him general manager of the company. In that role Vail, like a man who tastes combat for the first time, discovered his natural aptitude for industrial warfare. He applied himself vigorously, reorganizing the firm and putting the fight in Bell's employees, agents, and partners. In internal letters he called on the Bell side to give their all; for this battle, he believed, was the very test of their manhood.

"We have organized and introduced the business," he declared, "and we do not propose to have it taken from us by any corporation." To an agent who was wavering, Vail wrote, "we must organize companies with sufficient vitality to carry on a fight," for "it is simply useless to get a company started that will succumb to the first bit of opposition it may encounter."[19]

Vail's efforts surely helped morale, and some have credited them with preventing Bell's premature capitulation. But in truth the key to the fight was with Hubbard. Bell was overmatched in every area—finances, resources, technology—except one: the law, where it held its one all-important patent. And so, as the firm's eponymous founder lay in the hospital, Hubbard, an experienced patent attorney himself, retained a team of legal talent to launch Bell's only realistic chance of survival: a hard-hitting lawsuit for patent infringement. The papers were filed in September 1878. If Western Union was a figurative Goliath, the lawsuit was David's one slingshot stone.

The importance of Bell's lawsuit shows the central role that patent plays in the Cycle, and it is a role somewhat different than is usually understood by legal scholars. Patents are, by tradition, justified as rewards for invention. Owning a patent on the lightbulb, or a cure for baldness, means that only you (or your licensee) can profit from its sale. The attendant gains are meant to encourage investment in invention. But in the hands of an outside inventor, a patent serves a different function: as sort of corporate shield that can prevent a large industrial power from killing you off or seizing control of your company and the industry. In that oblique sense, a strong patent can sow the seeds of creative destruction.

The Bell patent is an example, perhaps the definitive example, of such a seeding patent. Had it not existed, there would never have been a telephone industry independent of the telegraph.

Yet it was hardly a foregone conclusion that Bell's patent would be its salvation. The validity of the license was somewhat in question: Elisha Gray, remember, had filed a similar patent, arguing, not without foundation, that Alexander Bell had stolen from his design the features that made the telephone actually work. Western Union, meanwhile, held various patents of its own relating to communication over wires, as well as to all of Edison's improvements to the telephone, which rights Bell was probably infringing. Western Union had the further advantage of the deep pockets required to wage a long legal battle. They could well have starved Bell out of existence or forced Bell to license its patent— also an effective death sentence, albeit at least a compensated one.

So how did puny Bell prevail against the mighty Western Union? If the story were a film or novel, one would have to charge the author with abuse of deus ex machina. For right at Bell's darkest hour it was saved by an unlikely and unexpected cavalry charge. Western Union came under attack from the financier Jay Gould, "King of the Robber Barons," who had been quietly acquring stock and preparing a hostile takeover. Now fighting for its own independence, Western Union was forced to look upon its tussle over the telephone as a lesser skirmish, one it no longer had the luxury of fighting.

Thanks to Jay Gould's blindsiding attack, and good old-fashioned corporate ineptitude on its own part, Western Union broke down and gave up on its imperial plans. Instead of dominating a business it could have bought for $100,000, the company entered into negotiations with Vail, who struck a tough bargain. Western agreed to abandon telephony forever, in exchange for 20 percent of rental income on the Edison telephone and a promise from Bell never to enter the telegraph market or offer competition to the Associated Press.[20]

Historians and business school professors have ever since puzzled over how a behemoth like Western Union could have submitted to such a raw deal so easily. One is tempted to fall back on the cliché "the harder they fall," but there were plenty of factors that made a difference.

Perhaps Western Union's leadership, without the benefit of Schumpeter's work (he was just about to be born), never fully understood that the telephone was not just a new and promising market but an existential threat. Such things can be difficult to see. Who, in the 1960s, would have imagined the computer industry would one day threaten the music industry? While it may seem obvious to us, Western Union might not have fully realized that the telephone would actually replace, not just complement, the telegraph. Recall that telephone technology was at the time both primitive and a luxury. For that reason, it is possible that Western Union thought it wasn't such a big deal to let Bell establish a phone service, imagining it was simply letting Bell run a complementary but unrelated monopoly.

Horace Coons, the communications chronicler, writing in 1939, lends some support to this idea. He attributes Western Union's retreat to its realization that staying in telephony would likely mean competing with Bell on an ongoing basis. As he wrote, "no one in the communications field was fond of the idea of competition. They had all experienced competition and they did not like it. . . . Both the telephone and the telegraph monopolies offered magnificent opportunities, [but] were not worth very much unless they were opportunities to be monopolies."[21]

For the purposes of our story, however, it is more significant to contemplate the counterfactual outcome. We all recognize how much a nation is shaped by its literal wars, yet a nation's large-scale industrial wars also inform its identity to a degree we don't always acknowledge. An America that had entered the twentieth century with Western Union as its single wire monopolist—a decidedly different arrangement from the one that came to be and one that would shape not just our telephone communications, but, as we shall see, radio and television broadcasting and ultimately the Internet—would likely have been, culturally, politically, economically, in innumerable ways great and small, an America significantly different from the one we know.

Instead, Bell, now grandly styled the National Bell Telephone Company, was left with the telephone market and began to lay the foundations of what is called the First Bell Monopoly. It was, however, far from what we'd recognize today as the telephone system. The First Bell Monopoly was a service for the rich, operating mainly in major cities in the East, with limited long distance capacity. The idea of a mass telephone service connecting everyone to everyone else was still decades away.

Meanwhile, in 1884, the Bell Company put Vail in charge of a new subsidiary meant to build its "long lines." Vail named the subsidiary the American Telephone and Telegraph Company—AT&T for short—a name that, one way or another, has figured centrally in the story of American communications ever since.

CHAPTER 2

Radio Dreams

One July afternoon in 1921, J. Andrew White paused before speaking the words that would make him the first sportscaster in history. White, an amateur boxing fan who worked for the Radio Corporation of America, stood ringside in Jersey City, surrounded by more than ninety thousand spectators. The boxing ring was but a tiny white square in a teeming sea of humanity. Everyone was waiting for the "fight of the century" to begin.[1]

In the ring the fighters looked mismatched. The larger was Jack Dempsey, the "Manassa Mauler," the reigning heavyweight champion,[2] who had grown widely unpopular for refusing to serve in World War I. Georges Carpentier, his opponent, had entered the ring to the strains of "La Marseillaise" and deafening cheers. The French war hero was obviously the crowd favorite.

In White's hand was something unexpected: a telephone. It was fitted with an extremely long wire that ran out of the stadium and all the way to Hoboken, New Jersey, to a giant radio transmitter. To that transmitter was attached a giant antenna, some six hundred feet long, strung between a clock tower and a nearby building. The telephone White was holding served as the microphone, and the rickety apparatus to which it was connected would, with a bit of luck, broadcast the fight to hundreds of thousands of listeners packed for the day into "radio halls" in sixty-one cities.

What was planned now sounds quite ordinary, but at the time it was revolutionary: using the technology of radio to reach a mass audience. Today we take it for granted that the TV or radio audience for some performance or sporting event is larger than the live audience, but before 1921 such a situation had never occurred. This fight, in fact, would mark the first time that more people would experience an event remotely than locally. That is, if everything went according to plan.

The idea to broadcast the fight came from a young man named Julius Hopp, manager of concerts for Madison Square Garden as well as an amateur radio enthusiast. He wanted to experiment with an application of radio technology that heretofore only hobbyists had played with—something they called "radio broadcasting."

Hopp could not do it alone. He found important backing, financial and technical, at the Radio Company of America (RCA), predominantly a military contractor, including its vice president, Andrew White, and more important, David Sarnoff, an ambitious young executive and enigmatic personality who would figure centrally in the history of radio. A Russian Jew who had immigrated as a youth, Sarnoff had an eye for promising ideas, coupled with a less admirable tendency to claim them as his own. Having managed to funnel several thousand dollars of RCA money to Hopp, he and White focused their combined energies on the Dempsey broadcast.[3]

The scale of the effort was unprecedented. But to be absolutely clear: Sarnoff, White, and Hopp were in no sense inventing radio broadcasting. They were, rather, trying to bring to the mainstream an idea that amateurs had been fiddling with for years. Just as email had been around since the late 1960s, though reaching the general public only in the 1990s, broadcasting in some form had been occurring since as early as 1912, and perhaps even earlier.

It was amateurs, some of them teenagers, who pioneered broadcasting. They operated rudimentary radio stations, listening in to radio signals from ships at sea, chatting with fellow amateurs. They began to use the word "broadcast," which in contemporary dictionaries was defined as a seeding technique: "Cast or dispersed in all directions, as seed from the hand in sowing; widely diffused."[4] The hobbyists imagined that radio, which had existed primarily as a means of two-way communication, could be applied to a more social form of networking, as we might say today. And the amateur needed no special equipment: it was enough simply to buy a standard radio kit. As *The Book of Wireless* (1916) explains, "any boy can own a real wireless station, if he really wants to."[5]

If the amateur pioneers had a leader, it was the inventor Lee De Forest, who by 1916 was running his own radio station, 2XG, in the Bronx.[6] He broadcast the results of the 1916 presidential election, and also music and talk for an hour or so each day. *QST Magazine,* the publication of the America Radio Relay League, reported in 1919 of De Forest's station, "we feel it is conservative to estimate that our nightly audience is in excess of one thousand people."[7]

Back in Jersey City, as the bout began, Dempsey ran at Carpentier, punching hard (you can watch the bout on the Internet), and while Carpentier puts up a spirited fight, the larger Dempsey clearly dominates. In the second round, Carpentier breaks his thumb, yet fights on. By round four, Dempsey is insuperable, landing blows to the body and head, seemingly at will, as the Frenchman stoops forward, barely able to stand. Then, in White's words: "Seven . . . eight . . . nine . . . ten! Carpentier is out! Jack Dempsey is still the world's champion!"

The broadcasters were in fact lucky it was over in just four rounds, for soon thereafter, their equipment blew up. Still it had held together long enough for more than three hundred thousand listeners to hear the fight in the radio halls. As *Wireless Age* put it: "Instantly, through the ears of an expectant public, a world event had been 'pictured' in all its thrilling details. . . . A daring idea had become a fact."[8]

What is so interesting about the Dempsey broadcast is that it revealed an emerging medium to be essentially up for grabs. It was in retrospect one of those moments when an amateur or hobbyist's idea was about to emerge from relative obscurity, with the same force, one might say, as Dempsey's blows raining down on Carpentier. And while not the cause of the extraordinary radio boom to follow, the Dempsey fight, which had taken so many ears by surprise, was in some sense its herald. While records are spotty, the number of broadcasting stations jumped from 5 in 1921 to 525 in 1923, and by the end of 1924, over 2 million broadcast-capable radio sets had been sold.[9]

Early radio was, before the Internet, the greatest open medium in the twentieth century, and perhaps the most important example since the early days of newspaper of what an open, unrestricted communications economy looks like. Having begun among some oddballs as a novelty aimed at bringing one's voice and other sounds to strangers via the airwaves, broadcasting was suddenly in the reach of just about anyone, and very soon all sorts of ideas as to what shape it should take, from the rather banal to the most utopian, were in contention.

The Open Age of American Radio

When in the course of human affairs things go wrong, the root cause is often described as some failure to communicate, whether it be between husband and wife, a general and a front-line commander, a pilot and a radio controller, or among several nations. Better communications, it is believed, lead to better mutual understanding, perhaps a recognition of a shared humanity, and the avoidance of needless disaster. Perhaps it is for this reason that the advent of every new technology of communication always brings with it a hope for ameliorating all the ills of society.

The arrival of mass broadcasting inspired, in the United States and around the world, an extraordinary faith in its potential as the benefactor, perhaps even a savior, of mankind. And while the reason may not be readily apparent, such belief is crucial to understanding the long cycles in the development of information media. For it is not just the profit motive that drives the opening up of a medium—there is typically a potent mix of both entrepreneurial and humanitarian motives.

Those who grew up in the late twentieth century have known the latter sort of idealism mainly as it manifests itself on the Internet in grand collaborative projects such as the blogosphere or Wikipedia and also in such controversial undertakings as Google's digitization of great libraries. This impulse is part of what has attracted thinkers like Lawrence Lessig, originally a constitutional theorist, to Internet studies, examining the anthropological and psychological consequences of complete openness and the promise it holds. Scholars such as Harvard's Yochai Benkler, Eben Moglen, and many others have devoted considerable attention to understanding what moves men and women to produce and share information for the sake of some abstract good.

Of course the human urge to speak, create, build things, and otherwise express oneself for its own sake, without expectation of financial reward, is hardly new. In an age that has radically commoditized content, it is well to remember that Homer had no expectation of royalties. Nor has the fact of payment for many types of information—books, newspapers, music—extinguished the will to communicate unremunerated. Well before the Internet, in a world without paid downloads, before even commercial television, the same urge to tinker and to connect with others for the pure good of it gave birth to what we now call broadcasting and practically defined the medium in its early years. In the magazines of the 1910s you can feel the excitement of reaching strangers by radio, the connection with thousands and the sheer wonder at the technology. What you don't hear is any expectation of cashing in.

Here is Lee De Forest addressing young people on the joys of the wireless:

> If you haven't a hobby—get one. Ride it. Your interest and zest in life will triple. You will find common ground with others—a joy in getting together, in exchange of ideas—which only hobbyists can know.
>
> Wireless is of all hobbies the most interesting. It offers the widest limits, the keenest fascination, either for intense competition with others, near and far, or for quiet study and pure enjoyment in the still night hours as you welcome friendly visitors from the whole wide world.[10]

What exactly were the hopes for radio? In the United States, where broadcasting began, many dreamed it could cure the alienating effects of a remote federal government. "Look at a map of the United States and try to conjure up a picture of what home radio will eventually mean," wrote *Scientific American*'s editor Waldemar Kaempffert in 1924.[11]

> All these disconnected communities and houses will be united through radio as they were never united by the telegraph and the telephone. The President of the United States delivers important messages in every home, not in cold, impersonal type, but in living speech; he is transformed from what is almost a political abstraction, a personification of the republic's dignity and power, into a kindly father, talking to his children.

There was even, perhaps unexpectedly for an electronic medium, hope for the elevation of verbal discourse. "There is no doubt whatever that radio broadcasting will tend to improve the caliber of speeches delivered at the average political meeting," read a column from the 1920s in *Radio Broadcast*.[12] "The flowery nonsense and wild rhetorical excursions of the soap box spellbinder are probably a thing of the past if a microphone is being used. The radio listener, curled comfortably in his favorite chair is likely to criticize the vituperations of the vote pleader quite severely. Woe be unto the candidate who depends for public favor upon wild rantings and tearings of hair."

There was even the hope for a more cultured society. "A man need merely light the filaments of his receiving set and the world's greatest artists will perform for him," said Alfred N. Goldsmith, the director of research at RCA, in 1922.[13]

> Whatever he most desires—whether it be opera, concert, or song, sporting news or jazz, the radio telephone will supply it. And with it, he will be lifted to greater appreciation. We can be certain that a new national cultural appreciation will result. . . . The people's University of the Air will have a greater student body than all of our universities put together.

All of these early aspirations partake of the idealistic expectation that a great social interconnectedness via the airwaves would perforce ennoble the individual, freeing him from his baser unmediated impulses and thus enhancing the fellowship of mankind. Such an intuition, of course, is not limited to communications technologies; it is a tenet of many religions that the distance between the individual and his fellows is an unnatural source of suffering, to be overcome. Perhaps this is why some were prepared to ascribe the miraculous potential of the new medium not to human cleverness but to Providence. "Radio proves the truth of the omnipotence of the Almighty," wrote *Radio Dealer* editor Mark Caspar in 1922.[14] "When the Bible tells us God is omnipresent and sees all we do and knows all our thoughts—we can now better realize that if we, mere humans, can 'listen in' and hear people talk all over the earth with a radio set, a foot or two long, what power must we ascribe to the Almighty? Can we longer doubt his omnipresence and omnipotence? Behold, the all-seeing eye!"

The power of an open technology like radio broadcasting to inspire hope for mankind by creating a virtual community is the more remarkable considering that radio was yet far from reaching its full potential as a communications medium. In fact, what it seemed to promise was, if anything, more thrilling than the present wonders. In De Forest's words, radio "is the coming Science, is moving ahead faster, possibly, than any other."[15] He urged young men to "take up Radio work because it offers a means of entertainment second to no other; gives useful instruction that can be made to produce tangible results later on; keeps everyone interested; enables you to get the news of the world by wireless and provides a pastime and hobby that will get the busy man's mind into other channels."

One must stress that it was not merely technological wizardry that set people dreaming: it was also the openness of the industry then rising up. The barriers to entry were low. Radio in the 1920s was a two-way medium accessible to most any hobbyist, and for a larger sum any club or other institution could launch a small broadcast station. Compare the present moment: radio is hardly our most vital medium, yet it is hard if not impossible to get a radio license, and to broadcast without one is a federal felony. In 1920, De Forest advised, "Obtaining the license is a very simple matter and costs nothing." As we shall see, radio becomes the clearest example of a technology that has grown into a feebler, rather than a stronger, facilitator of public discourse, the vaunted vitalities of talk radio notwithstanding.

But let us not exaggerate the "purity" of early radio: its founders and commercial partners had a variety of motives, not excluding profit. In the early 1920s, publications such as *Radio News* published lists of all the radio stations in operation, with their frequencies and what one might expect to hear on them—a forerunner of the once hugely profitable *TV Guide*.

Such listings show that many early stations were run by radio manufacturers such as Westinghouse, the pioneer of the ready-to-plug-in model, and RCA, both of which had an obvious interest in promoting the medium. Still many stations were run by amateurs, "radio clubs," universities, churches, hotels, poultry farms, newspapers, the U.S. Army and Navy; one was run by the Excelsior Motorcycle Company of Seattle.

The choices were dizzying. "A list of all that can be heard with a radio receiver anywhere within three hundred miles of Greater New

York would fill a book," explained one publisher of listings. "At any hour of the day or night, with any type of apparatus, adjusted to receive waves of any length, the listener will hear something of interest." A whole class of stations arose—for instance, just to broadcast jazz, which was otherwise inaccessible to most middle-class fans outside the urban centers where the art developed.[16]

As few recordings of radio in the 1920s survive, however, one must not romanticize the medium by supposing a quality of offerings to rival the diversity. Station schedules extended but a few hours a day. Content was limited to whatever broadcasters could wangle, whether starving musicians, gramophone recordings, or opinionated talkers. Yet we can imagine the wonder of simply tuning in, never knowing quite what we might hear—surfing the untamed world of the dial.

By its nature, early American radio was local, and hence the roots of "localism" in broadcasting. With an average range of thirty miles or so, an amateur radio station in, say, Seattle was not likely to have a national listenership. Stations that could reach the far corners of the country did not yet exist. The outer limit was represented by an event like the Dempsey-Carpentier fight, a sensation with a maximum signal range of two hundred miles. And so with no means to connect to other stations, and limited broadcast wattage, radio stations made a virtue of the necessity to be local. No baseball game or concert taking place nearby was too small to be a broadcast event. A local pastor could always count on his sermon being heard by more individuals than those sitting in the pews before him. There was no such thing as national radio, public or private. And for as long as such limitations persisted, so did the idealism surrounding radio. Even David Sarnoff, the future president of RCA, remarked, "I regard radio broadcasting as a sort of cleansing instrument for the mind, just as the bathtub is for the body."[17]

THE IDEALS OF BRITISH BROADCASTING

In 1922, John Reith, the youngest son of a Scottish minister, was appointed general manager of the newly formed British Broadcasting Company. At age thirty-three, he had no relevant experience—though admittedly individuals with credentials in broadcasting were few at the time—and so his selection was something of a mystery, even to him.

As Reith wrote in his diary, "I am profoundly grateful to God for His goodness in this manner. It is all His doing."[18]

Reith used the favor of Providence to build a distinct and lasting model of public broadcasting, and the early BBC represents a road not taken relative to radio broadcasting in America, one that would abandon the structural openness so stirring of utopian sentiment and yet in some sense more faithfully cultivate the improving ideals of public service. "The Policy of the Company," wrote Reith in 1924, is "to bring the best of everything into the greatest number of homes." In tune with Victorian convictions about human perfectibility, radio was employed as a means of moral uplift, of shaping character, and generally of presenting the finest in human achievement and aspiration. And it was this way from the beginning. Reith presided over the medium as a monopoly from its very inception, with no open period of broadcast pluralism, the thrilling free-for-all that had sprung up in America. His power was absolute yet governed by the British imperative of self-restraint.

Reith's intentions were as evident on the surface as at the core of his efforts. He opened a London studio in Savoy Hill, its appointments more suggestive of a gentlemen's refuge than the utilitarianism one might expect. Gale Pedrick, a BBC script editor, remarked: "Next to the House of Commons, Savoy Hill was quite the most pleasant club in London. There were coal fires, and visitors were welcomed by a most distinguished looking gentleman who would conduct them to a cosy private room and offer whisky-and-soda." Beginning in 1926, all announcers were required to wear dinner clothes during broadcasts, ostensibly to put any similarly clad performers at ease, and generally to preserve the decorum of the enterprise.[19]

In his 1924 book *Broadcast over Britain,* Reith gave definitive expression to his view of radio as a supremely dignified business.[20] The medium, he wrote, must not become "mere entertainment," catering to the "imagined wants" of the listener. There must be, he insisted, "no concessions to the vulgar." He believed that anything one might take for popular demand was but the contrivance of the broadcasters themselves. It was a view rather like the one expressed by Reith's contemporary Walter Lippmann in *The Phantom Public.*[21] As Reith would later put it, "He who prides himself on giving what he thinks the public wants is often creating a fictitious demand for lower standards which he will then satisfy."[22]

The mission of improving general sensibilities naturally led to cultural and educational programming, including lectures on important topics by learned men, though avoiding controversy. There was to be an element of what we might call self-help through building "knowledge, experience and character, perhaps even in the face of obstacles, though never proclaiming a competitive creed nor advertising a panacea."[23] Admittedly, the ban on provocation could limit the educational objective—even a talk on women's rights, for example, could be too touchy. Asa Briggs describes how George Bernard Shaw, invited to give a talk in 1924, was warned not to discuss politics or religion. "Politics and religion," he replied, "are the only things I talk about."[24]

Not that all restriction issued from Reith's own Victorian reticence. His vision of "a more intelligent and enlightened electorate" was sometimes limited by government pressure.[25] The BBC, though initially a private enterprise, was since its inception under the tight scrutiny of the government, with which Reith's relations were ever poor. In his diary Reith would vilify Winston Churchill as a "cur," "coward," "loathsome cad," and "blasted thug."[26] Unfortunately for Reith, his colleagues were hardly so stirred up by the prime minister and were perfectly content to toe the party line. As one BBC manager put it, "we do not wish to have the Broadcasting stations used for propaganda which will excite one section of the population and be very distasteful to another."[27] Hence the norms of British broadcasting continued to conform to those of polite dinner conversation, avoiding anything that might upset or inflame.

Perhaps the most famous of these norms is the one respecting "spoken English." Among the mandates of the BBC as custodian of the public trust was to save the King's English from corruption. (BBC English is still a recognizable norm of sorts, though now accommodative of popular usage to a degree that might well have horrified the founders.) Questions concerning "debatable language" were addressed by a particularly impressive advisory committee that included Rudyard Kipling, George Bernard Shaw, and the poet Robert Bridges, which met three times a year.[28] The committee could take credit for eliminating expressions like "broadcasted" and "listen in" from standard usage.

Reith's dream of lifting up the masses has an undeniable element of condescension about it. He had little curiosity about what the common people were interested in, nor, it must be said, was he especially fond of them. "I do not love, or even like, my neighbour," he once disclosed in a letter; "in fact I dislike him more and more as he hooliganizes about the roads with open exhaust . . . glorifying in being a damnable curse to the whole community."[29] This sentiment posed for him a certain crisis of faith. "I believe profoundly in the Christian ethic, but I am a very poor practitioner; I have said also that, to such extent as loving one's neighbour is an essential criterion of admission to the company of the elect, I absolutely fail to qualify."

Though he had largely succeeded in making the BBC conform to his vision, he was never content with his progress, and felt himself underappreciated. When he was knighted in 1927, but to no specific chivalric order, he wrote in his diary, "an ordinary knighthood is almost an insult. The PM has never comprehended the importance of our work." His dissatisfaction would persist into the 1940s, when he was created Baron Reith of Stonehaven. "I do not care two hoots or one hoot about honours, and often wish I had never taken one. What I do care about is the injustice of not being given or offered them."

Reith may have continued to harbor his grudges against the British government, but in a way, his legacy is indebted to that institution. As we have said, the BBC, however closely watched by Whitehall, did not come into being as a government organ but as a private company formed by a collective of radio manufacturers. Only later, in 1927, would it come under more direct public supervision, as a Crown Corporation—that is to say, a corporation owned by the king.

In this way, the BBC would for decades be spared the great controversy over advertising, which would consume and ultimately shape American radio. The BBC, as Reith tells in his memoir, "is not out to make money for the sake of making money."[30] The company's sustaining revenues came from the sale of licenses to receive broadcasts (ten shillings) and, in the early days, a royalty fee added to the price of radio sets. As for the American revenue model, the first parliamentary committee to consider radio banned advertising on the basis that it might "lower the standard"—though no explanation was given of how mention of tinned meat might have this effect.[31]

And so, this is radio broadcasting in the 1920s: On one side of the Atlantic, in the geographically vast United States, isolated clusters of local and mostly amateur operators, inspired by the enthusiasm of the

hobbyist and a somewhat vague though earnest idea of national betterment. In Britain, a private monopoly, with national reach, arguably
elitist but unquestionably and systematically dedicated to bringing "the
best of everything" to the general public. In either setting, the medium
would never be more hopeful or high-minded.

CHAPTER 3

Mr. Vail Is a Big Man

Located in northeastern New Mexico, the Johnson Mesa is a vast grassy plateau, some 8,600 feet above sea level and fifteen miles from the nearest big town. The grass is thick and runs to the horizon, broken only by the occasional barn, weathered and gray, and a single stone church, long abandoned.

On this mesa, in the fall of 1904, a farmer named Edmund Burch was stringing galvanized wire between lines of barbed wire fence. Using little more than the wire and his own hands, Burch, one of a handful of struggling homesteaders, was building an elementary telephone network to connect his farm with those of his mesa neighbors.[1]

Today, one rarely thinks of wiring one's own telephone network. But Burch was part of a movement of telephone self-connectors, the telecom DIYers of the first decade of the twentieth century. Bell had no immediate interest in wiring places like the Johnson Mesa, which lay a grueling two-thousand-foot climb from the nearest town. So Burch urged his fellow mesa farmers to forget waiting for Bell and "do it ourselves." "The farmer with the telephone," he asserted, "is with the times."

Burch had been inspired by, among things, a story in *Scientific American* entitled "A Cheap Telephone System for Farmers"; published in 1900, it told how an Indiana man had connected the farms in his area with nothing but galvanized wire and barbed wire. A no less inspiring piece in *Rural New Yorker* said, "Build strictly first-class as far as possible, but build your own lines by all means in some way. Use barbed wire only when absolutely necessary." Burke took its counsel on quality control. By 1904 he had a direct line to two neighbors and was demonstrating to the community that same magic that Mr. Bell had discovered thirty years earlier.[2]

Before long, Burch had founded the Mesa Telephone Company to wire the entire settlement, and in so doing became one of hundreds of such small telephone companies springing up across the nation under such names as the Swedish-American Telephone Company, the Home Telephone Company, or the People's Telephone Company.[3] His fellows in the cause styled themselves "the Independents." They were, by their own description, "an uprising of the people," a social movement dedicated to "American Industrial Independence."[4]

Though a very small group, the Independents would mount the first great challenge to the First Bell Monopoly. After the expiration of Bell's patent in 1894, hundreds of independent firms had cropped up to provide telephone service. The age of the Independents was the "open" phase in American telephony, characterized by a vision very different from Bell's. While now forgotten or unknown to most, that vision would profoundly change how Americans communicated.

In contrast to the Independents' clientele, that of the First Bell Monopoly consisted of businesses and rich individuals living in large East Coast cities; Bell was in no hurry to broaden the coverage of its network. In fact, Bell's business model was altogether too stagnant for Theodore Vail's taste. Like the Independents, Vail could sense the potential power of a national network, but if he yearned for industrial greatness, Bell's shareholders were monotonously interested in dividends alone. That conflict came to a head in 1887 after Bell, rather than plowing profits into expansion, announced a particularly fat dividend. Writing that his position at the company had become "embarrassing and unpleasant," a dispirited Vail retired to South America in search of adventure.[5]

The Independents, rooted in the farms and small towns of the West, were innovators, but of a conceptual kind, not the technical kind à la Alexander Bell. They saw a different world, in which the telephone was made cheaper and more common, a tool of mass communications, and an aid in daily life. They intuited that the telephone's paramount value was not as a better version of the telegraph or a more efficient means of commerce, but as the first social technology. As one farmer captured it in 1904, "With a telephone in the house comes a new companionship, new life, new possibilities, new relationships, and attachments for the old farm by both old and young."[6]

Typically, the rural telephone systems were giant party lines, allowing a whole community to chat with or listen to one another. Obviously there was no privacy, but there were benefits to communal telephony other than secure person-to-person communications. Farmers would use the telephone lines to carry their own musical performances. The historian Ronald Kline has described the telephone parties that were all the rage in some areas, with groups assembling to hear, as it were, a phoned-in concert. "The opening of the new telephone line at Ten Mile," reported the *Macon Democrat,* a Missouri newspaper in 1904, "was celebrated with gramophone, violin, banjo, french harp, guitar and organ Friday night."[7]

And so, while the Bell Company may have invented the telephone, it clearly didn't perceive the full spectrum of its uses. This is such a common affliction that we might name it "founder's myopia." Again and again in the development of technology, full appreciation of an invention's potential importance falls to others—not necessarily techni- cal geniuses themselves—who develop it in ways that the inventor never dreamed of. The phenomenon is hardly mystical: the inventor, after all, is but one person, with his own blind spots, while there are millions, if not billions, of others with eyes to see new uses that had been right under the inventor's nose. We shall see the story repeated throughout this book. For now, suffice it to say that it was simple farmers in the early 1900s who pioneered the use of the phone line for broadcasting long before the rise of radio broadcasting in the 1920s. Burch's Mesa Telephone Company offered its customers daily broadcasts of weather, train wrecks, and murders, the interval of programming announced by ten short rings. As Kline writes, "Every evening at a designated time, usually seven p.m., an operator would call all farms on a line and give the time, weather and market reports, newspaper headlines and local news, 'with a spicing of gossip.'"

·　　·　　·

In the theory of competition that applies to information industries, as to all others, we speak of *barriers to entry:* the obstacles that a newcomer must overcome to get into the game. But barriers in an information industry, trafficking as it does in expressive content, can represent more than a restraint on commercial aspirations; they can, depending on how crucially the information medium figures in a society's communications, also restrain free speech. If we want to define how "open" any industry is, we should start with a number: the *cost of entry.* By this we simply mean the monetary cost of getting into the business with a reasonable shot at reaching customers. Is it in the neighborhood of $100? $10,000? Or more like $1 billion? Whatever the magnitude, that number, most definitively, is what determines whether an industry is open or closed.

In the first decade of the twenty-first century, for instance, if you wanted to start a competitive mobile phone service, to take on AT&T, Verizon, and the rest, the price of entry—for a spectrum license, towers, and other necessities—was somewhere north of $10 billion. It's not the sort of expense most of us would take on in the pursuit of a hobby. Entry costs of such magnitude are not atypical. Thus, for most of the twentieth and twenty-first centuries, the phone market has been effectively closed. Starting a completely new phone service based on new wires was financially infeasible, as the costs of entry were enough to daunt even the most deep-pocketed firms, let alone rural cooperatives.

But for a brief time in the 1890s, all this was different. While we don't know exact present-value costs, they were low enough that farmers like Edmund Burch, as well as small-town entrepreneurs and rural cooperatives, could effectively compete with Bell. In this way, many towns ended up with two telephone systems, during what we call the "dual service" era.

Why was market entry cheaper? To begin with, telephony was at the time a decidedly low-tech affair, as evident from Burch's reliance on simple galvanized wire. In cities, you could generally just run wire aloft on poles, avoiding the costs of burrowing to reach homes. And in the countryside, it was even simpler: farmers like Burch in New Mexico could nail wires to farm fences, creating what they called "squirrel lines," and attach phones at the ends. It was telephony in the *Green Acres* style. These simple logistics, taken together with the absence of licensing costs, made things easy for motivated self-starters.

The economics of switching also made it possible for independent phone systems to compete with Bell. In today's automated world, the larger the network, the better it is, because you can reach more people in more places. But when human operators ("What number, please?"— the "telephone girls" to whom we shall return in another chapter) were needed to physically connect one phone line to another, a larger network meant a slower switching system, prone to bottlenecks and breakdowns. That weakness allowed the Independents room to start with just a few customers. In some places Bell had never even offered phone service, allowing the Independents the advantage of being first to market.

Bell's initial response to the Independents was simply to ignore or dismiss them. The trade journal *Telephony* reprinted stories of farmer telephones in its humor section, an attitude Bell could afford to share as long as the farmer lines and the Independents operated in communities Bell didn't want to serve. But as the Independents built more lines, and formed their own associations, they gradually grew to threaten Bell's control over the American telephone, until, by the turn of the century, the Bell companies undertook a campaign against forces they now called "the Opposition." It would be a rough campaign by any standard of industrial warfare. AT&T as a matter of course refused to make any connections between the Bell system and the Independents, but it would go much further to protect its monopoly. Relying on its profits in stronger markets, Bell would dramatically undercut the rates of local independent telephone companies in any contested area, a tactic known as *predatory pricing*. Sabotage of equipment was not unheard of, and it was practiced by both sides. Paul Latzke, a supporter of the Independents, wrote, "there has been wholesale bribery, systematic wrecking, and, at times, violence, almost anarchy."[8] According to one account, Bell would rip out wires and phones, and "in truly medieval fashion, pile the instruments in the street and burn them, as a horrible example for the future."*[9]

With a tendency toward moralizing bombast, the Independents complained to anyone who would listen. As one Independent wrote in *Sound Waves,* a "monthly magazine devoted to the interests of independent Telephony," "those who have watched the peculiar and heathen ways of the Bell monopoly know that it is, without doubt or question,

---

*There is a difference between creative destruction and merely destructive destruction.

the most conscienceless organization in the United States, compared to which the gigantic Standard Oil trust is a mere kindergarten of devious financial and industrial devices."[10]

However "devious," the Bell strategy was ultimately ineffective. No amount of unwiring could alter the fact that the Independents were meeting a demand for cheaper telephone service. By the early 1900s, Bell's dominance was beginning to erode, the company soon to be pinned like Gulliver by hundreds of Lilliputians. As the Independents grew profitable and more secure, even their rhetoric brightened with a new confidence: "The days of prosperity of the Bell companies are gone, never to return. The public has learned to appreciate good telephone service and courteous treatment, and will not again submit to the extortions and antiquated methods of the Boston trust."[11] By 1907, Paul Latzke would publish a small book called *A Fight with an Octopus,* in which he revealed the Independents had 3 million phones to Bell's 2.5 million, and total dominance in the West. Latzke further predicted that the "final battle" would take place in New York City: "The Bell people have made Manhattan Island their Gibraltar. Its defense will be a spectacle well worth watching . . . the greatest industrial battle of the age."[12]

FROM REPUBLIC TO EMPIRE

Sometime in the early 1900s, Vail, then living in Buenos Aires, was invited to Jekyll Island, South Carolina, to play cards with a man known to him only by reputation. During his visit, or shortly afterward, the man told Vail of a plan, then secret. He and a group of other financiers aimed to gain control of the Bell company, wishing not only to reestablish its former dominance but to build the greatest wire monopoly the world had ever seen. And he wanted Vail in charge. Vail knew that this man was to be taken seriously—for this seasonal resident of Jekyll Island was none other than J. P. Morgan, one of the greatest monopolists of that era or any other.[13]

In 1907, after gaining Vail's assent, Morgan set his plan in motion. In a lightning-fast series of financial maneuvers, he took control of Bell, forcing out the Boston owners. Vail's title would be president of American Telephone and Telegraph (AT&T), now the holding company for the entire Bell system. Rather like Steve Jobs's storied coming back to Apple, Vail's return to Bell, at age sixty-two, would change everything.

With Vail again in place, there began a great transition in telephony from an open, competitive phase to the Second Bell Monopoly, Bell's true imperial age of dominance in wire communications, which would last for most of the twentieth century. It was during this transition that the orthodoxy of centralized power in communications took its mature form. In exile, Vail had never ceased nursing dreams of empire, but now with Morgan's undreamed-of support, he was free to think big, even by his own outsized standards. The new slogan he was able to announce upon his arrival said it all:

ONE SYSTEM, ONE POLICY, UNIVERSAL SERVICE

The terminology is important to understand: it meant "unrivaled," not "for all." This was not "universal" as in, say, universal health care, but more nearly in the sense of the universal church. It was, as the historian Milton Mueller explains, universal service as an alternative to options, and as such it was a call for the elimination of all heretical hookups and the grand unification of telephony.[14]

To Vail and Morgan, building redundant phone lines between any two points was as senselessly wasteful as building twenty duplicate rail tracks between two cities, as sometimes happened in the nineteenth century. Why have twenty lines of varying standards where there could instead be one track of highest quality? They also accepted the other lesson of the railroads: without a single master, systemic chaos would undercut efficiency. Vail thought the "opposition" phone companies would stoop to any cut rate, and cut-rate service, just to be in the game. Using the formidable capacity of the Morgans to absorb loss, he undercut the price cutters.[15]

Vail's philosophy is well expounded in AT&T's annual reports, spirited and personal meditations on both AT&T and the responsibilities of a powerful public corporation. It was in these early years that Vail created what would remain the Bell ideology until the system's twentieth-century breakup. And interestingly, Vail had much ideological affinity with his foes, the Independents. He saw the merits in an ever expanding system, and he truly believed in the telephone as a "public utility" ultimately meant to serve every American. His reports, though obviously intended for general consumption beyond the ranks of share-holders, do nevertheless portray an earnest and sincere vision of the public good. Where he disagreed with the Independents was simply over the fact of their existence.[16]

As for J. P. Morgan, he was a mostly silent partner, and his name only rarely shows up in histories of the telephone. Yet Morgan's financing was absolutely crucial to the realization of Vail's vision and Bell's resur-rection as a monopoly. Whether Morgan shared any of Vail's sentiments about the public duties of a corporation we cannot know. But he cer-tainly concurred in his enthusiasm for monopoly as the optimal busi-ness model. Indeed, as we shall see over and over again, the shift from an open industrial phase to a closed market usually begins when capital interests spy the potential for vastly increased profit through monopoly, or when they demand greater security for their investments. Vail's access to Morgan's capital made his vision of the Bell system possible, but it also came with significant strings attached.

THE TAKEOVER

In 1909, at Morgan's direction and using his money, Vail seized a con-trolling interest in Western Union, Bell's childhood tormentor, making himself president of both. AT&T now controlled all instantaneous long distance communications in the United States. As the so-called long lines—those connecting one locality with another—were the scarcest part of the communications infrastructure at the time, to possess them exclusively was the greatest power. A combined AT&T and Western Union now shared customers, offices, and operations, creating a true monopoly in distance communications.[17]

With the support of Morgan, Vail began to take a softer line against the Independents, who were local. Where the old Bell had followed a scorched-earth policy, Vail now sought integration and consolidation. Former rivals were invited not to die but to join him—to rule over com-munications together, we might say, as father and son.

Part of Bell's new strategy was to abandon a tactic that had done so much in the 1890s to decimate the Independents: refusal of network connection. Vail's approach was now more subtle and complex: he used connectivity as a carrot rather than a stick; and it proved, together with merger and acquisition, an irresistible way to dominate the market. The story holds a powerful lesson for any independent business facing a much superior foe, a lesson as important in the 2010s as in the 1910s.

Vail's agreements offered the Independents membership in the Bell system, but they required the adoption of Bell's standards and Bell's equipment and imposed special fees for use of Bell's long distance lines, though with no promise of connecting a call to any non-Bell subscriber.[18] Vail's offers were, then, essentially the ultimatums that Genghis Khan made famous: join the network and share the wealth, or face annihilation. But Vail needn't have looked so far back for a role model; in his own time, John D. Rockefeller had pioneered the "purchase or perish" model to build Standard Oil.

The Independents tried to warn one another off the connection agreements with Bell. As one wrote in a bulletin: "You cannot serve two masters. You must choose between the people and a greedy corporation."[19] But even the relatively strong found resistance unsustainable and were forced to join. As for the relatively weak, they might simply be bought outright, sometimes through agents of Morgan who kept secret their affiliation with Bell. In 1911, Edmund Burch's Mesa Telephone Company was one of those to give up and sell out to the Bell company. What happened to Burch himself is a mystery, but his lines, and the mesa itself, were abandoned by the 1920s.[20]

Did the Independents ever have a chance? Not without their own long distance network. Without long lines the Independents were limited in what, ultimately, they could offer the customer. It was the AT&T long lines that connected Bell telephones, and they made the difference between a national network and a neighborhood of virtual cans and strings.

The Independents weren't stupid. There were some Independent long distance companies, though none individually or in any simple combination formed a network with the reach of AT&T's long lines. There were also efforts to build alternative nationwide long distance networks as early as 1899. That year, a group of financiers from Philadelphia known as the "Traction Kings" allied with others to form the "Telephone, Telegraph, and Cable Company of America." It announced that "the main object of this company will be the extension and perfection of long distance telephone service throughout the country, and in a secondary way the lessening of the rates."[21]

Here were the progenitors of MCI and Sprint, and the beginnings of long distance competition. Unfortunately, before it had even gotten under way, all the backers suddenly pulled out for reasons that remain mysterious. According to an FCC investigator's report decades later, in 1936, the pressure on the Traction Kings had come from J. P. Morgan himself, whose designs on a telephone monopoly were by then already formed. Indeed, as the FCC documented, no rival national long distance network could get financing in the United States or abroad. And so, in the absence of capacity, coordination, and cash, no real challenger to AT&T long lines would appear until the 1970s, some sixty years later. That was J. P. Morgan's lasting legacy.

Even putting aside the Morgan factor, however, Vail's strategy shows how selective openness can be even more treacherous for would-be competitors to navigate than a completely closed system. The option of being invited to dinner very effectively softens the fear of becoming dinner. It is the same logic Microsoft would follow in the 1990s, when its Windows operating system was similarly run as a partially open system. Like AT&T, Microsoft invited its enemies to connect, to take advantage of an open platform, hoping they wouldn't notice or worry that the platform came with a spring trap. For as with Bell, once having made one's bargain with Microsoft, there was no going back.

Antitrust

With "One Company, One System," Vail made explicit his vision of a communications monopoly. It cannot, however, have pleased Bell's attorneys to labor under a slogan expressing clear intent to flout the antitrust laws.

In the 1910s, laws such as the Sherman Act, the broadest antitrust statute, were still fairly recent efforts to contain the trusts that had grown to dominate American industries such as oil, steel, and the railroads. The law prohibited "agreements in restraint of trade" and punished a monopolist who abused its power. The Roosevelt and Taft administrations had made clear the bite of these laws in the early 1900s, culminating in the Justice Department's 1909 prosecution of Standard Oil and John D. Rockefeller for acts not so different from Bell's campaign against the Independents. The ensuing verdict would break Standard Oil into thirty-five pieces.

It was the year before that verdict when Taft administration officials came knocking, to begin what must by then have seemed the inevitable investigation and lawsuit against AT&T over its consolidation of the telephone industry. But from AT&T's first meeting with Justice, we see for the first time something that will occur again and again in the history of communications, the state's calculated exercise of discretion over whether to bless or destroy the monopoly power, deciding in effect what industry it will allow to be dominated. Theodore Vail will prove himself a high priest at winning the blessing of the state for monopoly dominance.

The threat posed by the Justice Department's case was hardly trivial. Just as Bell came under investigation, Thomas Edison's movie trust (the subject of a later chapter) was also under federal attack and would be dissolved in 1915. There was every reason to think the same would happen to the Bell system. But also at that very moment, Vail executed his most ingenious and surprising maneuver.

In a manner nearly unimaginable today, Vail turned to the government, agreed to restrain himself, and asked to be regulated. Bell agreed to operate pursuant to government-set rates, asking in exchange only that any price regulations be "just and fair." Imagine Microsoft in the 1990s asking the states and the Clinton Justice Department to determine the price of installing Windows, or Google today requesting federal guidelines for its search engine. Having spun much rhetoric about Bell as a public trust, Vail now seemed to be putting his money where his mouth was.

With this conciliatory if not quite prostrate attitude, AT&T was able to settle the lawsuit in 1913, acceding to a consent decree named the "Kingsbury Commitment" after Bell's vice president. Under the settlement, Bell made one big concession: it agreed to sell Western Union. It also agreed to permit Independents to retain their independence while enjoying access to its long distance services, and to refrain from acquiring further Independents in over one thousand markets.[22]

While the Independents may have regarded the Kingsbury Commitment as salvation—a "gift from Santa Claus Bell," in the words of one—the deal was not, in fact, all it may initially have seemed. True, by divesting itself of Western Union, Bell was giving up the dream of a complete monopoly over wire communications, but actually the telegraph was fast becoming a dinosaur anyway. True, Independents suddenly could hook up with Bell's long distance lines, but there is little evidence that many of them actually did. Superficially a victory for openness and competition, in time the Kingsbury Commitment would prove the insidious death knell of both.

The trick of the Kingsbury Commitment was to make relatively painless concessions that preempted more severe actions, just as an inoculation confers immunity by a exposing one's system to a much less virulent form of the pathogen. By offering to renounce hegemony in a dying industry and make available a service relatively few could still exploit, Bell spared itself the brunt—and the one truly meaningful remedy—of most antitrust proceedings: a breakup of the firm. With the government satisfied, and even Woodrow Wilson hailing it as an act of business statesmanship, Kingsbury's greatest achievement was to free Bell to consolidate the industry unmolested.[23]

The jujitsu of Vail's anti-antitrust strategy of the 1910s remains an apt lesson to any aspiring monopolist. The key was earnest profession of a good no one could dispute: making America the best-connected nation on earth by bringing the wonder of the telephone into every American home. Appropriating the most appealing rhetoric of the Independents, and arguing persuasively that the Bell system could get the job done more effectively, Vail turned his monopoly into a patriotic cause.

There is a long-running debate in the field of antitrust theory as to what should matter when judging the conduct of a monopolist. Robert Bork, the onetime federal judge and notoriously rejected Supreme Court candidate, is famous for arguing that the corporation's intent, whether malign or beneficent, should be irrelevant.[24] Yet as Bork himself knew, for most of the history of antitrust, attitude is everything, even if market efficiencies are supposed to matter most.

This was something Vail seemed to understand intuitively: that antitrust, perhaps all law, is ultimately pliable by perceptions of right and wrong, good and evil. He understood that the public and government would rise up against unfairness and greed, though not necessarily against size in and of itself. Had Goliath not cursed David by his gods, David might have kept his sling in his pocket. Vail heralded AT&T as the coming of enlightened monopoly, a public utility of the future. He promised to do no evil. And the government bought it.

To Be a Common Carrier and Friend of the State

From his handling of Bell's antitrust problems emerges a central tenet of Vail's thinking: the enlightened monopoly should do good as it does well, serving the public in close cooperation with the state. Vail's view of his firm as the handmaid of government, the telephone as a public utility, is at once the most sympathetic and scariest element of his vision. Vail saw no harm in, and indeed believed in, giants, so long as they be friendly giants. He believed power should be beneficently concentrated, and that with great power came great responsibility.

Vail's most meaningful concession—in principle if not in practice—was agreeing to serve as a *common carrier.** That pledge, in contrast to Western Union's original modus operandi, meant that Bell would refrain from picking winners in other sectors of the economy or public life—any area that privileged access to the growing reach of communications could influence. Despite being a monopoly, Bell was committing itself to noninterference and making itself equally open to all users of its service—that is, universal in the sense its initial claim of being universal had belied.† This is the essence of common carriage, a concept that may seem esoteric, but is as fundamental to free communications over wires and frequencies as the First Amendment is to free expression. The phrase itself is old, dating to fifteenth-century England and born of the need to reconcile the fact that in England, private entities were running what in most countries were public functions, such as roads, ferries, and so on.

Bell's dedication to common carriage was a promise to serve any customer willing to pay, charge fixed rates, and carry his or her traffic without discrimination. It made Bell's telephone service offer rather what a taxi service is meant to provide in most cities—a meaningful similarity, since the concept has its origins in transport.

---

*Technically, Congress declared telephony and the telegraph a common carrier in the Mann Elkins Act of 1910. But more important was Vail's embrace of the role.

†The alternative phrase for a common carrier is a "public calling," and the latter may capture more of the original meaning.

At the heart of common carriage is the idea that certain businesses are either so intimately connected, even essential, to the public good, or so inherently powerful—imagine the water or electric utilities—that they must be compelled to conduct their affairs in a nondiscriminatory way. As a simple example, if a man operates the only ferry over to town, that simple boatman is in a position of great power over other sectors of the economy, even the sovereign authorities. If, for example, he decided to charge one butcher more than another to carry his goods, this operator could bankrupt the one who didn't enjoy his favor. The boatman is thus deemed to bear responsibilities beyond those of most ordinary businesses.

The big question—now often the multi-billion-dollar question—is how to decide, as a matter of policy, what businesses should be considered common carriers with special duties to the public (as Bell positioned itself), which companies should be run by government (as the Post Office has been since Franklin founded it), and which should be "ordinary services" left mostly to forces of the free market.* In the Anglo-American common law tradition, one asks how essential or necessary the service is—how much other industries depend on it. Those industries that supply the means of trade in information, goods, or cash are more obviously vital even than, say, a country's sole producer of sugar.

Practically, this focus has led to four basic industries being identified as "public callings": telecommunications, banking, energy, and transportation. Each plays a certain essential role in the workings of the nation and the economy, and thus these are the industries that have attracted regulation as common carriers, or infrastructure.

Vail himself offers as apt a description as anyone of the common law orthodoxy:

> For the protection of the community, of individual life and health, there are some necessities that should be provided for all at the expense of all, such as roads, pure water, and sanitary systems for concentrated population, and reasonably comprehensive mail service. The determination between services that should be operated by the government

---

*Opponents of regulation in the twentieth century pushed the idea that only true monopolies ought to be considered public callings or common carriers. On the other hand, in the original English view, an industry need not be monopolized to be essential.

and those which should be left to private enterprise under proper control should be governed by the degree of necessity to the community as a whole as distinct from personal or individual advantage.[25]

So if we regard the Kingsbury Commitment as having sanctioned the most lucrative monopoly in history, it also made good on the essential goals of common carriage. Bell did, eventually, wire every home in the United States, and it provided decades of reliable service. But it should also be obvious to anyone—one need by no means be a raving libertarian—that there are some substantial dangers implicit in aligning the immense power of the state with the greatest of information monopolists.*

Vail died in 1920 at age seventy-four, shortly after resigning as AT&T's president, but by that time his life's work was done. The Bell system had uncontested domination of American telephony, and long distance communication was unified according to his vision. In 1921, Congress passed the Graham Act, recognizing AT&T's monopoly and removing any remaining obstacles to integration. The idea of an open, competitive system had lost out to AT&T's conception of an enlightened, licensed, and regulated monopoly. In this form, AT&T would remain in charge until the 1980s, and in not substantially different form it would return in the new millennium. As Milton Mueller writes, Vail had completed the "political and ideological victory of the regulated monopoly paradigm, advanced under the banner of universal service."[26] Vail's biographer adds, "the great work he created remains, never to come to an end so long as men buy and sell in the market place and social life endures."

What to make of Vail's legacy? Outside official Bell histories, Vail remains a controversial figure for being such a staunch and vocal monopolist. A man who takes a highly diverse and competitive industry and eradicates all competitors is an unlikely hero beyond his own company. Even among the hardest of the hard-grabbing moguls, he has few peers. And so the temptation to paint him as a villain is strong.

Yet if there is ever a logic and a benefit to dictatorship, industrial or otherwise, the verdict on the particular regime must, as Plato suggested,

---

*The technical term for such a system is "corporatism"; in its extreme manifestations it is called "facism."

inevitably depend on how one holds dominion. In this way the implacable megalomaniac Vail might well be redeemed by his sense of great
power's great responsibilities and his avowed dedication to the public
good, to which he always gave far more than lip service. He never pre-
tended that Bell had no choice in how the business was run; he simply
insisted that the non-free-market arrangement yielded higher dividends
for all. He accepted the duties of common carriage, as well as regulated
prices, but in return for monopoly's security and peace of mind. Proportionately, he probably delivered less profit for shareholders than Wall
Street might expect today. Vail was acutely aware of how important the
telephone network would be to the nation, and there is no evidence that
he ever put AT&T's profitability ahead of its obligation to serve. He
presents us therefore with a challenging figure: an unabashed monopolist, but a benign one, who lived up to his own ideals of enlightened
despotism. The fault in this arrangement therefore lay not so much with
Theodore Vail as with the men who would succeed him.

CHAPTER 4

The Time Is Not Ripe for Feature Films

In 1912, a small mustachioed man named Adolph Zukor sat patiently outside the office of the most powerful figure in American film: Jeremiah Kennedy, president of the Edison Motion Picture Patents Company of New Jersey. A Jewish immigrant whose accent indicated the Hungarian village he had left behind at age sixteen, Zukor was the owner of a small movie theater in New York's Union Square, and he had already demonstrated a remarkable power of determination in being granted this meeting. He had an ambitious plan to change American film, but he needed a license from Kennedy's Edison Company to execute it. He would find himself waiting alone outside that office for three hours.[1]

At the time, neither New York nor anyplace else in America was the global capital of the film industry. That was Paris, from which two firms, Pathé-Frères and Gaumont, ruled the world, with the Pathé studios alone distributing twice as many films in the United States as all American studios combined. From 1908 up until the Great War, French dominance gave birth to the first "feature"-length films (longer than twenty minutes), the invention of the newsreel, and the enduring genres of comedy, chase, and melodrama. French directors were the first to put famous stage actors before the camera and enlist well-known composers to write scores. The grandest theater in the world was the Palais

Gaumont on rue Caunlaincourt, which sat 3,400 even before it was expanded to accommodate 6,000.[2]

Meanwhile, despite a substantial role in inventing motion picture technology, the United States was a cinematic backwater. Film was popular, but it remained a novelty, shown in combination with live comedy routines, dancing monkeys, and other vaudeville acts. What American films existed were short—many no longer than a few minutes—with rudimentary plots and no recognizable performers.

French-style film had not yet crossed the Atlantic, and Zukor's plan was to bring the European experience to the United States. It was not as obvious a vision as it might seem: The American theaters, called "nickelodeons," though wildly popular, had a reputation for unpleasantness. As a 1910 article in *Moving Picture World* described the experience:

> I would have been more comfortable on board a cattle train than where I sat. There were five hundred smells combined in one. One young lady fainted and had to be carried out of the theater. I can forgive that, all right, as people with sensitive noses should not go slumming. But what is hardest to swallow is that the tastes of this seething mass of human cattle are the tastes that have dominated.[3]

But Zukor saw no reason that the medium should be an affair for the "seething masses" alone. In 1912, the immigrant who'd made a tidy sum as a furrier saw a perfect pilot project to elevate the U.S. market: *Queen Elizabeth,** a film starring the French actress Sarah Bernhardt, who was exceptionally popular in the United States. So certain was Zukor of this opportunity that he had laid out a small fortune, $18,000, for the American rights to the film. Certain, and grandiose: as he later told journalists, "We believe that we are doing a sort of missionary work for the higher art—that we are aiding in the cultivation of a taste for better things."[4]

But why the meeting with Kennedy of the Edison Company? While it may sound odd today, to screen *Queen Elizabeth,* even in his own theater, Zukor needed a license from the Edison Company. Edison was the leader of the "Film Trust," a cartel of ten firms that, at the time, owned every important American patent on motion picture technology. Using

---

*The French title is *Les Amours de la Reine Élisabeth.*

the power of its patents to decide the availability of films to theaters, the Trust was the de facto arbiter of what films would be shown in the United States.

In many ways the state of film art in America was simply the Trust's vision, under which only the short, the uncontroversial, and the uncomplicated were granted the right of production. As for stars or artistic credits, they were banned. Thus *Queen Elizabeth,* which ran forty minutes and included a marquee performance, while a typical European production, was way outside the bounds for American film.

Kennedy finally admitted Zukor to his office. In his autobiography, Zukor wrote that he wasn't offended by the wait, because at the time he wasn't important enough to be offended. Kennedy listened politely to his proposal, but would do no more. "The time is not ripe for features," Kennedy said, "if it ever will be."[5]

Stuck with a giant investment, Zukor had little choice. He became, in the words of the film scholar James Forsher, "one more outlaw."[6] Zukor began to travel a path that would lead slowly but inexorably to the creation of Hollywood, and all it has meant for America and the world. If Kennedy was the most powerful man in American film in 1912, he little knew that the furrier seated in his office would soon, as the future president of Paramount Pictures, succeed him. *Queen Elizabeth* fulfilled Zukor's hopes, and it anchored a whole new business model premised on "famous players," or what we call bankable stars. Zukor would make for a different sort of despot than Kennedy, implacably driven, assuming a Don Vito Corleone–like status in early Hollywood, mixing favors and intimidation in equal measure. But in 1912 he was simply the ambitious owner of a small theater in Union Square. How he rose to the commanding heights of film is, as we shall see, essentially the story of the industry in America.

ORIGINS OF THE EDISON TRUST

Unless you are a film historian, you probably don't know who invented the movie, at least not the way you know who invented the telephone or the lightbulb. Such ignorance is usually a sign that the inventor was somehow bought out or suppressed, or failed to found his own industry in the manner of Alexander Bell. The American film industry is, rather, an instance of the Kronos effect: most of film's inventors were co-opted by the reigning power of the entertainment industry, such as it was, namely the phonograph. As a consequence, if the American film industry can be said to have had a founder, it would be none other than the godfather of the gramophone, Thomas Edison.

How about the inventors of film technology? In France, a man named Louis Lumière invented a working camera and projector in 1895, though he would soon abandon building an industry around them in favor of seeking out new inventions. Meanwhile, as so often happens, the very year that Lumière invented his projector, in the United States a man named Charles Francis Jenkins, together with a collaborator, invented another one he dubbed the "Phantoscope." By September, the pair had set up a rudimentary movie theater at the Cotton States exhibition, in Atlanta, Georgia.

Like Lumière, Jenkins did not found the film industry, though in his case it was not for want of interest. Rather, it was his partner's decision to sell out to Edison that scuttled his hopes. Edison immediately entered the market with the "Vitascope," essentially the Phantoscope by a different name. Eventually Jenkins had little choice but to sell his own patent interest in the first motion picture projector—for $2,500. "It's the same old story," he would say, years later; "the inventor gets the experience, and the capitalist gets the invention."[7]

Control of the Phantoscope empowered Edison, but not enough to ensure dominance of the emerging industry. Another company, Biograph, soon came out with its own camera, and the industry would be consumed for almost a decade by litigation over conflicting patents.

By 1908, the primary litigants decided to settle their differences by forming the Motion Pictures Patent Company, in the offices of which we first meet Zukor. The Film Trust, as it would be more commonly known, comprised the largest film producers (Edison, Biograph, and others) and the leading manufacturer of film stock, Eastman Kodak. In the name of avoiding "ruinous" competition, this cartel pooled sixteen key patents, blocked most film imports, and fixed prices at every step of filmmaking and exhibition. There was, for instance, a set price per foot of film that distributors would pay producers, another price (originally $2 per week) that exhibitors paid for the use of patented Trust-owned projectors, and so on. So long as its affiliates paid the set rates, a healthy profit was more or less guaranteed. And with all related patents pooled, the Trust was also able to end the acrimonious infringement lawsuits among its members.

Shortly after its formation, the Trust held a series of meetings to introduce its new rules to the rest of the American film industry, most importantly to the "exchanges," as the key distributors were then called, and the major theater owners. In 1909, at one of those meetings, in the Imperial Hotel in New York, sat Carl Laemmle, a small elflike man, barely five feet tall. Laemmle was from Germany, and like Zukor and many other Jewish immigrants of that era, he had made his money in the garment trade before switching to theaters in 1906 or so. Now, Laemmle was attending the meeting at the Imperial as a major film distributor of the Midwest.[8]

What Laemmle heard in that meeting did not sit well with him. Only Trust members would be permitted to make films or import them into the United States, the penalty of doing so unsanctioned being a patent infringement lawsuit. Every theater owner had to hold a license to exhibit films, the licensing fee being $2 a week. And any distributor or theater that broke the rules would be subject to an immediate boycott, denied any access to films. As John Drinkwater, Laemmle's biographer, describes one meeting, "the audience was not invited to express opinions; it was merely ordered to submit."[9]

To go it alone against the Trust was a most daunting prospect. Cooperation, obviously the path of least resistance, seemed much more sensible, and as a major distributor, Laemmle was large enough to have been well rewarded for his assent. While we'll never know for sure, according to his biographer it was pure outrage that goaded Laemmle to undertake industrial combat instead. "He was convinced on the spot that the Trust was in every way an evil thing, menacing the whole future development of the industry," says Drinkwater. To Laemmle, "the whole character of this new tyranny was corrupt and demoralizing—so he believed, and the belief was not captious but a deep, a passionate conviction."[10]

And so on April 24, 1909, Laemmle became the first to openly and publicly challenge the Trust, declaring himself "an Independent."[11] His mission was a long shot at best. And he was risking the simple starvation of being denied films, the sine qua non of his business. He was also inviting patent lawsuits and all manner of personal attacks the Trust might mount. It was in many ways a suicidal path for a moderately successful immigrant businessman.

Laemmle's bold decision, like Zukor's earlier, presents an interesting example of dynamics we have observed before. We have seen how important outsiders are to industrial innovation: they alone have the will or interest to challenge the dominant industry. And we have seen the power of considerations beyond wealth or security—factors outside the motivations of the ideal rational economic actor—in inspiring action to transform an industry. Laemmle's instinctive loathing of the Trust's domination, his desire to be free, would have a deep and lasting effect on American film.

In his 1909 declaration of independence, Laemmle exhorted his "fellow fighters" to denounce the "film octopus," boldly if not quite rationally arguing that the defeat of the Trust was inevitable. The "Independents," he said, "as sure as water runs down hill, will win this fight with flying colours." He called others to join what amounted to a campaign of civil disobedience, including a refusal to pay the $2 per week to "smoke your own pipe." And he made a personal pledge he had no obvious way of honoring: to supply films to any who joined his cause, an "ironclad promise to give you the best Films and the best service at all times in spite of Hades itself."[12]

Unfortunately for him, most of Laemmle's peers, lacking the appetite to fight the Trust, either accepted the rules or gave up the business. In 1910, the Trust began to consolidate the film exchanges by systematically buying them out, acquiring, according to Upton Sinclair, 119 of the 120 major exchanges.[13] Among those deciding to throw in the towel were three brothers, Jack, Sam, and Harry Warner. Harry Warner planned to become a grocer, and so, following an alternative course of history, Warner Bros. might today be a supermarket chain.[14]

Laemmle, however, did have a few allies, among them very useful friends overseas. In 1909, a group of French, Italian, British, and German producers formed the International Projecting and Producing Company, whose goal was to challenge the American Trust that was blocking their imports. And so European productions, of which Zukor's *Queen Elizabeth* was only the first, began to give the Trust's films a run for their money.[15]

But Laemmle's most important supporter was the sole exchange owner who'd refused to sell out, the owner of the Greater New York

Film Rental Company, one Wilhelm Fuchs (later William Fox). Fox was another Jewish immigrant, albeit with a much harder life story than even Zukor's. A destitute childhood on New York's Lower East Side, selling candy and stove polish to support his family, had left him with a dead arm and a paranoid sensibility.

The name Fox continues to loom large in American media, whether as Twentieth Century–Fox, or Fox News, or Fox Broadcasting. And here is the proverbial source of the Nile: an angry rebel with socialist leanings who refused to knuckle under. When Fox rejected their offer, the Trust not only stripped him of his license, but publicly accused him of renting their films to a brothel in Hoboken. But their hope of crushing or embarrassing Fox into submission was a major miscalculation. The Trust's assault energized the man, and with Laemmle, he would emerge as their fiercest opponent in the New York area.

There was a third principal player on the Independent side, a Westerner named William W. Hodkinson, a theater patron turned owner (in Ogden, Utah), and later an exchange man. Hodkinson was a rarity among the key Independents, being neither a Jew nor a New Yorker. He had initially joined the Trust, helping to run the General Film Exchange in Salt Lake City. But he was also something of an idealist, his motto being "Better Pictures, Higher Admissions, Longer Runs, for a Better Audience." Having failed to bring the film powers around to his thinking, he would quit the Trust in 1913, citing its "non-progressive attitude." He would go on to found Paramount Pictures as a competitor to the General Film Exchange, the new firm's enduring mountain logo originating as a doodle from his own hand.[16]

For all their bravado, however, the Independents obviously faced many daunting practical challenges, first among them the Trust's film embargo. Laemmle had promised prospective coconspirators "the best Films" even while his break with the Trust cost him his access to the only films available. However improbable, the only conceivable options were to violate the Trust's ban on imports, or to create their own competing supply. The former—Zukor's course—held its own limitations, and so, not without misgivings, Laemmle and Fox became film producers. Thus was the Hollywood studio born, not out of choice, let alone glamour, but of brutal necessity.

The studio Laemmle opened near Union Square soon began making films as quickly and cheaply as it could, relying on French sources for raw film stock (since Eastman Kodak was part of the Trust). He named his firm the Independent Motion Picture Company (IMP); it would later be known as Universal Studios.[17] Fox soon followed, creating Fox Features, whose first production would be *Life's Shop Window*. At this point they crossed paths with Zukor; fresh from his rebuff by Kennedy, he joined in 1912, making his star-centered films in the European style.[18]

With the rise of insurgent producers allied with Paramount, the industrial warfare reached a new level of intensity. To distribute films illicitly was one thing; but to produce them was to attack the very heart of the Trust's legal monopoly. Merely to operate a camera without a license was to violate patents owned by the Trust. Beginning in 1910, the Trust commenced a scorched-earth legal campaign meant to make an example of Laemmle. Over three years, their lawyers would sue him 289 times. Laemmle's biographer describes the Trust's strategy thus:

> Injunction suits—let there be injunction suits in large numbers, let them flock in from all quarters, let the federal courts and the state courts buzz with them. Scour the country for infringements, set spies on every independent camera, projecting machine, reel of film, that could be found. Let actions breed and multiply . . .[19]

Rarely has an essential tension between free expression and intellectual property been laid so bare, made so explicit, as it was in the Trust's patent suits. The Trust, using its economic power and the patent laws, was able to harness the power of the state in the attempt to destroy its budding competition and their new type of films, leaving them only one choice. Now and again, in the course of the Cycle, a little lawbreaking will prove a useful thing.

THE ORIGINAL WEST COAST–EAST COAST FEUD

As the historian Lewis Jacobs writes, "Independents fled from New York, the center of production, to Cuba, Florida, San Francisco, and Los Angeles. . . . The safest refuge was Los Angeles, from which it was only a hop-skip-and-jump to the Mexican border and escape from injunctions and subpoenas."[20] Whatever it stands for today, Hollywood was once a place for industry outlaws on the lam.

Often, though, film history, written mostly by cinéastes, can tend to romanticize the great move west as something akin to the von Trapp family's escape over the Swiss Alps. Here, for instance, is Maurice Bardèche's account:

> For all their audacity and their ruses, the outlaws faced defeat when salvation suddenly opened before them. . . . They quickly gathered together their cameras, their painted scenery and their make-up boxes and set forth on an exodus to the West. . . . Here were sunny skies which made elaborate studio buildings unnecessary. A few planks, some trees, a bungalow to sleep in, a café for leisure moments were sufficient. If detectives showed up, they could pile actors, scenery and cameras into a car and disappear across the border for a few days.[21]

Less romantically, Tino Balio points out that the Trust itself had been producing films in Los Angeles before the Independents got there.[22] Suffice it to say by the mid-1910s the suburb of Hollywood was clearly the new home for the Independents, with seventy independent production companies, including Fox and Universal, located in the vicinity.

We should pause to ask: What exactly was at stake in this cross-country feud? The Trust was a cartel intent on monopolizing the industry. Economics textbooks portray the harm of monopoly as its tendency to restrict supply and set high prices. But in the case of the Trust, the goal was to make a cheap product, and so the effect was to depress prices—can there be any harm in that? Yes, and in fact this is a case where the greater harm of monopoly reveals itself to be not economic but expressive. The Trust's rules controlled not just costs, but the very nature of what film, as a creative medium, could be. In an information industry the cost of monopoly must not be measured in dollars alone, but also in its effect on the economy of ideas and images, the restraint of which can ultimately amount to censorship.

This is not free speech idealism for its own sake. As we've seen, the ban on most imports kept Americans from enjoying or participating in the developments in film taking place in Europe. The severe domestic limits on length (in general, films were to run ten minutes and could rarely go over twenty) made complex filmmaking difficult, if not impossible. The ban on "stars" was also a hindrance. Hoping to prevent the problem of "celebrity"—an actor's gaining a following that might lead to unreasonable salary demands and higher costs—the Trust unwittingly neutralized the incentive for film acting to develop as an art or a serious profession.

Last and perhaps most damaging was the Trust's arrogation to itself of the role of official censor. The Trust simply did not allow films it deemed inappropriate to be made or exhibited. It took its cues from the National Board of Censorship, a private organization formed in 1909 to review films for immorality. In this judgment, its view was expansive. Not only lewdness could be banned; even scenes that, for instance, made burglary seem easy could violate the imperative of moral uplift.[23] And so the National Board and the Trust, private institutions outside the reach of any constitutional scrutiny or accountability, were in effect America's film censors.

Not that the Independents, though rebels, were ideological crusaders either. They wanted to break open the film industry for their own reasons of commerce, not as agents of free speech. The open era of film was not, as that of radio was, launched by idealistic amateurs. But whatever the motive behind Laemmle and Fox's instinct to fight the Trust, the effect was to blow open a new and incredibly powerful medium of expression and one with greater economic potential than had been allowed before. The film industry, once cracked, would be an extreme example of how an open industrial market and an open economy of ideas can overlap entirely.

As the battle between the Independents and the Trust wore on, the Trust, perhaps increasingly desperate, began taking the law into their own hands. In came the private enforcers, on the theory that "though an injunction will not stop a man from making films, a broken camera will."[24]

THE OUTCOME

In 1912 it was by no means clear whether Hollywood or the East Coast Trust would dominate the future of American film (nor, for that matter, whether American film would be dominant in relation to European). There was no reason to bet on Hollywood. As the historian Paul Starr writes, "The Trust consisted almost entirely of Anglo-Protestant businessmen, and their central figure, Thomas Edison, was an American legend, while the Independents were nearly all socially marginal Jewish immigrants, originally without significant financial backing or political connections."[25] The Edison Company and the Trust had their patents, plenty of money, control of the theaters and distribution, and all of the advantages that go with being a "first mover."

And yet as the 1910s progressed, the Trust grew ever weaker, and the Independents grew stronger. Why?

We can make the matter more perplexing by comparing the struggle in the film industry to what was happening concurrently in the telephone industry. Both featured a group of "Independents" opposing a would-be monopolist, in one case the Edison Trust, in the other, AT&T. And yet we see opposite outcomes: AT&T would bury the farmers and their barbed wire, going on to rule American telephony for decades, while Tinseltown would reduce film's East Coast origins to the subject of a trivia question.

While there was no one key to the film Independents' victory, we might say they won by a process of funded innovation—by guessing right about what the next step in film could be and attracting capital to their guess. Rather than the endless pulp offered by the Trust, the Independents imported big European pictures (such as *Queen Elizabeth*) or produced films of similar ambition and complexity, creating the demand their product was fulfilling.

In *An Empire of Their Own: How the Jews Invented Hollywood,* Neal Gabler makes this point a different way, by comparing WASP and Jewish cultural sensibilities at the time. For the former, movies "would always be novelties." These "aging WASPs," he writes, "were increasingly losing touch with the predominantly young, urban, ethnic audience—the audience from which the Jewish exchangemen and theatre owners had themselves recently risen."[26]

Hollywood's entrepreneurs, moreover, were adept at gaining Wall Street financing, at a time when the idea of a bank funding a cultural product was unheard of. In contrast, the Trust was slow to turn to banking, doggedly relying on its system of fixed prices.[27] But the set fees paid to producers meant an upper limit on budgets, limiting production flexibility and ensuring that a Trust film would never be as unusual or eye-catching a confection as independent or foreign films. As Zukor once put it: "what they were making belonged entirely to technicians. What I was talking about—that was show business."[28]

We might say, more simply, that the Trust, unlike AT&T, did not have the House of Morgan behind it, and perhaps that is all that need be said. For as we shall see, the history of American culture is as often a story of financing as of artistic merit. The Trust overrelied on its prices, the patent law, and lawyers, whereas AT&T relied on its financial power, a much more dependable asset.

As a general rule, cartels try to stay away from courts, just as a fugitive, even one wrongly accused, advisedly steers clear of the police station. Oddly enough, the Trust spent most of its time in court, and that is where it met its final fate. For in an act of industrial jujitsu, in 1912 Fox and Laemmle both filed antitrust actions against the Trust in their defense against the patent lawsuits.

The Trust was in name and fact a tempting target for the antitrust laws, as it made no secret of being a price-fixing cartel, and as James Grimmelmann writes, "the Ninth Circle of antitrust hell is reserved for price fixers."[29] The countersuits gained the attention of the Taft administration, which began its own investigation.

The Trust's defense against price-fixing charges was fascinating and colossally unpersuasive. In court, they openly admitted their purpose of dominating the film industry. But they argued that their existence was necessary both to "improve the art" of cinema and to perform censorship on behalf of the government, fulfilling a "neglected function of the State." The Trust proposed, in effect, that it was due an exemption from the law because, as a private regulator of free speech, it was performing a public service.

The claim to be a surrogate censor probably seemed less bizarre in that jurisprudential and cultural climate than in our present one. Nevertheless, the theory failed to impress the courts, and the Trust was unable to strike a lifesaving deal. Not that there was much to save by then, the Trust's ranks and coffers having been depleted. In 1915, a federal district court finally ordered the tattered Trust be dissolved.[30] The American film industry was, for the first time, an open industry.

As American film opened up, it took off in directions few could have imagined. An industry famous for its lack of imagination entered an era of astonishing creative breadth, soon to challenge Europe as it never could before. The sheer volume of producers and exhibitors now working meant that every genre could be explored to its outer limits, and the demand was there to meet supply. Four thousand, two hundred and twenty-nine films were reviewed by the industry press in 1914 alone (an average of more than eleven new films every day). Specialty films proliferated for every niche market perceived: for blacks, Jews, and Irish, for socialist, racist, anarchist, trade unionist and antilabor. As the film historian Steven Ross writes, "the relatively inexpensive costs of production and the constant demand for films allowed producers to indulge their political sentiments, or those of their directors and writers."[31] Film in the late 1910s through the 1920s was consequently an astonishingly diverse and fecund medium—"as diverse as human thought," to borrow the description a Supreme Court opinion of the 1990s would use for the Internet.[32]

The beginning of the First World War in Europe gave the Americans a wide-open path to global supremacy. The European film industry, like other aspects of the culture, would never fully recover from the Great War, and Paris lost its place as the world capital of film. Once-mighty Pathé was sold off in pieces. George Méliès, the most famous director of the early 1900s, met a harsh fate. With his studios commandeered by the French army, Méliès, desperate for cash, sold his entire film archive to a junk merchant who melted it down to make footwear. In the 1920s, Méliès was discovered selling candy and toys in a booth at the Montparnasse train station.

What happened to the Edison Trust? All its members quickly passed into obscurity or were bought out, with the exception of Eastman Kodak, which had already left the Trust by the time of its collapse. By the 1920s, the cartel that a decade before had ruled American film, seemingly invincible, had been completely eliminated.

The founders of Hollywood, for the most part, went on to riches and fame, including the very first rebels, Laemmle, Fox, and Hodkinson. Their studios—Universal, Twentieth Century–Fox, Warner Bros., and Paramount—continue to dominate American film. But as we shall see, more important still than any of these would be that man waiting outside that office in New York City, Adolph Zukor. It was he, above all, who would manifest that rare trait Schumpeter described as "the dream and the will to found a private kingdom."

CHAPTER 5

Centralize All Radio Activities

It is inconceivable," said Herbert Hoover, secretary of commerce, at the first national radio conference in 1922, "that we should allow so great a possibility for service, for news, for entertainment, for education, and for vital commercial purposes to be drowned in advertising chatter."[1] Hoover's remarks reflected the accepted wisdom of the times: that advertising on radio was unacceptable. That is to say, they reflected what radio broadcasting was in the early 1920s: a decentralized industry founded on a rather idealized notion of an emergent technology, the technological utopia of its time.

Hoover would convene several more such meetings in Washington, D.C., to create a form of self-rule for the broadcast industry. He believed not in law, command, or controls, but rather in what he called "voluntarism."[2] That ideal inescapably implied meetings to build consensus on shared norms in a friendly environment.

According to a report of the first conference, all agreed that "direct advertising in radio broadcasting service [should] be absolutely prohibited." J. C. McQuiston, the head of publicity for radio manufacturer Westinghouse, spoke for many when he wrote that advertising "would ruin the radio business, for nobody would stand for it."[3]

Yet despite Hoover, and the idealism of radio's dreamers, other forces had designs of their own on the future of the medium. Listeners who

"Advertising by Radio Cannot Be Done; It Would Ruin the Radio Business, for Nobody Would Stand for It."

were tuned in to New York's WEAF at about 5:15 p.m. on Monday, August 28, 1922, heard this:

> Let me enjoin upon you as you value your health and your hopes and your home happiness, get away from the solid masses of brick, where the meager opening admitting a slant of sunlight is mockingly called a light shaft, and where children grow up starved for a run over a patch of grass and the sight of a tree.[4]

This, the world's first major radio advertisement, was a promotion for a housing development named Hawthorne Court. In format rather like what we'd now call an infomercial, the spot urged listeners to leave Manhattan for the leafy comforts of Queens. It was also the opening shot in what would become the battle to redefine radio and ultimately to make it a closed medium.

WEAF was the flagship station for AT&T, the telephone monopolist. More than Hoover or any other individual or entity, AT&T, it turns out, would define American broadcasting and entertainment in its inception. Indeed, while NBC sometimes calls itself "America's First Network," Bell actually got there first; by 1924, its National Broadcasting System (NBS) comprised sixteen stations reaching 65 percent of the American homes with radios.[5] To a degree few understand, the mighty broadcast networks, CBS, ABC, and NBC, that would dominate American domestic life in the twentieth century were all ideological descendants of the Bell system.

AT&T had a unique advantage in early radio broadcasting: monopoly ownership of the nation's only practical means of moving sound around the nation, namely, its long distance network. The network built for carrying telephone traffic was perfectly suited to carrying radio programs as well.* As an unanticipated dividend of Vail's adroitness, AT&T was the only company in a position to form an entity the world had never seen before: a broadcast *network*. The value of a network, as opposed to a mere station, is in the power to harness economies of scale. Even in the early 1920s, producing one show for sixteen stations meant that AT&T could pool the revenues from sixteen different audiences to create a single, higher quality product. The network is what made possible the production of broadcast news and entertainment as we would recognize it. The NBS network also made it possible for American presidents, beginning with Calvin Coolidge, to give speeches reaching the entire nation at the same time, the form of political address that would reach its apotheosis with Roosevelt's "fireside chats."

But we are getting ahead of ourselves. The development of AT&T's network, the National Broadcasting System, immeasurably important as it was, was preceded by another Bell first: advertising. Advertising is a force with few peers in the cultural history of the twentieth century, but its significance in the 1920s was to create a new and more sustainable business model for a radio station. Selling radio sets—the old revenue model—was a good if limited business, for ultimately few households would need more than one radio every few years. But advertising revenues could expand indefinitely—or so it seemed then.

Advertising, in time, proved almost a license to print money, and the effects on broadcasting of the revenue model it introduced can scarcely be overstated. It gave AT&T, and later the rest of the industry, an irresistible incentive not just to broadcast more but to control and centralize the medium. To see why, compare the older model: When revenues came from the sale of radio sets, it was desirable to have as many people broadcasting as possible—nonprofits, churches, and other noncommercial entities. The more broadcasters, the more inducement for the con-

---

*The national telegraph wire network was also still in existence, but it was of poor quality; efforts to use it to carry radio transmission were a failure.

sumer to buy a radio, and the more income for the industry. But once advertisements were introduced, radio became a zero-sum game for the attention of listeners. Each station wanted the largest possible audience listening to *its* programming and *its* advertisements. In this way advertising made rivals of onetime friends, commercial and nonprofit radio.

At first AT&T denied any interest in advertising, simply describing its place in the radio business in terms that had saved its telephone hegemony: "common carrier" of the airwaves. As the firm prepared to operate WEAF at 660 AM, it issued an announcement: "Anyone desiring to use these facilities for radio broadcasting should make arrangements with Mr. Drake, general commercial manager."[6] As with the telephone network, for a fee anyone could get on the AT&T radio network and broadcast as they liked. In some sense, the common carriage concept provided cover and plausible deniability of any change in modus operandi: AT&T wasn't advertising—its customers were.

Such caution, too, informed the types of advertising AT&T initially allowed. It barred any mentions of price, or other possibly jarring details such as the color of a package or the location of a store. In consequence, ridiculous as it may sound, many of the first advertisements took a form more educational than commercial. Gillette's first radio ad, for example, was a lecture on the history of beards.[7] In time, NBS would also develop the idea of sponsored programs and acts, among the first the *A&P Gypsies* and the *Eveready Hour*.[8] And so it was NBS that originated "entertainment that sells," and NBS that pioneered radio programming aimed at turning citizens into consumers—the basic formula that has dominated American radio and television for more than eight decades.

Within a few years, the rest of the radio industry was feverishly trying to imitate AT&T's model—no surprise, considering how obvious and overwhelming its advantages were. Advertising and sponsorship gave radio stations a sustainable financial base—real money to pay speakers and musicians, who had formerly worked for free, with all the limitations on quality that that arrangement implies. But there was only so much the competition could accomplish without AT&T's long distance network.

When a utopian, open medium such as radio had been begins to close up, sinister forces may seem to be at work. There is sometimes truth to that impression, an extreme instance being the Third Reich's creation of a centralized broadcast system for propaganda. But just as

often, the closing is driven by a hunger for quality and scale—the desire
to improve, even perfect the medium and realize its full potential, which
is limited by openness, for all its virtues. It was the *Eveready Hour* that
led the way toward broadcast fare of higher quality and polish.[9]

Let there be no doubt that AT&T had a typically clear idea of what
the structure of the radio industry should be. The company saw no
reason not to apply Vail's winning ideals again, envisioning a vibrant,
high-minded radio monopoly to go with its telephone monopoly. As
A. H. Griswold, an AT&T executive, disclosed in a speech in 1923 with
all the can-do hubris of that corporate culture:

> We have been very careful, up to the present time, not to state to the
> public in any way, through the press or in any of our talks, the idea
> that the Bell System desires to monopolize broadcasting; but the fact
> remains that it is a telephone job, that we are telephone people, that we
> can do it better than anybody else, and it seems to me that the clear,
> logical conclusion that must be reached is that, sooner or later, in one
> form or another, we have got to do the job.[10]

To close the loop entirely, AT&T set about designing its own radio sets,
presenting President Coolidge with one of its handsomer models.[11] In a
final stroke, such as to this day inspires heated debate over network neutrality, AT&T's new radios were engineered to receive only AT&T broadcast frequencies—and, not surprisingly, only AT&T programming.*

RADIO RESISTANCE

By the mid-1920s it seemed likely, if not certain, that AT&T would
dominate the radio industry. The firm held the all-important long lines,
and its president, Walter Gifford, was aggressive and, in the mold of his
predecessors, fond of conquest. The only thing standing in his way was
a company that the U.S. government had already sanctioned to monop-
olize radio, just as AT&T had been granted a warrant to rule telephony.

---

*Actually, the motivations for the exclusivity were complex. One reason, obviously, was to
favor AT&T's stations. But AT&T had also joined a radio patent pool in the 1910s that arguably prohibited it from manufacturing radio sets; the fixed frequencies were seen as grounds
for an exception to this prohibition.

And so the clash that was shaping up for the future of broadcasting would be, if not quite one of fellow titans, substantially different from the war Bell had earlier waged against its Lilliputian rivals in telephony.

We first encountered the Radio Corporation of America ringside at a boxing match in 1921, but this strange creature needs a better introduction via a few historic analogues. Structurally the RCA was rather like the BBC, a national champion; but unlike the British company, it was neither established nor sustained with public duties. Rather, in 1919, the RCA was formed mainly in response to the navy's insistence that all vital radio technologies be held by an American firm, in the interests of national security.[12] And so RCA was fashioned out of the existing American Marconi Company to pool and exploit the rights to use more than two thousand patents owned by General Electric, United Fruit, Westinghouse, and AT&T. In consideration for the licenses, General Electric was made majority owner of RCA, but AT&T and Westinghouse also had substantial stakes. Hence one of the odder features of the contest for broadcasting: AT&T was in a battle with its own property.

RCA's general in the battle with AT&T was David Sarnoff, a genius of industrial combat also present at that boxing broadcast in 1921, who shall play a recurring role in this drama.[13] Sarnoff was in midcareer, a rising star within RCA, when he was suddenly presented with a chance to take on AT&T and become the defining mogul of American broadcasting. The metaphor has been used before, but Sarnoff loved to imagine himself David confronting Goliath. For his part, AT&T's Gifford at first refused outright to negotiate with Sarnoff, whom he is said to have declared an "abrasive Jew."[14]

While AT&T was a phone company first and foremost, it was also the larger and more aggressive of the two champions, and it seemed to hold a decisive advantage: ownership of *the* network, the nation's only quality long distance lines. Against AT&T, RCA would face some of the same problems of access and interconnection that had doomed the telephone Independents in the 1910s.

Let us pause to imagine what things might look like if Sarnoff had not been able to find a way to achieve what seemed impossible and AT&T had won the battle for radio. Imagine that nearly every radio station and every radio set in America was AT&T's, along with every telephone and wire. The power the phone company would have had over American culture and communications is beyond comparison in the annals of democracy, comparable in structure only to what the fascist and Communist regimes in Europe were creating.

But back to the story. Sarnoff needed a network to compete with AT&T, but there was no obvious way to get one. Despite its common carrier pledges, Bell denied any rival radio station access to its wires. According to one Sarnoff biographer, a Bell executive told him, "Transmission by wire is ours. Stay out of it."[15] RCA did experiment with leasing parts of the (lower quality) telegraph network to carry programming, but the result was "a loud buzz."

And so Sarnoff conceived a tactical shift. As mentioned earlier, AT&T was prohibited from manufacturing radio sets: it had signed an agreement with RCA that said it has "no license . . . to make, lease or sell wireless telephone receiving apparatus except as a part of or for direct use in connection with transmitting apparatus made by it."[16]

On the arguable ambiguity of that bit of legalese Sarnoff decided to stake his company's future. Under the terms of the agreement, he brought a secret binding arbitration proceeding against AT&T, contending that its new radio sets violated the conditions of RCA's license pool.[17] He was either lucky or a more astute reader of contractual language than other executives, for after hearings, the arbitrator found not only that AT&T was violating the patent agreement by manufacturing radio sets, but that its broadcasting activities were illegal as well.

Unfortunately for AT&T, the arbitrator's ruling coincided with their losing another crucial suit over the patent for the vacuum tube, without which they could no longer manufacture radio receivers or transmitting equipment. Topping off these woes, at about the same time, the Federal Trade Commission, a new agency created to enforce the antitrust laws, launched an investigation into the radio industry.

Obviously, AT&T had a lot to lose from another brush with any antitrust enforcers. Nonetheless, unwilling to concede defeat, the firm struck back with a report claiming violations of the statute on the part of RCA. The absurdity was lost on no one: AT&T, the state-sanctioned telephone monopoly, was accusing another state-created monopoly, RCA, of being an illegal trust, with the transparent aim of blocking RCA from entering a market RCA had been created—with the express cooperation of AT&T—to exploit!.

This was not Adam Smith's vision of competition, nor even Schumpeter's, but rather American industrial policy gone amok. And all of this maneuvering that could have so altered American communications and culture transpired behind the scrim of corporate confidentiality, not to be made public until scholarly investigation decades later.

Despite its initial bravado, sometime in 1926, AT&T would lose its belly for the radio fight. The reasons have never been fully elucidated, but it is clear that by this time the opponents were caught in something of a "prisoner's dilemma." One or the other side could have tried to gain the upper hand by going public with its charges, inviting the possibility of a long and costly federal lawsuit. Or they could strike some kind of deal in secret. They chose the latter option. The two firms decided to work together on a new national broadcasting service, based on Bell's NBS. AT&T would sell its network and stations to RCA, preserving its long distance networking, while RCA took care of everything else. Though the settlement was in both firms' interests, there is no question but that AT&T had blinked and that the deal was a major victory for Sarnoff, who, just as Bell had done in scaring off Western Union in the 1870s, used the law to prevent AT&T from dominating radio completely.

So while radio was supposedly developing in the United States without direct government intervention, contrary to the British model, in fact it was a case of two government-sponsored champions dueling over the same industrial prize in a decidedly unbloody bout. For in the end, AT&T's National Broadcasting System never died, but simply morphed a bit. Walter Gifford and David Sarnoff, finally on speaking terms, relaunched the entity under an almost imperceptibly different new name:

ANNOUNCING THE NATIONAL BROADCASTING COMPANY, INC. . . . . The purpose of that company will be to provide the best programs available for broadcasting in the United States. . . .The Radio Corporation of America is not in any sense seeking a monopoly of the air. . . . It is seeking, however, to provide machinery which will insure a national distribution of national programs, and a wider distribution of programs of the highest quality.[18]

NBC had been born, and with it a new ideal of American broadcasting.

A New American Model

"Commercialism is the heart of the broadcasting industry in the United States," wrote Henry Lafount, a commissioner of the Federal Radio Commission, in 1931.[19] By the 1930s, times had indeed changed in American radio. What was once a wide-open medium, mostly the province of amateur hobbyists, was now poised to become big business, dominated by a Radio Trust; what was once an unregulated technology would now come under the strict command and control of a federal agency.

The rise of the AT&T/NBC model led directly to this transformation. Through most of the 1920s, the regulation of American radio had been light, with Hoover's vision of voluntary virtue resting on the hope that goodwill made formal rules unnecessary. With networking and advertising the new keys to financial viability, however, the larger broadcasters and manufacturers of radio sets had no use for government evenhandedness. They wanted, rather, a government policy that would aggressively favor commercial broadcasting. It may seem surprising to regard Hoover as a naïve idealist, but in this context that's what he was. The companies that had, while small and dependent on the sale of radio sets, been perfectly happy with Hoover's rule now launched an attack on his authority. In 1926, Eugene McDonald, president of both the National Association of Broadcasters and Zenith Corporation, accused the president of "one-man control of radio" and called Hoover a "supreme czar." Deliberately flouting Hoover's rules, McDonald began using frequencies reserved for Canadians, provoking a potential fight with the British Empire. Hoover had no choice but to order him to stop, but McDonald sued to challenge Hoover's right to do so, and a federal district court found that Hoover, all along, had lacked any authority to assign radio frequencies.[20]

It was in the wake of Hoover's defeat that, in 1927, Congress saw the need to create the Federal Radio Commission, a congressional agency of enormous importance in our broader narrative, as the only body dedicated to the problems of communications in the United States. Unfortunately, the FRC was tainted from the beginning, its policy closely wedded to the interests of NBC and the navy. Congress's prime concern in forming it seems to have been denying Hoover, already a promising aspirant to the presidency, too much power over broadcasting; hence the creation of an independent commission, as opposed to authority in the Commerce Department.[21] They might equally have sought to set up a commission with a public service mandate like the BBC's, or even one to preserve the diversity of radio broadcasting. But they did neither, instituting instead only a new bureaucracy widely seen as captured ab initio.[22]

After a period of difficulty finding staff, the FRC, founded to favor "general" broadcasting, almost immediately set about jackbooting its way where Hoover had trod so lightly. Hoover had pictured himself a careful gardener, trying to cultivate commercial, educational, and other nonprofit radio stations on the same dial; the FRC saw its mission as more to plow up the radio dial, making way for a bigger and better radio of the future. In effecting its program of clearing the airwaves, the agency relied on a new distinction between so-called general public service stations and propaganda stations. These were, in effect, synonyms for "large" and "small" respectively, but it was apparently easier to assault the underdog if one could label him a propagandist, even though the term's present pejorative sense would not take hold until used to describe communications in Nazi Germany. In any case, by whatever name, the FRC favored the large, networked stations affiliated with NBC (and later CBS). Because the large operators had superior equipment, and fuller and more varied schedules, the FRC could claim not implausibly that they better served the public.[23]

As the commission would soon announce, "There is not room in the broadcast band for every school of thought, religious, political, social, and economic, each to have its separate broadcasting station, its mouthpiece in the ether."[24] So declared the commission in the course of shutting down a well-known station in Kansas famous for its medical quackery.

What is immediately striking about this pronouncement is how much it reads like a calculated antithesis of the First Amendment. Less visceral analysis reveals it to be based on a false technological premise. It is true that interference was a problem. Without *any* order to the radio dial, no station could be heard. But the FRC had a real choice of whether to back more low-power stations, or fewer high-power stations. There was, in fact, room on the broadcast band for every school of thought, if broadcast rights were confined to localities and lower-wattage transmitters. It was simply a matter of how one envisioned dividing up the ether.

The FRC's views closely aligned with those of RCA, NBC, and the rest of the industry, which now depended not on more radio stations, but on huge audiences for just a few stations. Government's mission had become to free up frequencies to make room for stations that could reach huge areas, or the whole nation at once—so called "clear channels." With its General Order No. 32, the FRC demanded that 164 smaller stations show cause why they ought not to be abolished.[25] The commission went further with General Order No. 40, which reset the entire radio dial, shuttering or reducing hundreds of small stations to create forty nationwide clear channels, and cramming the remaining six hundred channels into fifty leftover frequencies. Following No. 40, writes Robert McChesney, "U.S. broadcasting rapidly crystallized as a system dominated by two nationwide chains supported by commercial advertising." Commissioner Lafount described it as the "structure or very foundation" of American broadcasting, and indeed it was.[26]

And yet there was much to be said in defense of the new. The networks of the 1930s can be credited with creating a broad listenership for quality programming, such as the famous radio serials of the period. Reflecting the AT&T ideal of enlightened monopoly, perhaps, the networks also carried some sense of public service, with every station in theory a trustee of the public airwaves. So in addition to the entertainment designed to sell products, the networks broadcast "sustaining programs," money-losers run in the public interest. From this concept grew their news departments, also unprofitable but serving the public good.

By the mid-1930s, it was clear that the Cycle had turned with respect to radio, and the medium was completely transformed. The days of the freewheeling American dial were over. In fact, so it was around the world, as virtually every nation began to regulate radio, abandoning the decentralized way of the early American experiment, in many cases without ever having passed through it.

The most striking example, of course, were the Germans, who moved directly to the centralized radio model in the 1920s and by the 1930s had installed radio broadcasting as the centerpiece of the Nazi state's propaganda campaigns. Joseph Goebbels, Hitler's propaganda minister, saw the radio as a central instrument in achieving *volksgemeinschaft*, the unified national community. "A government that has determined to bring a nation together," as he put it, "has not only the right, but the duty, to subordinate all aspects of the nation to its goals, or at least ensure that they are supportive." For Goebbels, industrial structure was a critical part of making this happen. "Above all," wrote Goebbels, "it is necessary to clearly centralize all radio activities."[27]

The fate of open radio gives credence to the inevitability of the Cycle, yet we can also see how much of what happened was a matter of choice. There were some attractive features of early American radio worth preserving, and they could have been preserved given less heavy-handed support of the new paradigm. But the defenders of those virtues, Hoover and a few senators, lacked the political clout to prevent a wholesale flip from an open to a closed system. The American government ended up failing to affirm a considered vision of what broadcasting should be, only following and accommodating the evolution of business models. Once the industry had concluded that its profits could be maximized if more people listened to fewer stations, the government, acting as if the business of America were only business, did the industry's bidding, showing only the most feeble awareness of its consequences for the American ideal of free expression.

As the years went by, the founders of the commercial system would begin to credit themselves, and not the amateurs, for the creation of American radio. Sarnoff, as head of RCA and the founder of NBC, made himself the defining mogul of American radio. He began spinning vainglorious tales for reporters and historians that he had been first to envision radio broadcasting in 1914, that the Dempsey bout broadcast had been his idea, and that he had pioneered the national broadcasting network. The amateur hobbyists and inventors like Lee De Forest— even AT&T, for that matter—were brushed out of the official portrait, as Sarnoff proceeded like the ancient Chinese emperors who rewrote history as soon as they came to power, to prove they had had Heaven's mandate all along.[28]

CHAPTER 6

The Paramount Ideal

Since 1909, when it had opened boasting "the world's finest theatre pipe organ," Tally's Broadway had become Los Angeles's leading "first-run" theater, *the* place to see the latest and best.[1] The proprietor, Thomas Tally, was a true forerunner in the film industry, even credited by *The New York Times* with coining the term "motion picture." He had also been a stalwart ally of the Independents in their fight against the Trust, faithfully subscribing to W. W. Hodkinson's "Paramount Program" of thoughtfully selected Independent offerings. But for all Tally's loyalty, in 1916 a salesman from Paramount visited him with some most unwelcome news.

Things had changed at Paramount since founder Hodkinson's expulsion and the firm's merger with a group of producers. The new management was now offering terms very different from those Tally had known before. Hereafter, if he wanted Paramount's "star" films, he would be required to buy en bloc—a full year's worth of films, all from Paramount's production partners.[2] He would, moreover, be obliged to buy the films "blind," or without preview.

The agent represented Adolph Zukor, the onetime rebel Independent who that year had taken over Paramount, implementing the star system, whereby recognizable names became the essential asset in film. Among the essential assets he controlled, the greatest was Mary Pickford, the most popular actress of the 1910s and the anchor that made his block sales model feasible. "As long as we have Mary on the Program," one of Paramount's salesmen said, "we can wrap everything around her neck."[3] To get Mary, you had to buy the block.

While the Independents had just recently broken the chokehold of the Trust, Zukor was now showing every sign of wishing to reestablish empire, but with himself as the presiding mogul of American film. In some sense the latter objective was already in hand. Having engineered the takeover of Paramount, merging his Famous Players production studios with Paramount's distribution might, he was now president of the largest motion picture company in the United States, if not the world. While never public about his ambitions—indeed, he spoke out frequently about the "evil" of combinations—his actions left little room for doubt.

In 1917, Tally and some like-minded theater owners decided to defy Paramount. After a meeting in New York they announced the formation of the "First National Film Exhibition Circuit." The group comprised twenty-six major exhibitors, from San Francisco, Chicago, Philadelpia, Boston, New York, and other cities. At the meeting was Samuel "Roxy" Rothafel, manager of what was at the time the largest theater in the United States, the Strand on Broadway. First National's goal was simple: "to find means of repressing Zukor before he could acquire dictatorial power."[4]

The stage was set for another of the great industrial battles over an information medium, by now a familiar sort of contest, though in truth, no two will prove entirely alike when one considers the specific distribution of advantages and blind spots, the array of heroes and villains, or the distinctive consequences for American culture. In this instance, we have on the one side the Independent theater owners—a disparate group numbering in the thousands, not so formidable individually but, for now, holding collectively the preponderance of power in the industry. Their opponents, far fewer, were the new generation of big producers who had supplanted the Trust, including William Fox, Carl Laemmle, and, with his ascension at Paramount, the greatest of them all, Adolph Zukor.

At stake: not just control of the industry—Would it be open or closed to all but a handful of studios?—but also the character of the medium: Would it continue to be varied and independent, tailored to a variety of

sensibilities, or produced at previously unexampled scales and national-
ized, so to speak, for a single homogeneous audience?

AN IDEOLOGICAL CONFLICT

William W. Hodkinson had had his own strong ideas about the ideal
structure of the film industry.[5] Readers will remember he'd been the odd
man out, the gentile from Utah who became a key ally of the Jewish
immigrant insurgent leaders. Having started as a member of the Trust,
Hodkinson, with an ornery streak, came to believe strongly that every
"layer" of the film industry should remain separate—in other words,
that producers should focus exclusively on making films, exhibitors on
running theaters, and distributors on bringing the two together. Otherwise, he concluded, the quality of film would suffer: "The history of the
business has shown that the most successful pictures have been developed by individual efforts rather than by mass production."[6]

What Hodkinson opposed is what economists call *vertical integration*—the stacking, as it were, of the parts of an industry that perform
different functions (here, production, distribution, and exhibition) to
create a consolidated single entity. (The phenomenon is distinguished
from *horizontal integration,* the more common effort to dominate a single function, in the way that Bell progressively took over nearly every
telephone company, different firms that were doing the same thing but
in different markets.)

Hodkinson's Paramount, in its original form, was composed of eleven
distributors, collectively serving as America's first national distributor of

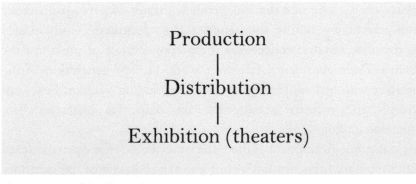

Production
|
Distribution
|
Exhibition (theaters)

The structure of the film industry

feature-length films, and as such a crucial link between the Independents and the market.[7] Paramount advanced funds to the producer in exchange for exclusive distribution rights, and all relations were contractual. By furnishing such security, Hodkinson thus hoped to encourage longer films of higher quality, according to his motto, "Better Pictures, Longer Runs."[8]

When the Trust began to crumble, Adolph Zukor soon became Hodkinson's ideological rival among the Independents. As allies against the Trust, they had been rather like Trotsky and Stalin—united merely for convenience of the revolution, the former preoccupied with a great vision, the latter with power. But Zukor's plan for the film industry was predicated on achieving a system of mass production not much different from that favored by other magnates of the time, such as Henry Ford. The central idea was to control as many parts of the business as possible. In steel, that meant owning the mines, the trains, and the mills. In film, it required owning the talent—stars, directors, and writers—as well as the studios, the distribution networks, and, ultimately, the theaters.

Hodkinson believed in what is sometimes called craft, or authorial filmmaking, wherein one creator did nearly everything, writing, directing, producing, and casting his own film.[9] He was, in fact, among the original backers of a tradition that we identify now with directors like the Coen brothers, Peter Jackson, Woody Allen, and Francis Ford Coppola. In contrast Zukor saw not craft but the latest methods of production as the true stock-in-trade. He would come to promote the "central producer" model, concentrating most of the decision-making authority in the producer rather than the director. With streamlined production and virtually guaranteed audiences, films could be grander and more elaborate than ever. It was a new idea for a cultural industry: there was no need to settle for the meager profits of the nineteenth-century model still ruling the stage; with the twentieth-century methods of production, one could have a balance sheet to match!

In the Hodkinson-Zukor divide emerges a recurrent contest over American culture. Zukor's ideology foresaw one mighty firm in control at every level, coordinated, integrated, and centralized, with one mogul at the hub, in this case himself. Where Hodkinson's revolt against the Edison Trust was abstractly conceived as a break with the machine, Zukor's ambitions, no less the product of his times, were more nearly a usurpation, a revolt to replace one machine with another. As the historian Lewis Jacob writes, "To be important a thing had to be big—and so the movie became one of the biggest things in American civilization."[10] Everywhere, businesses were growing into giant, consolidated operations, powered by Wall Street money. It was no giant leap to imagine film as a national monopoly, and Zukor planned to make it his own. For despite every disadvantage in youth he believed himself destined for greatness in the land that had received him. This meant success not in something insubstantial as culture but a respectably large American enterprise, such as he dreamed movies could become. Those born into the ruling class, like the founders of the Edison Trust, could afford to lord it over a cut-rate cash cow. Zukor's ambitious self-creation called for something grander.

In 1916, Zukor's proposal of an alternative machine called into question what the Independents had stood for in their heroic revolution: Throw off domination, or simply throw off East Coast think-small domination?[11] Hodkinson, for his part, was in no doubt, baldly refusing to give up on the "Paramount Ideals." He would to the bitter end continue to regard the curatorial model of funding and distribution as essential to the creative vitality of the industry. In a later interview, he would continue to insist on the distributor as a "neutral" middleman, rather than one "trying to pull the chestnuts out of the fire for some producer of unworthy pictures or giving exhibitors something for nothing."[12]

Among the agents Zukor dispatched to sell his vision was Benjamin Hampton, who would later write an important history of the period. Hampton recounted how he pressed Zukor's case to Hodkinson: "I said bluntly that the owners of Paramount were in business to make money and that his adherence to what he called 'Paramount ideals' would come to nothing." Zukor, he warned, would simply buy out the partners and force him to quit. Not that Hodkinson was unaware what he was up against; he simply didn't care. "I am right," he replied, "and if I'm put out of Paramount for being right, there will be another place for me in the industry." Finally, as Hampton relates, "he turned in his revolving chair and gazed at the pigeons wheeling above the marble portico of the Astor Library. Silence settled on the room. He could not change his point of view."[13]

Zukor stealthily made good Hampton's prediction, buying out a majority of Hodkinson's partners. On July 13, 1916, at a dramatic board meeting, Hodkinson was ousted from the presidency in a single vote. The new president of Paramount, Hiram Abrams, announced: "On behalf of Adolph Zukor, who has purchased my shares in Paramount, I call this meeting to order."[14]

With one of the original proponents of separating production and distribution gone, Zukor made himself head of the new combination, the first major integrated studio in America and now the largest film corporation in the world.*

We have now seen something of Zukor in action. It is worth studying the man a bit more in establishing a profile for the defining mogul archetype that figures so often in this narrative. He was, as we've said, a figure from central casting, and if we were making a film, he might be rejected for any role requiring a measure of nuance: an orphaned immigrant Jew of small stature but pugnacious, who spent his teenage years boxing larger foes. The newspapers would call him the "Napoleon of Motion Pictures," but what he sought was something more akin to boss-of-bosses status in the industry, dispensing decisive preferments and punishments in accordance with his personal code of honor. He liked to operate in secret, leaving others to wonder what he was up to until he sprang his plans. "I began to compare him with the many industrial and financial magnates whom I had met," wrote Hampton. "I soon decided that nothing like Zukor had yet appeared in America." The director Cecil B. DeMille recalls "the steel and iron, the indomitable bravery and driving determination, in that little man. . . . There would come a time when he would put his two clenched fists together and, slowly separating them, say to me, 'Cecil, I can break you like that.' "[15]

THE BLOCK AND THE CHAIN

With his takeover of Paramount, Zukor became a de facto leader of the Independents, arriving nearly where he wanted to be: in the place of prime mover of motion picture. What lay between Zukor and the

---

*The new combination, called the Famous Players–Lasky Corporation (Jesse Lasky being the one of eight partners to get a credit), would continue to distribute films under the Paramount Pictures trademark, until the entire firm came to be known as Paramount Pictures.

absolute control over film he craved? Not his fellow studio executives. True, these rebel comrades, including William Fox, head of Twentieth Century–Fox, Carl Laemmle at Universal, MGM, and the Warner brothers, were technically his competitors, but in effect they were still operating jointly as a cabal, their common roots on New York's Lower East Side having forged a lasting if unofficial (or at least undisclosed) unity. No, Zukor's real opponents, as we've said, were the theater owners, recently organized in the First National Exhibitors Circuit. At 3,600 strong, they were still a mostly disaggregated and diverse mob, but as they held the power to decide what would be exhibited and what wouldn't, they controlled most of the industry's revenues.[16]

Soon after organizing, Tally and the First National leadership demanded an end to block sales and blind booking, backing up the move with a boycott of Paramount, the primary purveyor of film blocks. By the summer of 1917, Zukor had blinked. His official statement in *Photoplay* magazine said: "After August 5, 1917, any theatre in America can secure Paramount Pictures and Paramount Stars just as it chooses to book them. . . . The Restrictions are Off." Flexing its newfound muscle, the theater coalition went further, making a startling announcement in 1918: First National had acquired film's two biggest cash cows, Charlie Chaplin and Mary Pickford, the latter "America's Sweetheart" and the beating heart of Zukor's business.

There was no great cleverness behind the coup, only more money and creative freedom. Tally's group offered Pickford and Chaplin each an unprecedented million-dollar contract, together with the right to make any films they wanted. Details of Chaplin's deal survive: eight two-reel films a year, at $125,000 per film, and his own studio on Sunset Boulevard.[17]

Zukor, despite his wealth, could not match these offers or the financial might of the theater owners. As Hampton relates, Zukor tried promising Pickford $250,000 simply to retire for five years, but she refused, saying, "Oh, I couldn't do that, Mr. Zukor. I love pictures, and I'm just a girl. I couldn't quit now."[18]

As was his custom when not taking action, Zukor would remain silent for some time after First National had stolen his stars and broken his block booking program. In hindsight, we can see that it was during this quiet interval that the entire plan for the Hollywood studio was hatched in Zukor's head. He came to understand that if his studio and the others wanted to control the entire industry, they would need to control distribution. And so he blazed the path that the rest of the studios would soon follow.

In 1919 Zukor made a bold move, issuing a $10 million stock offering, unheard of by the standards of the film business, and with those funds, as the 1920s began, he launched a direct assault on the exhibition industry. His plan was to acquire every theater he could, build new ones, and reintroduce block bookings. As was his wont, however, he disavowed being up to anything at all, insisting that Paramount "had no desire to enter into the exhibition business unless forced to do so because of a lack of proper theaters in a particular community, or because of our inability to obtain proper representation for its product."[19]

In this struggle between Zukor and the theaters we see the lineaments of a classic contest between a single, integrated firm and a disaggregated, loosely allied, industry; this opposition reveals what the economist Mancur Olson has described as the organizational advantages of a small group (like the Hollywood cartel) over a group comprising thousands (the theaters).[20] Zukor, as the leader among the studios, adopted the very soul of divide-and-conquer. He established a presence in large cities, but without trying to take on the powerful theaters in New York or Los Angeles. Instead, his campaign targeted the weakest links in the theater alliance, in small towns, particularly in the South, New England, and the Midwest.

In New England, Zukor joined forces with a man named Alfred Black, who "was so successful in persuading or frightening exhibitors that he soon had thirty or forty houses in New England."[21] For his Southern strategy Zukor relied on a team headed by a tough character named S. A. Lynch. As Hampton writes, "the methods of the energetic Lynch in dealing with theater owners were hardly gentle. His emissaries soon became known by such names as 'the wrecking crew' and the 'dynamite gang.'"[22]

Lynch crisscrossed the South like Sherman and presented the top theater in each town with a choice of either the chain or the block: either sell out and join the "chain," or else agree to exclusive or preferential showing of Paramount films—the "block." Those who would not comply were punished with a new theater built next door, or threats of worse. Here is how the Saenger Amusement Company, a Lousiana theater operator, described the experience in a full-page newspaper ad: "The methods they are using are as near Bolshevism as anything I know of. They hope to gain a hold for each tentacle of their octopus by threats and brute financial force."[23]

In this, the first use of the *chain* model to destroy independent theaters, we see the application of Henry Ford's credo of central organization to the point of sale.[24] The chain approach had been pioneered in the 1910s by firms like A&P and Woolworth's, decades before Walmart would use it to conquer retail to a degree previously unimaginable. But taking the chain concept to the film industry was an entirely radical innovation, one introducing consequences, cultural as well as economic, that continue to this very day.

As he began to build and acquire theaters, Zukor also infiltrated First National, and by 1921 he had, in addition to more than three hundred theaters, three seats on First National's board. There were rumors (later proved true) that he had bought out the members he supplanted. Soon the rest of the First National board were thoroughly "demoralized," as the historian Richard Koszarski writes, and "no longer sure which of their number was now in the enemy camp."[25]

As his campaign wore on, other facets of Zukor's character appeared. Meeting obstacles and resistance, denounced by the theater owners at every turn, he nonetheless refused to be cast as the villain, instead displaying a fascinating adroitness at placating and befriending his enemies. One memorable incident of 1921, for example, has Zukor arriving alone at the organization of independent producers. He apologized for the excesses of the Southern strategy. "Tears rolled down his cheeks," writes Hampton, "as he declared his lack of personal responsibility for the acts of oppression committed by the Black and the Lynch crews."[26] Perhaps deep down he had some misgivings, betraying those with whom he had fought the Trust. As recently as 1918 he had been fulminating that "the evil of producing and exhibiting coalitions is one of the gravest perils that has ever confronted the motion picture industry. If the business is to progress, it must advance on the basis of free and unhampered selection of product for exhibitors, large and small."[27]

Strange words from the man who founded the Hollywood cartel and destroyed independent film in America. If ambivalence ever troubled his mind, it never stayed his hand.

THE BLOCK

Zukor would continue for two more years before his campaign of theater buying and reintroduction of the block attracted federal attention.
In 1921 the Federal Trade Commission announced Complaint No. 835,
an investigation into the trade practices of Adolph Zukor's Paramount
Pictures.[28] Zukor, the complaint alleged, had conspired to "monopo-
lize the motion-picture industry, and to restrain, restrict and suppress
competition in the distribution of motion picture films. . . ."[29] The
gathering of evidence began, and the investigation would last for years,
centering on questions never answered to everyone's complete satisfaction, even to this day: Was block booking really such a bad thing? Why
did the exhibitors oppose it so vehemently?

Interestingly, it was W. W. Hodkinson, the deposed founder of Paramount Pictures, who had first introduced a prototype of block booking
through his "Paramount Program" in 1914—a move he would live to
repudiate, writing in an essay: "I am a Frankenstein. I created the thing
which has grown into an uncontrollable monster!"[30] And at the time he
used it, the practice was actually welcomed by theaters. But by the late
1910s, after Zukor's rise and thereafter, the independent theaters vocif-
erously rejected being forced to buy films they did not want. And the
studios had begun to insist upon the defense they would ever after cling
to: block booking was simply a form of bulk sales, such as any scaled-up
modern industry depends on. As large, modern operations, the studios
could not be expected to tailor their menus to the tastes of thousands of
independent theaters.

There was a crucial difference between Hodkinson's block sales and
those of Zukor et al.: under Hodkinson, the distributor chose the best
films he could find and sold them as a package. For years after Zukor
had combined distribution and production, however, theaters would
complain of the block as merely a device for coercing them to buy many
third-rate films just to get a few good ones.

At some level, what most galled the theater owners was not being
sold a bill of goods, literally, but the loss of discretion. "The exhibitors
demand that they be given a voice in the selection of entertainment
for their people," wrote one of them, a P. S. Harrison, in 1935.[31] The
theaters had lost not only a say in the business decision of what to carry

but also their cultural power to curate: to promote tastes and views, to fit their programs to local audiences. Here we see a rift that will appear in virtually every information industry, the fault line between the virtues of centralized and of decentralized decision making, between the imperative to produce at scales that justify production costs and the desire for variety.

And it was not exclusively a matter of taste. Men like Harrison were less vocal about the right to choose the films they would show than about the right *not* to show those they considered objectionable. "The question," he wrote in *The Christian Century*, "is whether the American people will continue vesting in a small clique of picture producers the right to control a medium which has so much influence upon the lives and minds of the people, particularly upon the minds of our young men and women."[32] In essence, then, some theater operators opposed block booking because it denied them the right to censor films for their audiences.

As we shall see, the Supreme Court would in 1948 and again in 1962 agree with Harrison and other independent exhibitors that block booking did indeed violate the antitrust laws.[33] How? By "add[ing]," the Court found, "to the monopoly of a single copyrighted picture that of another copyrighted picture which must be taken and exhibited in order to secure the first."[34] Most economists who have studied block booking since the 1960s, however, have tended to defend it as harmless and in some ways efficient. Most famously, in 1963 George Stigler, a Nobel Prize–winning star of the Chicago school of economics[35] disputed the idea that bundling could "extend" a monopoly, arguing it did not confer any advantage or leverage a firm holding copyrights didn't already have.[36] In 1983, the economist Benjamin Klein suggested as justification the avoidance of "oversearching"—the time and expense of bargaining over particular films, which he called "goods of uncertain and difficult to measure quality."[37]

Stigler and Klein might be right that block booking cannot, by itself, extend or expand the monopoly power of a particular copyright, and that selling by giant lots can be efficient for a studio supplying thousands of theaters. But what they neglect to consider are the potential consequences for the nature of the product itself.

Suppose the block sale is undertaken by an oligopoly of the top five producers? That in fact is what happened. There was no explicit agreement among the studios to sell by the block. But the absence of collusion doesn't change the impact of a practice if all subscribe to it. And so, beyond any effects on receipts, a consequence of the general adoption of block booking may have been to displace from prime venues any films not produced by the major studios—to fill, as it were, the shelf space with industry products, a practice I call *parallel exclusion.* In fact, this was one reason the Supreme Court mentioned for barring the practice in 1962: that exhibitors "forced by [the studios] to take unwanted films were denied access to films marketed by other distributors. . . ."[38]

Again we must confront the reality that cultural and information industries pose special problems for standard industrial analysis, complicating the rules of supply and demand by virtue of the product's less tangible forms of value. We might understand perfectly well how block booking and vertical integration reduced the costs of industrial production, while understanding nothing of what these innovations meant for film as a form of expression. Interestingly, when it comes to products like film, such inefficiencies as "higher search costs" might be a good thing, if the result is greater variety in what gets seen and heard. As the film critic Pauline Kael wrote in 1980, "there are certain kinds of business in which the public interest is more of a factor than it is in the manufacture of neckties."[39]

THE TURNING POINT

Few realize that 1926, before the triumph of the "talkies," was the turning point for American film. In that year Zukor scored his greatest victory—the one that finally broke the independent theater industry—with his takeover of Balaban and Katz, Chicago's most powerful independent theater chain, the Midwestern backbone of First National.[40] With the fall of Chicago, the fight was essentially over. Zukor now had direct control over more than a thousand theaters, including many of the most important.

Zukor's Publix Theater Corporation, a Paramount subsidiary, was now the first true national theater chain. At its height, it claimed total dominance in the South and Midwest, as well as considerable power everywhere else. Every day, 2.5 million customers came through Paramount's doors. A hobbled First National, meanwhile, retreated to film production before finally selling out to an ascendant Warner Bros. in 1928. By joining

Warner Bros., the nation's last great independent theaters had given up their war with Hollywood and effectively joined the system.[41]

In 1927, Zukor and his allies also managed effectively to contain the threat of the FTC investigation.[42] In 1926, after heavy lobbying by the film industry, its friend Abram Myers was appointed by Calvin Coolidge to head the commission. Myers's FTC concluded the block booking investigation and issued a rather weak reprimand of the industry. *Variety* labeled it a mere "gesture," and Zukor announced that he would ignore the decree. It was, in any event, reversed by the District of Columbia Court of Appeals in 1932.

The late 1920s would thus prove little more than a mop-up operation. Paramount, MGM, and Universal hunted down and destroyed most of the independent theaters, producers, and distribution companies. Warner Bros., once a small independent producer, managed to join the ranks of the major studios with the first truly successful sound film, *The Jazz Singer.* Among the others, only United Artists, formed by D. W. Griffith and a group of movie stars, would survive, going on to play, as we shall see, an important role in the film industry of the 1970s.

Meanwhile, Thomas Tally, founder of First National Exhibitors Circuit and onetime Zukor rival, abandoned the film industry. His time had passed: together with his son, he established a ranch outside Los Angeles. As for W. W. Hodkinson, he survived as an independent producer until 1929, nurturing the independence of directors like Cecil B. DeMille. Yet he, too, would quit the industry, moving to Central America, where he started an airline; but to his dying day in 1971 he would maintain that Hollywood had made a giant mistake when it followed "that character" Adolph Zukor.[43]

The rise of Hollywood and of the Zukor model is another definitive closing turn of the Cycle. In the course of a single decade, film went from one of the most open industries in the United States to one of the most controlled. The flip shows how abruptly industrial structure can change when the underlying commodity is information. For no sooner had the age of the independent theater owners ended than the openness of the film business had, too. And with the rise of the Hollywood studio—our most visible manifestation of mass-produced culture, then as now—began a rule that would last for decades.

Part II

Beneath the All-Seeing Eye

SIX YEARS BEFORE *Brave New World,* as we have seen, Aldous Huxley could already glimpse where the centralization and mechanization of culture was leading. He foresaw culture's future dominated by commerce. He also saw the prospect of global standardization. "In 3000 A.D." wrote Huxley, "one will doubtless be able to travel from Kansas City to Peking in a few hours. But if the civilization of these two places is the same, there will be no object in doing so."

By the late 1930s every one of the twentieth century's new information industries was fixed in its centralized imperial form. The glories of the new arrangements were evident. Hollywood film was in its golden age, turning out classics like *The Wizard of Oz* and *Gone with the Wind.* NBC and CBS, with help from New York's advertising agencies, had perfected the concept of "entertainment that sells," symbolized by the soap opera and other sponsored programs, like Texaco Star Theater. And the Bell system had become the paragon of a communications monopolist, best captured in the Bell slogan: "The System Is the Solution."

It was also a fact that each of the new media had at least some sense of public duty encoded, as it were, in its DNA. Bell served as a common carrier offering universal service. The networks ran their "sustaining" programs and news departments, under the watch of the FCC. And Hollywood, while a business, had also been motivated by the concept of film as an art form, and had created itself to produce better entertainment than the Edison Trust had seen fit to offer, more like what was available on the stage.[1]

Yet amid these glories of progress—perhaps even necessary to achieve them—there had also been created with respect to freedom of expression one of the least hospitable environments in American history. The 1920s, that heyday for small inventors and alternative voices, were decidedly over. "The times are not propitious for the recognition of great, rebellious, or unorthodox talent," wrote Lawrence Lessing in 1956. "Large impersonal forces are loose in the world, in this country as in more tyrannous parts of the globe, sweeping aside the individual of high merit in pursuit of some new, corporate, collective and conformist destiny."[2]

We turn now to what information empires mean for speech and innovation. Most who study these topics are obsessed with government's role in censorship and providing incentives to innovate. But the state's role, while significant, cannot compare to the power of industry to censor expression or squelch invention.

While the accomplishments we owe to the structures of the 1930s are undeniable, it is essential to understand what was repressed, blocked, or censored by the new system, if we are to understand what was—and is—really at stake.

CHAPTER 7

The Foreign Attachment

Henry Tuttle was, for much of his life, president of the Hush-A-Phone Corporation, manufacturer of a telephone silencer. Apart from Tuttle, Hush-A-Phone Inc. employed his secretary. The two of them worked alone out of a small office near Union Square in New York City. Hush-A-Phone's signature product was shaped like a scoop, and it fit around the speaking end of a receiver, so that no one could hear what the user was saying on the telephone. The company motto emblazoned on its letterhead stated the promise succinctly: "Makes your phone private as a booth."[1]

Advertisements for the cup ran frequently, usually in classified sections. This one from the October 14, 1940, edition of *The New York Times* is typical:

> PHONE TALK ANNOYS? HUSH-A-PHONE PREVENTS.
>
> DEMONSTRATION EITHER TYPE PHONE.
>
> HUSH-A-PHONE CORP., CHELSEA, 3–7202.

If the Hush-A-Phone never became a household necessity, Tuttle did a decent business, and by 1950 he would claim to have sold 125,000 units. But one day late in the 1940s, Henry Tuttle received alarming news. AT&T had launched a crackdown on the Hush-A-Phone and

Leo Beranek and the Hush-A-Phone

similar products, like the Jordaphone, a creaky precursor of the modern speakerphone, whose manufacturer had likewise been put on notice. Bell repairmen began warning customers that Hush-A-Phone use was a violation of a federal tariff and that, failing to cease and desist, they risked termination of their telephone service.[2]

Was AT&T merely blowing smoke? Not at all: the company was referring to a special rule that was part of their covenant with the federal government. It stated: *No equipment, apparatus, circuit or device not furnished by the telephone company shall be attached to or connected with the facilities furnished by the telephone company, whether physically, by induction, or otherwise.*

Tuttle hired an attorney, who petitioned the FCC for a modification of the rule and an injunction against AT&T's threats. In 1950 the FCC decided to hold a trial (officially a "public hearing") in Washington, D.C., to consider whether AT&T, the nation's regulated monopolist, could punish its customers for placing a plastic cup over their telephone mouthpiece.

The story of the Hush-A-Phone and its struggle with AT&T, for all its absurdist undertones, offers a window on the mind-set of the monopoly at its height, as well as a picture of the challenges facing even the least innovative innovator at that moment. As such, the case is an object lesson in the advantages and disadvantages of monopoly. For while it may seem a minor matter, the Hush-A-Phone affair raised fundamental questions about innovation in the age of information monopoly.

AT&T's crackdown wasn't the only challenge Tuttle faced in the 1940s. Over the years, as the telephone had assumed its "modern" design, the Hush-A-Phone, first conceived in the 1920s, was obliged to adapt. Tuttle sought solutions to his hurdles in academia, specifically at the Massachusetts Institute of Technology and Harvard University. In 1945, he queried Leo Beranek, then a young acoustics expert at MIT. The men would meet in New York, and Beranek, thinking the problem an interesting one, agreed to design an improved telephone silencer.

Tuttle was lucky; though not well known at the time, Beranek would soon emerge as one of his field's great authorities, going on to design the acoustics for the United Nations complex and Lincoln Center in New York, and for the Tokyo Opera City Concert Hall, as well as writing the classic textbook *Acoustics*. Rather more relevant to the Hush-A-Phone challenge, during World War II Beranek had worked with a team of scientists at Harvard on the problem of communications in the din of airborne cockpits. In both cases, as Beranek understood, the key to intelligibility was the middle-range frequencies. The silencer he designed for Tuttle would sacrifice the lower-range sounds—creating a slight boominess—in exchange for privacy and external silence. Once he'd developed a prototype according to these specifications, he applied for a U.S. patent and sent his plans over to Tuttle, who enthusiastically dispatched a contract under which Beranek would be paid twenty cents per unit.

Tuttle and Beranek didn't exactly consider themselves a threat to the Bell system. Indeed, once when I asked him if he ever viewed himself as a Bell competitor, Beranek just looked at me as if I were crazy. Rather, their modest aim as independent, outside inventors was a minor improvement to the telephone handset, and an ungainly one at that. So why was AT&T determined to run Hush-A-Phone out of business?

Caught in this seemingly trivial battle over a bauxite cup is a debate over the merits of two alternative models of innovation: centralized and decentralized. Representing the decentralized model was Hush-A-Phone, with Beranek operating, in effect, as Tuttle's system of innovation—

a lone inventor of sorts, qualified in acoustics but unaffiliated with Bell. Representing the centralized model was AT&T's already quasimythical Bell Labs, the entity established to ensure that AT&T, and AT&T alone, moved the phone system along its path into the future.

THE GREAT BELL LABS

In early 1934, Clarence Hickman, a Bell Labs engineer, had a secret machine, about six feet tall, standing in his office. It was a device without equal in the world, decades ahead of its time. If you called and there was no answer on the phone line to which Hickman's invention was connected, the machine would beep and a recording device would come on allowing the caller to leave a message.[3]

The genius at the heart of Hickman's secret proto–answering machine was not so much the concept—perceptive of social change as that was—but rather the technical principle that made it work and that would, eventually, transform the world: magnetic recording tape. Recall that before magnetic storage there was no way to store sound other than by pressing a record or making a piano roll. The new technology would not only usher in audiocassettes and videotapes, but when used with the silicon chip, make computer storage a reality. Indeed, from the 1980s onward, firms from Microsoft to Google, and by implication the whole world, would become utterly dependent on magnetic storage, otherwise known as the hard drive.

If any entity could have come up with advanced recording technology by the early 1930s it was Bell Labs. Founded in 1925 for the express purpose of improving telephony, they made good on their mission (saving AT&T billions with inventions as simple as plastic insulation for telephone wires) and then some: by the 1920s the laboratories had effectively developed a mind of their own, carrying their work beyond better telephones and into basic research to become the world's preeminent corporate-sponsored scientific body. It was a scientific Valhalla, hiring the best men (and later women) they could find and leaving them more or less free to pursue what interested them.

When scientists are given such freedom, they can do amazing things, and soon Bell's were doing cutting-edge work in fields as diverse as quantum physics and information theory. It was a Bell Labs employee named Clinton Davisson who would win a Nobel Prize in 1937 for demonstrating the wave nature of matter, an insight more typically credited to Einstein than to a telephone company employee. In total, Bell would collect seven Nobel Prizes, more than any other corporate laboratory, including one awarded in 1956 for its most famous invention, the transistor, which made the computer possible. Other, more obscure Bell creations are nevertheless dear to geeks, including Unix and the C programming language.

In short, Bell Labs has been a great force for good. It is, frankly, just the kind of phenomenon that makes one side with Theodore Vail about the blessings of a monopoly. For while AT&T was never formally required to run Bell Labs as a research laboratory, it did so out of exactly the sort of noblesse oblige that Vail espoused. AT&T ran Bell Labs not just for its corporate good but for the greater good as well. This is not to be naïve about the corporate profit motive: Bell Labs contributed to AT&T's bottom line far more than plastic wire insulation. Nevertheless, it's hard to see how funding theoretical quantum physics research would be of any immediate benefit to shareholder value. More to the point, it is hard to imagine a phone company today hiring someone to be their quantum physicist, with no rules and no boss.

For, in part, the privileges AT&T enjoyed as a government-sanctioned monopoly with government-set prices were understood as being offset by this contribution to basic scientific research, an activity with proportionately more direct government funding in most other countries. Put another way, in the United States, the higher consumer prices resulting from monopoly amounted, in effect, to a tax on Americans used to fund basic research. This unusual insinuation of a corporation between the government and its goal of advancing American science goes a long way to explain how AT&T, as it matured, became in effect almost a branch of government, charged with top-secret work in the national interest.[4]

For all the undeniable glory of Bell Labs, there emerge little cracks in the resplendent façade of corporatism for the public good. For however many its breakthroughs, there was one way in which the institution was very different from a university: when the interests of AT&T were at odds with the advancement of knowledge, there was no question as to which good prevailed. And so, interspersed between Bell Labs' public triumphs were its secret discoveries, the skeletons in the imperial closet of AT&T.

Let's return to Hickman's magnetic tape and the answering machine. What's interesting is that Hickman's invention in the 1930s would not be "discovered" until the 1990s. For soon after Hickman had demonstrated his invention, AT&T ordered the Labs to cease all research into magnetic storage, and Hickman's research was suppressed and concealed for more than sixty years, coming to light only only when the historian Mark Clark came across Hickman's laboratory notebook in the Bell archives.

"The impressive technical successes of Bell Labs' scientists and engineers," writes Clark, "were hidden by the upper management of both Bell Labs and AT&T." AT&T "refused to develop magnetic recording for consumer use and actively discouraged its development and use by others."[5] Eventually magnetic tape would come to America via imports of foreign technology, mainly German.

But why would company management bury such an important and commercially valuable discovery? What were they afraid of? The answer, rather surreal, is evident in the corporate memoranda, also unearthed by Clark, imposing the research ban. AT&T firmly believed that the answering machine, and its magnetic tapes, would lead the public to abandon the telephone.

More precisely, in Bell's imagination, the very knowledge that it was possible to record a conversation would "greatly restrict the use of the telephone," with catastrophic consequences for its business. Businessmen, for instance, the theory supposed, might fear the potential use of a recorded conversation to undo a written contract. Tape recorders would also inhibit discussing obscene or ethically dubious matters. In sum, the very possibility of magnetic recording, it was feared, would "change the whole nature of telephone conversations" and "render the telephone much less satisfactory and useful in the vast majority of cases in which it is employed."[6]

And so we see that the enlightened monopolist can occasionally prove a delusional paranoid. True, once magnetic recording arrived in America, there were a few, from Nixon to Lewinsky, whose sordid secrets would be exposed by it. But, amazingly enough, we all still use telephones. Such are the liabilities of being subject to the whim of even the most high-minded corporation: even the fantasy that the fate of the company could be at stake can have significant consequences. It was safer to shut down a thrilling line of research than to risk the Bell system.

This is the essential weakness of a centralized approach to innovation: the notion that it can be a planned and systematic process, best directed by a kind of central intelligence; that it is simply of matter of assembling all the best minds and putting them to work in unison. Were it so, the future could be planned and executed in a scientific manner.

Yes, Bell Labs was great. But AT&T, as an innovator, bore a serious genetic flaw: it could not originate technologies that might, by the remotest possibility, threaten the Bell system. In the language of innovation theory, the output of the Bell Labs was practically restricted to *sustaining inventions;* disruptive technologies, those that might even cast a shadow of uncertainty over the business model, were simply out of the question.

The recording machine is only one example of a technology that AT&T, out of such fears, would for years suppress or fail to market: fiber optics, mobile telephones, digital subscriber lines (DSL), facsimile machines, speakerphones—the list goes on and on. These technologies, ranging from novel to revolutionary, were simply too daring for Bell's comfort. Without a reliable sense of how they might affect the Bell system, AT&T and its heirs would deploy each with painfully slow caution, if at all.

Perhaps the response seems less neurotic if we consider how deep-seated can be the apprehension of the Kronos effect. Not for nothing would the Bell system prove itself among the best defended and most secure monopolies in corporate history. Whatever the opportunity inherent in new technology, there was always also a threat, one that prudence demanded be devoured at birth. Bell's own genesis had proved that bit of wisdom. In 1876, Alexander Bell had patented the machine that eventually dethroned and replaced what was then the nation's greatest corporation, Western Union. What charm of the new can possibly rival the instinct for self-preservation? Certainly not a plastic cup.

THE TRIAL OF INNOVATION

Thus did AT&T in deadly earnest go about hushing the Hush-A-Phone. At the two-week trial (technically a hearing), the company showed up with dozens of attorneys, including a top litigator from New York City, and no few expert witnesses. Legal representatives of each of the twenty-one regional Bells came as well, necessitating that extra seats be installed in the hearing room—bleachers for AT&T's lawyers. On Hush-A-Phone's side were Harry Tuttle, his lawyer, the acoustics professor Leo Beranek, and one expert witness, a man named J.C.R. Licklider.[7]

Bell's lawyers mounted a powerful assault against the cup and the cup-bearing company itself. The argument was that the Hush-A-Phone posed substantial harm to telephone service, and at the same time, that the company by that same name, in selling a useless device, was essentially perpetrating a fraud on the public. Bell led with a Bell Labs engineer, W. H. Martin, who set out to show that the Hush-A-Phone impaired telephone service. According to his tests, it created a "transmission loss" of 13 decibels and "receiving loss" of 20 decibels. The loss, he said, was greater "than the total of all the improvements which have been incorporated in the Bell System handsets and the accompanying station apparatus for a period of over 20 years."

The next Bell witness, AT&T vice president John Hanselman, testified more broadly in favor of Bell's ban on "foreign attachments," justifying it in terms of the public good and AT&T's stewardship of the telephone system. It was among the firm's duties to protect the consumer from such useless gimmicks, he maintained. And if there was any use to a Hush-A-Phone, he suggested, unembarrassed by his own casuistry, AT&T would have invented and marketed it. Nor was uselessness the gravest threat. Foreign attachments created outside Bell's design and control standards posed all manner of hazard, including power surges up the phone lines that might electrocute Bell repairmen and send them falling to their deaths. On cross-examination by Tuttle's lawyer, Hanselman finally conceded that such a calamity had never occurred. But, he insisted, there is always a first time.

Like the next Bell witness, Hanselman also testified that there was no demand for a voice silencer in any case. If there were, as a reporter summarized his perfect corporate smugness, "it would clearly be brought to his attention." The proof was that hardly anyone was actually using the Hush-A-Phone, and so the consequences of banning it were furthermore negligible. That there might be some good in the operation of the free market per se simply did not figure in his conceptual universe: "It would not be feasible," he told the FCC, "to allow customers to buy devices on the open market."

AT&T's counsel left no stone unturned, subsequent witnesses reaffirming the same points made by earlier ones, until finally, one made the startling charge that the Hush-A-Phone was unhygienic. As related in *Telecommunications Reports,* "Mr. Burden [stated] that he had sufficient experience as a plant man cleaning the cone-shaped transmitters formerly used by operators to realize that a receptacle such as the Hush-A-Phone would collect food particles, odors, and whatnot over a period of time."

With that, AT&T rested its case.

The theory of Hush-A-Phone's case was, in bald contradiction to AT&T's claims, that the phone silencer was indeed an effective and useful device, and one that AT&T didn't offer. As such, it was of no potential harm to consumers or to the telephone system or its representatives. Testifying first, Henry Tuttle presented the FCC with a list of reasons why customers might want a telephone silencer. Sanity was one, the reduction of office noise being "important to the mental health of employees," in his view. Privacy was another, a necessity for many professionals and businessmen, which claim he supported with a list of Hush-A-Phones in use in Washington, D.C., including a number in congressional committee rooms. In addition, Tuttle distributed a collection of testimonials called "Phone Conversations Overheard," a compilation of cautionary tales was included the woeful story of one man who'd been disinherited when his uncle overheard his nephew making unflattering remarks about him on the telephone. *If only he'd had a Hush-A-Phone . . .*

But the strongest part of Hush-A-Phone's case was the technical aspect. In addition to Beranek, who already enjoyed some eminence, Tuttle brought in, on his designer's advice, a friend of Beranek's from army days who was now a professor at Harvard, one J.C.R. Licklider. The pair of academics, who will reappear later in their better-known identities as the founders of the Internet, "gave the hearing," according to one reporter, "a somewhat ivy-covered atmosphere, taking the spectators and participants back to their college days."[8] But more importantly, they lent an air of unimpeachable authority. Beranek and Licklider, also an expert on acoustics, had run a battery of tests on the silencer, demonstrating that conversations conducted with it remained satisfactorily audible while effectively free from eavesdroppers. This proof did not prevent Bell's lawyers from launching a fierce and lengthy cross-examination, resulting in numbingly complex disagreements about the means of testing word articulation quality. Still, Licklider's report had provided actual data bearing on the question of intelligibility, whereas the counterargument depended on some abstract claim of "transmission loss."

It is worth pausing to observe that as Bell's lawyers squared off against Licklider and Beranek over technology, the world was witnessing, unbeknownst to anyone, even the combatants, the first of many engagements between AT&T and the Internet's founders—in effect, the Fort Sumter, so to speak, in the epic fight between those later to be called the "Netheads" (backers of the Internet) and the "Bellheads." Never mind that it was 1950 and they were arguing over a plastic cup sold in classified ads.

To close their case, the Hush-A-Phone team offered dramatic demonstration to rival O.J.'s bloody glove. Tuttle called his secretary and asked her to speak into a telephone receiver, first with, then without the Hush-A-Phone attached. In accordance with Licklider's findings, the device did indeed alter the acoustics of the telephone transmission, making it sound more "boomy." Yet, as was evident to all, the speech remained intelligible. The Hush-A-Phone, in other words, indubitably worked.

One Mind or Many?

Bell was right about one thing at least: the Hush-A-Phone wasn't terribly popular, and it showed few signs of catching on. To understand AT&T's all-out response as more than merely neurotic, then, one must see the device not for what it was but for what it represented: a threat to the system, and by extension to a sanctified method of innovation. The device itself did not so much effect as foreshadow an intolerable loss of control. It might have failed per se and yet encouraged people to attach all manner of other devices to the telephone, thus coming to see Bell's technological holy of holies as something anyone could tamper with. It might even lead to a future in which people purchased their own telephones!

Here, then, we come to the second weakness that afflicts centralized systems of innovation: the necessity, by definition, of placing all control in a few hands. This is not to say that doing so holds no benefit. To be sure, there is less "waste": instead of ten companies competing to develop a better telephone—reinventing the wheel, as it were, every time—society's resources can be synchronized in their pursuit of the common goal. There is no duplication of research, with many laboratories chasing the same invention. (That avoidance of redundancy in applying brain capital should sound familiar from Vail's philosophy of industrial organization: centralized innovation is the R&D sibling of monopoly, with the same type of claim to efficiency.) Yet if all resources for solving any problem are directed by a single, centralized intelligence, that mastermind has to be right in predicting the future if innovation is to proceed effectively. And that's the problem: monopoly presumes a prescience that humans are seldom capable of.

AT&T and other proponents of centralized innovation assumed the future of the telephone system to be not only knowable, but indeed known. As Beranek put it, "Bell had created the best telephone system in the world. They had the big laboratory. Their attitude was, we don't need you." Bell never missed a chance to assert its need to control every single working part of the system. As it enunciated the need in its legal briefs:

> It would be extremely difficult to furnish "good" telephone service if telephone users were free to attach to the equipment, or use with it, all of the numerous kinds of foreign attachments that are marketed by persons who have no responsibility for the quality of telephone service but are primarily interested in exploiting their products.[9]

Quality control, by such lights, depended on control of every other kind.

Unfortunately, it would be decades before innovation theorists challenged this Bell orthodoxy. In the 1980s, the economists Richard Nelson and Sidney Winter examined the record of human innovation and devised what is now called the "evolutionary" model of innovation. Their thesis implied that in fact innovation is much more a process of trial and error than theretofore imagined. General human ignorance about the future leads to a great many human errors. Furthermore, the human element always introduces an irrationality, even to the point of such paranoia as Bell evinced concerning magnetic recording. Thus if everything is entrusted to a single mind, its inevitable subjective distortions will distort, if not altogether disable, the innovation process. By contrast, Nelson and Winter argued, the most rapid or efficient innovation typically results when the widest range of variations are proposed and the invisible hand of competition, as proxy of the future, picks among them. It is rather like Darwin's idea of the relative fitness of individuals in determining the evolution of species, and like natural selection it depends on the power of accidents.[10]

Hush-A-Phone was a forerunner in this latter-day approach to innovation. Looking at the telephone and AT&T's network, Tuttle saw what we today would call an *innovation platform.* In other words, the Bell system was something that people could and should try to improve, with add-ons and new features. The innovation Tuttle envisioned was privacy, but other outsiders would imagine devices that could answer the phone and transmit images or other forms of data. But AT&T believed that if phone subscribers wanted privacy, they could cup their hands over the receiver.

The irony of the Hush-A-Phone affair is that no company should have understood the importance of outside invention better than Bell, whose eponymous founder was the very archetype of the outsider bearing long-shot ideas that turn out to shape the future. Yet by the 1950s AT&T had left the spirit of Alexander Bell behind. Or perhaps, more ominously, Bell understood that its inevitable downfall would come at the hands of one exactly like its founder, and the only strategy was to temporize. Eventually, a meal gets away even from Kronos.

Hush-A-Phone Decided

Bernard Strassburg, chief counsel to the FCC during the Hush-A-Phone trial, considered the result of the proceedings to be preordained. "In my view, Tuttle's prospects of winning his case before the FCC were, from the start and without court intervention, virtually nil," he would write. "It was the conviction of the FCC and its staff that they shared with the telephone company a common responsibility for efficient and economic public telephone service and that this responsibility could only be discharged by the carrier's control of all facilities that made up the network supplying that service."[11]

After the hearing in 1950, the FCC sat on the Hush-A-Phone case for *five years.* Federal agencies have some discretion about when they will decide things, and the FCC elected to stall, allowing AT&T to continue its ban on foreign attachments. It was not until late in 1955 that the FCC issued a brief decision.

AT&T, the federal agency decided, had been right: the Hush-A-Phone was indeed a danger to the telephone system and a nuisance to consumers—"deleterious to the telephone system and injures the service rendered by it," in the words of the report. More generally, the FCC held that "the unrestricted use of foreign attachments . . . may result in impairment to the quality and efficiency of telephone service, damage to telephone plant and facilities, or injury to telephone company personnel."[12] The preposterous image of the electrified repairman had apparently sunk in.

The decision was painful and expensive for Tuttle, who had financed the lawsuit himself and devoted years to the case to no avail. When he heard the news Beranek contacted Tuttle and forswore any further royalties due him for inventing the silencer. "I just felt bad for him." With little left to lose, Tuttle decided to go for broke and appeal the FCC's decision, again at his own expense. Arguments were heard, and one year later, in 1956, eight years after Tuttle's initial FCC filing, the D.C. court of appeals released its decision. The panel of federal judges, headed by David Bazelon, reversed the commission and vindicated Tuttle and the Hush-A-Phone.

In a scene reminiscent of the conclusion of *Lord of the Flies,* the D.C. court administered the judicial version of a reality check on Bell's tortured logic: "To say that a telephone subscriber may produce the result in question by cupping his hand and speaking into it," wrote Judge Bazelon, "but may not do so by using a device . . . is neither just nor reasonable."[13] The court also admonished the FCC for delaying its decision for five years. Finally, in one crucial phrase—one that in time would unravel AT&T, the phrase, in fact, that would result in its eventual breakup—Judge Bazelon affirmed that the subscriber has the "right reasonably to use his telephone in ways which are privately beneficial without being publicly detrimental."[14]

With Hush-A-Phone's modest victory, the door was cracked not only to every manner of ancillary device in the 1970s, but, as we shall see, to the collapse of Bell's once indomitable empire. Though not without a battle royal: if AT&T was willing to launch a thousand ships to ward off a plastic cup, one can imagine the force that would be brought to bear against a genuine rival in the form of MCI. But in 1956 that eventuality was as yet in the distant future.

Having won its case, Hush-A-Phone ran a series of advertisements proclaiming its device newly approved for use by federal tariff. Unfortunately, it could not keep up with Bell's own stately pace of product design, and when the phone company began to sell new handsets again, sometime in the 1960s, Hush-A-Phone folded. Such are the wages of stifling innovation: to this day, while the annoyance of mobile phone chatter, the banality of overheard conversations, has become a cliché, there is not a Hush-A-Phone or its equivalent to be found.

Hush-A-Phone's valiant founder died sometime in the 1970s, to be forgotten, apart from one great cultural reference. In the 1985 film *Brazil*, Robert De Niro plays a maverick repairman who does unauthorized repairs and leads a resistance movement against a totalitarian state. The hero and hope of that dystopia is named Harry Tuttle.

CHAPTER 8

The Legion of Decency

In 1915, a young Jesuit priest and professor of drama named Daniel Lord published an essay, denouncing George Bernard Shaw. "The bubbles he throws before the eyes may sparkle," wrote the Chicago-born Lord of the internationally acclaimed playwright and critic, "but they are as worthless as the trinkets for which the Indians bartered priceless territory." His work, declared Lord, is "devoid of that first of all necessary qualities, truth."[1]

So began another career of the Reverend Daniel Lord, as crusader against "filth" in the public sphere, a cause to which he would devote much of his life. Throughout the first half of the twentieth century, Lord would advance a memorable and distinctive vision of what the media, the old and the new, were for. Like John Reith at the BBC, Lord believed that the aim of any cultural product should be an improving one: to inform and to educate as well as to entertain. But Lord's specific version of improvement was much more austere. He considered it the very point of human communication to reinforce the received truth, never to challenge it.

Among his cultural interventions, Lord would edit the Catholic magazine *The Queen's Work* (the title referring to the Virgin Mary), for which he also wrote widely circulated essays opining on such controversial subjects as abortion, divorce, and anti-Semitism (he condemned all of these). As a Catholic essayist, Lord was just one in a long line. But his lasting contribution to American culture was distinctive. It was he who wrote and would later help to enforce the famous (or by some lights infamous) Production Code that specified what was acceptable in Hollywood film from 1934 until the 1960s. At the height of their powers, he and his allies gained effective control over film in the United States, practicing without any formal or official authority a censorship to rival that of any authoritarian regime.

THE PLOT AGAINST HOLLYWOOD

In the late 1920s, Lord was one of a small group of Catholic activists who envisioned a new type of activism directed at the burgeoning film industry. His comrades included Martin Quigley, the publisher of a Chicago-based film industry trade journal, and William Hays, a charismatic president of the Motion Picture Producers and Distributors of America, who would be in effect the industry's face in this campaign against the industry. Hays would appear on the cover of *Time* magazine in 1926 as the "Polychromatic Pollyanna," the personification of the Production Code that Lord had written but that was often referred to as the "Hays Code."[2]

Another member was Joseph Breen, who apart from Lord would prove the most important figure in the private censorship of Hollywood. Breen, who originally worked for Hays in public relations, was described by his biographer, Thomas Doherty, as "Victorian Irish," characterized "neither by leprechaun charm nor whisky-soaked gloom, but by a sober vigilance over the self and a brisk readiness to perform the same service for others, solicited or not." That Breen did not hold Hollywood or its moguls in high esteem, at least initially, is clear. In private letters from 1932, he complains that "these Jews seem to think of nothing but money making and sexual indulgence." As for the industry's lower orders: "People whose daily morals would not be tolerated in the toilet of a pest house hold the good jobs out here and wax fat on it." If Lord was the legislative branch of the Production Code, Breen was its executive, the enforcer. It was he, for example, who would try, though unsuccessfully, to change Rhett Butler's famous line at the end of *Gone with the Wind,* "Frankly, my dear, I don't give a damn" (Breen's suggestion: "Frankly, I don't care").[3]

Convinced that Hollywood was corrupting Americans, these three men, each of them connected to Hollywood in one way or another, all of them Catholics, were passionately determined that something be done. What's most interesting about the Lord-Quigley-Breen approach is their view that more effective censorship of the industry could be achieved directly, with the support of the church, than by government regulation. Decrying what they called the "janitor" model, they deemed it unsatisfactory to try to clean up a mess already made; far better to stop vulgar films from being produced in the first place. And far better to accomplish this by private intimidation than by public policy.

The Catholic enforcers subscribed to a principle that legal scholars would come to call "prior restraint"–the regulation of films *before* production. Quigley had personally insisted on this strategy based on his experiences in 1920s Chicago, where despite the Church's lobbying, the board of censors would permit an offensive film to be made and released. Quigley knew that even in a town run by Catholics, the industry, flush with cash, could always bribe politicians or the police to get its films through.

Indeed, the Catholic enforcers were astute to realize that in a democracy, official censorship could never be as effective as private. Officials, if they could be prevailed upon to act at all, would act after the fact, and then it would be up to law enforcement, an inherently imperfect enterprise, to make the rules stick. The three men reasoned that if they wanted their Christian values—which were, after all, the values of America traditionally—to be upheld, they would have to find a way of restraining the film industry themselves. More directly, Breen would declare himself the one man "who could cram decent ethics down the throats of the Jews."*[4]

Pre-Code

Hollywood films in the early 1930s are still among the edgier in the history of American mass entertainment, and thanks to the introduction of sound, their departure from what had come before seemed even starker

---

* Despite such rhetoric, Breen's biographer insists that by the 1940s Breen was a committed and vocal opponent of anti-Semitism.

at the time. The sensibility was embodied by Mae West, a Zukor star, in her films *She Done Him Wrong* and *I'm No Angel*. In both, West plays a mature, liberated sexpot—essentially the same character as *Sex and the City*'s Samantha Jones, but with a Brooklyn accent. It is in those films that West speaks her most remembered lines, such as "Is that a gun in your pocket, or are you just happy to see me?" and "When I'm good, I'm very good. When I'm bad, I'm better."

Here is how Mick LaSalle, a film critic and authority on pre-Code Hollywood, describes the films of the early 1930s:

> They celebrate independence and initiative, whether the protagonist is honest or crooked. They prefer the individual to the collective and are deeply cynical about all organized power, such as the government, the police, the church, big business and the legal system. Anything that gets in the way of freedom, including sexual freedom, they tend to be against. In the same way, anybody who tells somebody what to do is usually the villain.[5]

These might sound like traditionally American values, too, but not to the likes of Father Daniel Lord. After the second of Mae West's films was released, he wrote an angry letter to the trade association threatening a "day of reckoning."[6] When Paramount announced a third Mae West vehicle, in 1934, that day came.

"I wish to join the Legion of Decency, which condemns vile and unwholesome moving pictures. . . . I hereby promise to remain away from all motion pictures except those which do not offend decency and Christian morality."[7]

So reads the membership pledge of the body that became a crucial part of the Catholic Church's offensive against the film industry in the wake of *I'm No Angel*. Catholic parishioners across the United States were invited to join the Legion of Decency; Protestants and Jews were welcome, too. It wasn't the first mass morality movement in the United States, but it was perhaps the best subscribed: at its height, in 1934, the Legion claimed 11 million members.

The Catholic boycott, or at least the threat of one, was the Legion's most lethal weapon against Hollywood. The aim was to eliminate attendance at films judged immoral, but at Breen's urging, some Catholic authorities urged a boycott of all motion pictures. The intensified version of the strategy, Breen believed, would assure the intimidation of not only the studio executives but lower-level employees, such as a district manager of Warner Bros. whom he described as the "kike Jew of the very lowest type."[8]

The Legion of Decency's project was no less fascinating for being so venomous: a Catholic movement designed to discipline Jewish producers on behalf of a Protestant majority. But perhaps most impressive is just how effective it was. Some of this may be put down to good timing. At the same moment, also under pressure from the Church, the incoming Roosevelt administration threatened to get involved, and compounding Hollywood's headaches, that very year, a series of academic studies suggested that films were dangerous to children.[9]

Facing boycotts and possible federal action, late in 1934 the industry cartel agreed to abide by the Production Code drafted by Daniel Lord. (Hollywood had actually acceded to the Code earlier but had tried initially to ignore it.) Breen would be head of a new "Production Code Administration" with personal authority to review all treatments and scripts. He would also oversee the issuance of the Production Code Seal of Approval. A theater in the trade association showing a film that lacked the Seal was committing a violation of the Code, one of many punishable by a $25,000 fine.[10]

STRUCTURE

Adolph Zukor had no censorial instincts per se. He built the Hollywood studio system, but to make bigger, better films, and to guarantee a good return on his investors' dollar. His foremost concern was that his films sold, and in pursuit of that goal he was no prude. In fact, Paramount's scandalous movies were the main inspiration for the Catholic backlash.

But even though motivated chiefly by dollars and cents, Zukor had created, however incidentally, an industrial structure very amenable to speech control. In fact, had Zukor and his cohorts at Warner Bros., Universal, and Fox not wiped out the independent producers, distributors, and theaters, the rule of the Production Code would not even have been possible. One can hardly imagine Lord, Quigley, and Breen gaining anything like the leverage they managed over Hollywood if industry power were still dispersed among thousands of producers, distributors, and theater owners. As it happened, the transformation in Hollywood content was so abrupt as to be shocking, and it demonstrates a substantial vulnerability of highly centralized systems: while designed for stability's sake, they are in fact susceptible to extremely drastic disruption.

What did the Code require? Today we remember it most vividly for the rule that put married couples in twin beds, but there was much more to it. The Code was not just a litany of "thou shalt nots," but rather a body of received ideas articulated by Lord as to what film should be. It was a Manichaean notion of the right and the wrong, the good and the evil. Film should, in all cases, reaffirm these distinctions, never complicate them.

The Code did not ban treatment of controversial subjects. For example, corruption might figure in a perfectly acceptable plot, but it had to appear in suitably limited form that would not corrupt the general morals. An individual judge or policeman could be dishonest, but not the whole judicial system. A man might be unfaithful to his wife, but the institution of marriage itself could not be presented as a sham or otherwise assailed. In Lord's terms, the ideal film was one wherein "virtue was virtue and vice was vice, and nobody in the audience had the slightest doubt when to applaud and when to hiss."[11]

The basic conception was captured in the Code's three principles:

> No picture shall be produced which will lower the moral standards of those who see it. Hence the sympathy of the audience should never be thrown to the side of crime, wrongdoing, evil or sin.
>
> Correct standards of life, subject only to the requirements of drama and entertainment, shall be presented.
>
> Law, natural or human, shall not be ridiculed, nor shall sympathy be created for its violation.[12]

Gregory Black, a historian, described the artistic toll imposed by these seemingly simple requirements: "If the movie industry was to satisfy the reformers it had to give up—not simply clean up—films that dealt with social and moral issues," for the Code insisted that "films must uphold, not question or challenge, the basic values of society." They should be "twentieth century morality plays that illustrated proper behavior to the masses." But the Code was not limited to these main directives, in the face of which one might find interpretive latitude; included was a much longer list of painfully specific applications.

On the topic of dance in films, the Production Code set standards that might have satisfied the Taliban:

DANCES
1. Dances suggesting or representing sexual actions or indecent passions are forbidden.
2. Dances which emphasize indecent movements are to be regarded as obscene.

And in case it might be unclear what else might constitute "obscenity":

OBSCENITY
Obscenity in word, gesture, reference, song, joke, or by suggestion (even when likely to be understood only by part of the audience) is forbidden.

The effect on film of the new self-enforced system was immediate and stark. As Mick LaSalle writes, "the difference between pre-Codes and films made during the Code is so dramatic that, once one becomes familiar with pre-Codes, it becomes possible to tell, sometimes within five minutes, whether a 1934 film was released early or late in the year." Even Betty Boop, the squeaky-voiced animated flapper, was converted into a "maiden aunt" with lower hemlines.[13]

SPEECH

The story of Daniel Lord and the Legion of Decency goes to a central contention of this book: in the United States, it is industrial structure that determines the limits of free speech.

This may sound odd to some. The expression "free speech" may more immediately call to mind the First Amendment to the U.S. Constitution, or equivalent codifications in other countries: a right whose exercise is granted by law and seemingly not at the mercy of other human institutions, particularly commercial ones. But the story of the Code makes quite clear that the reality is very much otherwise. For the First Amendment guarantees, among other things, only that "*Congress* shall make no law . . . abridging the Freedom of Speech." Even if the judiciary had been jammed full of civil libertarians and card-carrying members of the ACLU, it wouldn't have made the least difference. The Lord-Breen system of censorship had nothing to do with the First Amendment or the courts, for it had nothing to do with any law subject to judicial review. The Legion of Decency was an entity wholly independent of government, its power over an industry deriving from that industry's own self-imposed structure. The Constitution may protect us from government's limiting our freedom of speech, but it has nothing to say about anything we might do to limit one another's. There is nothing in the Constitution that stops Church A from compelling Mr. B to pipe down.

A central metaphor in the national discourse about free speech was introduced in the 1920s by Justice Oliver Wendell Holmes, when he wrote that "the best test of truth is the power of the thought to get itself accepted in the competition of the market, and that truth is the only ground upon which their wishes safely can be carried out," referred to commonly as the concept of a "marketplace of ideas." It captures the idea of a figurative market, where anyone with a tongue to speak and ears to hear might freely peddle and receive opinions, creeds, and various other forms of self-expression. The hope is that in such a domain, the truth will win out.[14]

But what if the figurative "marketplace of ideas" is lodged in the actual and less lofty markets for products of communication and culture, and these markets are closed, or so costly to enter as to admit only a few? If making yourself heard cannot be practically accomplished in an actual public square but rather depends upon some medium, and upon that medium is built an industry restricting access to it, there is no free market for speech. Seen this way, the Hays Code was a barrier to trade in the marketplace of ideas. And even without them, the higher the costs of entry, the fewer will be the ideas that can vie for attention.

The trick is that a concentrated industry need not be censorial by nature in order for its structure to produce a chilling effect on the freedom of expression. The individual actors may very well have such a predilection, but they are not the real problem. The problem is that a "speech industry"—as we might term any information industry—once centralized, becomes an easy target for external independent actors with strong reasons of their own for limiting speech. And these reasons may have nothing to do with the industry per se. The Catholic Church of the 1930s obviously wasn't a film studio and had no aspirations in the film business. But having its own motivations for preventing certain forms of expression it deemed objectionable, it found the means to do so in the very design of America's foremost industry of cultural and expressive production. It was the combination of the Church and the Hollywood studio system that produced one of the most dramatic regimes of censorship in American history.

One Man

By the mid-1930s it had become clear that the possibility of making any given film in America depended on the discretion of one man, to the point that in 1936 *Liberty* magazine could credibly opine that Joseph Breen "probably has more influence in standardizing world thinking than Mussolini, Hitler, or Stalin." Yet the film industry, the Catholic Church, and the White House all embraced the system, proclaiming a new day for American cinema. Eleanor Roosevelt lauded the new code in a national radio address: "I am extremely happy," she said, that "the film industry has appointed a censor within its own ranks."[15]

Not that the Code was foolproof. From 1934 onward, there were times when Joseph Breen accepted a film, only for it to be condemned as unwatchable by local chapters of Legions of Decency in other parts of the country.[16] And as Breen spent more time in Hollywood, his sensibility mellowed into something more forgiving. But there can be no doubt about the effect of the Code regime on the nature of American filmmaking. "The pre-Code era didn't fade," writes LaSalle. "It was ended in full bloom and with the finality of an axe coming down."[17]

It is true that Zukor and his colleagues at other studios would in these years oversee many of Hollywood's finest and most beloved productions, including *Philadelphia Story* and *It's a Wonderful Life* (though admittedly the cult of the latter was not contemporary, arising decades later in a cultural climate desperate for moral assurances). But in exchange, they accepted a regime that demanded—indeed, defined—the so-called Hollywood ending and its abiding respect for authority. Serious challenges, explicit or implicit, to such institutions as marriage, government, the courts, or the Church are virtually nonexistent in the films of the 1930s, '40s, and '50s. In other words, there was no place for the expression of remotely subversive views or anything that questioned the status quo. Social critique may have existed in the culture, but it was not promulgated through the culture's mightiest megaphone.

If Congress today wanted to pass a law with the same effect as the Code, it would be struck down in an instant. Yet the Production Code, perhaps the strictest abridgement of speech in U.S. history, worked for two reasons. The right—by reason of patent, not constitutionality—even to make a film in the first place was in the hands of but a dozen men. And those dozen, by reason of commercial imperatives, made themselves subject to the veto of one man, Joseph Breen. We cannot claim to understand the life of free speech in America without understanding how that happened.

CHAPTER 9

FM Radio

In the year 1934, Edwin Armstrong was at the height of his career in every sense. From a laboratory at the very top of the new Empire State Building in New York, with an antenna attached to the spire, Professor Armstrong was testing a new radio technology, firing radio waves at receivers dozens of miles away. The signal was clear and clean, producing an audio fidelity beyond anything heard before. "Actual measurements," writes Lawrence Lessing, a radio historian, "showed clear reception out to at least three horizons—or a distance of about 80 miles." Armstrong called his invention "frequency modulation" radio, or FM for short.[1]

The man who had put Armstrong in the Empire State Building was David Sarnoff, whom we met ringside at the Dempsey fight fourteen years earlier. Sarnoff was then an ambitious young executive obsessed with new technologies such as radio broadcasting. By now—as president of the Radio Company of America, the nation's most important radio manufacturer and parent company of NBC, the preeminent broadcast network—Sarnoff was the single most powerful man in American broadcasting, the defining mogul of an information industry, whose importance was comparable only to that of Theodore Vail, Adolph Zukor, or a very few others. As such, he was also a symbol of the mass culture redefining American sensibilities.

As for Professor Edwin Armstrong of Columbia, he was a different archetype, one of the lone inventors who had long sustained the American approach to innovation. Working from his attic in Yonkers, he had already won fame developing three fundamental radio technologies—the regenerative circuit, the superregenerative circuit, and the superheterodyne receiver—the first of them while still an undergraduate. He was also something of a lively character, commuting to Columbia on a bright red Indian motorcycle. In the early 1920s, he'd become smitten with Sarnoff's secretary and bought a French sports car, driving her away in it. Another time, in manic glee, he climbed the four-hundred-foot radio tower on the roof of radio station WJZ in midtown New York. Above all, Armstrong was an inveterate radio idealist, for whom the technology would always seem magical and of unlimited potential.[2]

As fellow early radio enthusiasts, Armstrong and Sarnoff had bonded on February 1, 1914, having spent the whole night picking up radio signals using Armstrong's technologies. For years Armstrong would send Sarnoff a telegram on the anniversary of that day. Armstrong, in fact, owed his substantial fortune to Sarnoff, who in the 1920s had persuaded RCA to buy Armstrong's patents. In the early days, it was possible, then, for both an inventor and a businessman to inhabit the camp of radio utopianism.

Sometime in the 1920s, Sarnoff had tasked Armstrong to design a "little black box" that might remove the static and distortions of AM radio. Armstrong had taken the direction, and now, nearly a decade later, he delivered, though his solution was not a "black box" but a new technology for transmitting radio waves. Duly impressed with the results, Sarnoff installed his old friend at RCA's expense in the Empire State Building, the tallest building in the world, ordering that Armstrong be provided with anything he needed to perfect the technology.[3]

Yet as time passed, and the encouraging test results continued to mount, Armstrong noticed something strange. He had given Sarnoff an exclusive look at FM both out of personal loyalty and because the sale of his last patent to RCA had granted the firm an option on his next one. And what he had produced went beyond what had been asked; indeed, his innovation would give RCA the opportunity to take radio to its full potential. It would never occur to him that the Radio Corporation of America would want to do anything less. And yet the better the

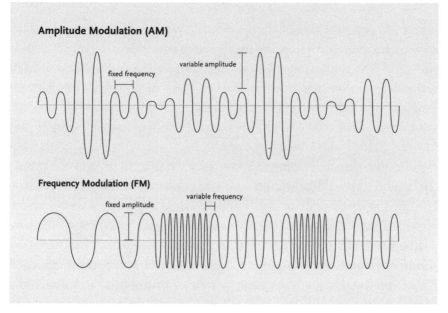

AM and FM compared. Modulating amplitude, as opposed to frequency, required far more power and created far more static.

tests proved FM to be, the more distant and evasive RCA's management seemed to become.

As months passed, Armstrong grew impatient for an answer. Was RCA interested in the commercial development of FM radio or not? Still receiving only vague demur, one day late in 1934 he confronted Sarnoff in Rockefeller Center.

"Why are you pushing this so hard?" asked Sarnoff.

"There is a Depression on," said Armstrong. "The radio industry needs something to put life in it. I think this is it."

"Yes," said Sarnoff, "but this is no ordinary invention. This is a revolution."[4]

That was all Sarnoff would say. Armstrong again pressed him, but Sarnoff changed the subject. Armstrong returned to his lab and carried on with his tests, until he was politely asked to remove all of his equipment from the Empire State Building.[5] From that day onward, over the next decade, in ways both subtle and not, Sarnoff and the rest of the AM radio industry quietly campaigned to relegate FM radio to irrelevancy.

Three important waves of innovation followed the great consolidation of broadcasting in the 1920s: mechanical television, electronic television, and FM radio transmission. And despite the importance of each technology, what is so striking is that none managed to produce an independent industry capable of challenging the dominant Radio Trust, comprising primarily RCA, NBC, and NBC's industrial allies, CBS, General Electric, and Westinghouse.

Over two decades, from 1926 to 1946, each invention, and the inventor behind it, would fall victim to David Sarnoff and the radio industry. Sarnoff's story is perhaps this book's most compelling parable of the Kronos effect, and what bears most attention is the power of his particular methods. For while Vail and Zukor consolidated industries by means of financial pressure and corporate acquisition, Sarnoff managed his empire by using government to restrict inventions, and hence the future. In pursuit of this aim, the Radio Trust perfected the art of controlling both public discourse and federal regulation to the detriment of any would-be rivals. They grew to recognize a truth that had eluded Western Union in the 1880s: the best antidote to the disruptive power of innovation is overregulation. That is to say, the industry learned how to secure the enactment of seemingly innocuous and sensible regulations that nonetheless spelled doom for any rival. In fact, a careful review of the 1930s makes clear that the early FCC was among the most useful tools of domination that industry has ever invented.

FM Radio

Why wasn't Sarnoff, once a radio idealist himself, interested in FM? What Sarnoff wanted when he asked for a "black box" was something that would improve AM radio. But the creative muse has its own agenda, and what Armstrong wound up developing was no improvement on AM, but a replacement for it. FM radio was, in other words, a quintessential disruptive innovation. And whoever Sarnoff may once have been, by the 1930s he was a man who could ill afford to embrace a disruption of the AM radio economy, and in particular, the NBC system.

Now one must appreciate just how important a development FM

was, both technologically and in respect of its potential to revolutionize radio broadcasting. Most of us know that FM is capable of a far higher fidelity in sound reproduction than AM. It's a matter of less static: in the 1930s, when it appeared, FM was capable of signal-to-noise ratios of 100:1 or better, as compared to 30:1 for AM stations.[6] But perhaps even more innovative was FM's capability to broadcast using far less power than AM, thus obviating the high-wattage stations controlled by a few broadcasters. If managed correctly, Armstrong realized, the adoption of FM could clear the way for many more broadcasters than the large networks dominating the AM dial. Let out of its box, there was even some possibility that FM could make the FCC less relevant, as America returned to the lightly regulated, decentralized industry that was the original 1920s vision of the medium.

Beyond radio, FM technology held another promise: as a potential competitor to AT&T's long distance lines—the most powerful communications network in the world. In the late 1930s, Armstrong experimented with small mountaintop relay stations to connect FM stations in different cities. If FM could connect radio stations, networks like NBC would potentially no longer need AT&T's long lines to ferry their programming around the nation.[7]

Finally, Armstrong also believed that FM had applications for much more than just broadcasting music and news. At one eye-popping demonstration at his Empire State Building laboratory in 1934, he showed RCA that FM could carry a facsimile reproduction of *The New York Times,* and telegraph messages as well—a form of wireless fax! In other words, Armstrong foresaw not just better radio, but a multipurpose communications technology. While it may sound astonishing, even in our own times the full potential of FM radio remains untapped.*

You might think that the possibility of more radio stations with less interference would be generally recognized as an unalloyed good. More radio stations means more choices for the consumer and more opportunities for speakers, musicians, and other performers. But by this point the radio industry, supported by the federal government,[8] had invested

---

*As of 2010, the Federal Communications Commission was interested in licensing many more low-power FM stations; yet Congress, under pressure from broadcasters, passed the Radio Broadcasting Preservation Act of 2000, which made licensing of low-power stations much more difficult.

heavily in the status quo of fewer stations. Radio's business model, as we've seen, was essentially "entertainment that sells"—shows produced by advertisers, with revenues dependent on maximizing one's share of the listenership. Hence, the fewer options the better. Even the preeminent radio manufacturer RCA was not immune to this logic. More stations might mean more radios sold, but as owner of NBC, RCA now viewed the network's interests as synonymous with its own. And so we see another instance of how vertical integration of an industry creates a vested interest in limiting free expression. Profit is tied not to the proliferation of many voices but to the propagation of a few—to the mass production of speech, as it were.

The campaign against FM radio in the 1930s and 1940s is a study in rhetoric as a weapon of industrial warfare. While it changed over time, the strategy pursued by Sarnoff and his allies was to belittle FM, talk it down, and generally promote a conventional wisdom that favored the AM industry. It shows, as we shall see, that perhaps the most effective way to gain power over the future is to dictate popular assumptions. RCA and other broadcasters did so by focusing on the promise of a new medium: television. The technology of FM was rarely mentioned in the radio industry's endless promotion of the latest and greatest. When mentioned at all, FM, lauded as theoretically interesting, was also minimized as a mostly unproven technology, experimental and of marginal utility.

The message had a strong effect on two audiences, the first of these being the federal government. The government can act only on the basis of what it understands to be established fact. Much of what is called lobbying must actually be recognized as a campaign to establish, as conventional wisdom, the "right" facts, whether pertaining to climate change, the advantages of charter schools, or the ideal technology for broadcasting. Much of the work of Washington lobbyists is simply an effort to control the conversation surrounding an issue, and new technologies are no exception.

The facts, as the FCC was given to understand them, were that FM might eventually be a useful improvement on AM, but that its time had not yet come. And so, for six years after its invention, the FCC banned commercial FM broadcasting and limited experimental FM to a single narrow high-frequency band. In contrast to the early, unregulated days of AM radio, there was no way for an FM station even to get started without breaking the law. And even if one were so bold, with no radio manufacturer selling FM sets to consumers, there were no listeners. And without listeners, there was no industry.

Investors were the other target of the industry's indoctrination program. Compounding a general scarcity of investment capital in the 1930s, rules banning commercial FM made it all the harder for would-be stations to attract investors. In fact, the very first FM radio stations were nonprofit ventures, financed by Columbia University at Armstrong's urging, a precursor of Stanford's funding of the Google search engine in the 1990s.

Only thanks to Armstrong's tireless advocacy did the technology not die, and indeed begin to gain acceptance, including, eventually, allocation of spectrum for commercial use. He had a flair for the dramatic, and his demonstrations of FM's wonders made headlines, undermining the efforts of the industry to suggest there wasn't much to it. In 1935, he shocked the Institute of Radio Engineers with a low-power FM broadcast from Yonkers of remarkably high fidelity. "A glass of water was poured before the microphone in Yonkers; it sounded like a glass of water being poured and not, as in the 'sound effects' on ordinary radio, like a waterfall." By such enthusiastic efforts Armstrong managed to convince a few like-minded people to found FM stations, and by 1941, yielding to the technological reality, the FCC allocated spectrum between 50 and 60 MHz.[9]

But, as mentioned, the disinformation campaign's trump card was television, and the Radio Trust's insistence that it, not FM, was the future. Many standard histories in fact ascribe the slow development of FM to the rise of television. Of course, there is no denying that television was the greater leap. But insofar as television never has supplanted radio, the Radio Trust was presenting a false choice. There was no reason for the federal government not to allow the development of both a new FM industry and a television industry. Except in the industry's tale spinning, it never was a matter of one or the other.

The Second World War put the development of all consumer technologies on hold, even though FM did see adoption by the army and navy, to whom Armstong extended a free license to his patents. FM's best chance to become what some called "radio's second chance" would not

come until immediately after the war, when the coast was finally clear
for commercial use, and the new technology gained the grudging support of even RCA/NBC (or at least parts of RCA), which knew its utility all along and now wanted its share.[10] Yet it can be no coincidence
that in 1945 the FCC would also move to enact new rules, ostensibly for
the benefit of the FM industry, yet much favored by the AM dial, above
all NBC and CBS, the broadcast duopoly.

That year, the FCC announced that the FM frequency band would
be moved from its original 50–60 MHz home to the now familiar 88.5–
108 MHz range. The shift came with a few additional rules. At the networks' urging, FM radio stations owned by AM stations were required
to carry the exact same programming—so-called simulcasting—for
the supposed benefit of the American consumer. And all FM networks
were obliged to use AT&T's network for long-range broadcasting,
precluded from developing their own long distance lines or carrying
long-range broadcasts. New limits were put on the maximum wattage
of FM stations, which already required little power, thus neutralizing
the advantage in range that FM stations naturally had over AM.[11]

The rules stoked controversy and drew heavy opposition from the
infant FM industry and its allies. The FCC nevertheless defended them
in terms reasonable on their face, but specious under their technical
particulars. The band relocation and the wattage limits, for instance,
apart from clearing bandwidth for television, were officially justified
as protecting FM broadcasts from interference, particularly something
called "skywave interference" that critics believed to be a bogeyman, but
whose existence could never conclusively be disproved.

In its defense, the FCC had by now conceded that FM was the superior technology and described its rules as making possible a migration
from AM. There was a defensible objective in preparing an orderly
future for FM, in contrast to AM's undesirably chaotic early days. But
even if one assumes benign intentions, federal planning is never a good
midwife for a new industry. And less charitably, one must allow that the
whole concept of a migration was designed so that existing AM station
owners would dominate the FM dial. Indeed, by 1949, 85 percent of
the "new" FM station licenses had been extended to AM station owners, who tended to duplicate their AM programming on FM. In other
words, the FCC managed to expand capacity while quashing the pos-

sibility of new programming and new voices. FCC, in short, would countenance FM only so long as it posed no threat to the powers of the existing industry.[12]

Generally, the 1945 FCC order was bad news for the infant FM industry, and the independent elements of AM radio. Its new wattage guidelines made obsolete 400,000 FM radio sets that consumers had already bought and required every FM station to buy a new transmitter. Officially, the FCC estimated that the effects of the order would set FM back four months. But despite the exertions of Armstrong and the fledgling industry to stay afloat, FM would not recover from these blows for decades. The exciting technology developed under the auspices of the industry in 1934, its technical superiority notwithstanding, was effectively dead by 1952.

In 1940, Armstrong had predicted that FM would supplant AM in five years. Actually, it would take until the 1970s for it to catch on, and until the 1980s to reach the popularity of AM.[13] It cannot be denied that the emergence of television at the same time took some of the wind out of FM's sails. It is likewise true that wholly replacing one technology with another does take time. But the studied efforts of the AM industry and the complicit restrictions imposed by the FCC, even if occasionally well-meaning, certainly retarded FM, preventing it from developing into anything more than AM in stereo. Its other remarkable capabilities—especially the potential for many new low-cost radio stations or long distance relays—remain unexploited even to this day. In Lessing's description of what happened, "the vast concentration of economic power that marked the field of mass communications . . . had rolled over FM and crushed it into a shape less threatening to their monopolistic pattern of operations."[14]

Professor Armstrong would never see his greatest invention reach mass acceptance. But what finally crushed him had less to do with the burdens of championing FM's struggle than with his more personal battle against RCA and David Sarnoff. RCA had begun changing its tune about FM in 1946, when it saw the technology's value for providing sound to television broadcasting. Sarnoff, understanding well the merits of Armstrong's invention, decided to put FM receivers in RCA's television sets. But instead of seeking the partnership of his old friend, Sarnoff and his firm decided simply to use Armstrong's technology and wait for him to sue.

RCA's official position was that in-house engineers had invented and patented a "different" FM. In reality, there was no such thing, and Armstrong, like Alexander Bell in the 1870s, was forced into litigation that would last the rest of his life. Unfortunately, though he was extremely stubborn, Armstrong was facing an opponent that was ready for him. RCA's lawyers from the firm of Cahill, Gordon did what New York law firms specialize in: turning lawsuits into wars of attrition. It was a fight of man against corporation, with discovery and pretrial motions artfully employed to delay justice until the size of Armstrong's legal bills alone threatened to determine the contest. The inventor's deposition alone lasted over a year.[15]

It was during this litigation that Sarnoff and Armstrong came face-to-face for what was almost certainly the last time. Sarnoff was deposed by his own lawyers with Armstrong at the plaintiff's table. Then Armstrong's lawyer asked Sarnoff how the men knew each other, and Sarnoff with perfect sangfroid replied, "We were close friends. I hope we still are." Later, when RCA's lawyer asked him who had invented FM radio, Sarnoff expanded on the party line: "I will go further and I will say that the RCA and the NBC," he testified, "have done more to develop FM than anybody in this country, including Armstrong."

The denial that Armstrong had in fact invented FM was, if not the last straw, then very nearly the last. A man's decision to end his life is inevitably a complex one and cannot easily be blamed on any one person or event. Yet it is clear that by the 1950s, whatever faith Armstrong had had in the fairness of the system—of business, of justice, or of life itself—was being chipped away by the travails of FM, which he had made his own. He was nearly bankrupted, as the lawsuit with RCA consumed the once great fortune from his earlier patents. His marriage (to Sarnoff's onetime secretary) had also fallen to pieces over his refusal to settle. In 1954 it all became too much too bear. On February 1, forty years to the day from the night he and Sarnoff had spent searching for radio signals, he wrote a final note, dressed neatly, and walked out the window of his thirteenth-floor Manhattan apartment.

The story of FM radio gives some taste of what television's inventors were in for, and a sense of David Sarnoff's genius in the art of industrial combat. He saw not just that FM could replace AM, but that television would more generally replace radio, and by implication, destroy the Radio Corporation of America. He was a man who proved it is possible to defy both Joseph Schumpeter's doctrine of creative destruction and, as he turned RCA into a television company, the adage that you can't teach old dogs new tricks.

CHAPTER 10

Now We Add Sight to Sound

Before them sat a crudely built machine made of wire and wood, incorporating an old tea chest and a bicycle lamp. At its center was giant spinning disk made of wire and cardboard. This was the scene in January 1926, in London's Soho district, as a rumpled bespectacled man named John Logie Baird received a group of reporters and scientists in his cramped attic laboratory.[1]

That Baird had persuaded the press and, even more, scientists from the Royal Institution of Great Britain to come see his contraption was impressive. For Baird was an eccentric inventor with limited formal training—rather reminiscent of "Doc" in *Back to the Future*. Up to that point, his greatest invention had been a type of hosiery designed to absorb dampness, known as the "Baird undersock." Less successful was his follow-up effort, pneumatic footwear (a crude precursor to Nike's "Air" shoes) that had an unfortunate tendency to explode underfoot.[2]

Baird's latest was indubitably a great leap beyond inflatable footwear: the world's first working television. Much as Alexander Bell had coaxed voice out of a wire, Baird, fifty years later, had discovered how images, too, could be sent over a filament with the help of a giant spinning disk.* Lack of training in electronics—as with Bell—sometimes proved

---

*Known as a Niptow disk.

The world's first television

an advantage, for he would try things others would deem ridiculous. To improve television scanning, for instance, he once got hold of a human eyeball, hoping to discern some applicable secret of operation. ("Nothing was gained from the experiment," he would concede in his journal. "It was gruesome and a waste of time.")[3]

Back in his attic laboratory, at Baird's command, the ghostly image of a face appeared on a screen. As the London *Times* reported, "the image as transmitted was faint and often blurred, but substantiated a claim . . . it is possible to transmit and reproduce instantly the details of movement, and such things as the play of expression on the face."[4]

As with so many technologies, television resulted from several simultaneous inventions. Almost immediately after Baird introduced his prototype, an American, Charles Francis Jenkins, invited the American press to see his television in Washington, D.C. Jenkins's "radiovisor" was a handsome machine made of polished wood, and the American press came away favorably impressed, *The New York Times* promptly declaring Jenkins the "Father of Television." Attentive readers may remember meeting Jenkins more than thirty years earlier, when, still in his twenties, he had coinvented the first American motion picture projector, only to see his partner sell out to Thomas Edison. ("The inventor gets the experience, and the capitalist gets the invention.")[5]

Unfortunately for both men, neither Baird's nor Jenkins's prototype worked all that well. Hence the appearance of the third great independent inventor. In September 1928, just two years after Baird and Jenkins had demonstrated their devices, a twenty-two-year-old San Francisco resident named Philo Farnsworth was showing the press a remarkably high-resolution clip of Mary Pickford combing her hair on the screen of his own contraption.* The "young genius," as the *San Francisco Chronicle* described him, had invented the world's first electronic, as opposed to mechanical, television, having replaced Baird's spinning disk with a cathode ray gun, for which he would secure a broad patent. "It is a queer looking little image," said the *Chronicle* of the Farnsworth electronic television, but "perfection is now a matter of engineering."[6]

Among these three men there was nearly every necessary ingredient for yet another saga of disruptive innovation in the great Anglo-American tradition. We have an existing technology, radio, already backed by a large, concentrated industry, vulnerable to being supplanted by fast-moving entrepreneurs and a giant technical leap. All that was wanting was capitalization, and perhaps the right individual to help these inventors hang on to what they had created.

Yet, as sometimes happens, this is exactly the point when the natural course of the narrative breaks, when the needle, as it were, is yanked from its groove disrupting the song of Schumpeterian disruption. Enter the force that perennially has bedeviled the smooth operation of Schumpeter's basic theory: the reluctance of the obsolete to go gently. Despite its vulnerabilities, the radio industry was deeply unimpressed by predictions of its demise, indeed would ferociously resist usurpation.

If any man was equal to yanking the needle, to turning the tide of history toward his own purposes, it was radio's emperor, David Sarnoff, president of RCA and the founder of NBC. Sarnoff could see plainly that television was coming—there was no burying *this* innovation—and so he determined that when it did appear, television would be under the firm control of his company and his industry. Television was not to

---

*As Farnsworth took the Pickford image from the film *The Taming of the Shrew*, the first public demonstration of electronic television was almost certainly an infringement of copyright. Later Farnsworth would use for his demonstrations Walt Disney's talking cartoon *Steamboat Willie*, also presumably without a license.

evolve, in response to ambient forces, à la Darwin; rather, it would be created as though by the God of the Old Testament, in the image of the creator, in this case radio. For Sarnoff, the paramount objective was that television not pose a threat to radio's hold on the attention of Americans in their homes, their attention to the advertising that was now the industry's lifeblood. In this purpose, he was endowed with superhuman means: the technical and financial resources of an industrial leviathan. And as if that were not enough, he would, as in other circumstances, also enlist the aid of the federal government in his struggle to survive the onslaught of creative destruction.

To say that television shaped American popular culture and social norms in the twentieth century, to the point of virtually creating them, is to state the obvious. If for no other reason, then, it is worth understanding just how it began.

In the early 1930s, it was by no means obvious what television would be. It might have emerged as an independent new industry to challenge the NBC-CBS duopoly and its advertising models. It might for some time have languished as an industry while thriving as an open, amateur medium, a hotbed of diverse content and points of view, such as radio was in the 1920s or Internet video in the early years of the twenty-first century. Or it might arise as something more like today's cable television or Hollywood, a producer of more elaborate programming not dependent on advertising. It was, in any case, never preordained that television would be the lackey of AM radio.

The story of television's founding presents a stark contrast to that of the heroic inventor-founder of American mythology. It is a dispiriting case of what happens when Kronos finishes everyone on his plate. The direct cultural consequences would be profound: two (later three) networks defining the medium that would define America, offering programming aimed at the masses, homogeneous in sensibility, broadly drawn and unprovocative by design, according to the imperatives of "entertainment that sells."

An Industry Is Born

Within a year of 1926, Jenkins's and Baird's primitive televisions had sparked a contest to found an industry. As *The New York Times* wrote,

"One of the strangest and most exciting races that the world ever witnessed is now in progress. Eight inventors, working individually and
in teams, are reaching out to clutch a prize as rich as any ever won by
Edison, Bell or Marconi." It very much seemed, then, as if another of
those great American (or Anglo-American) sagas of invention and progress was under way. Who, indeed, would be the next Bell or Edison?
In the United States, Jenkins was the clear front-runner. In the summer of 1928, Jenkins opened the world's first television station, W3XK,
in Washington, D.C. With programming five days a week, he soon
bragged of over 25,000 viewers, though the claim is hard to evaluate. By
1929 he'd opened a second station, in New Jersey, and now with two had
planted the seeds of the first television network.[7]

Jenkins's success soon began to attract competition from larger firms.
Among others, units of both General Electric and AT&T developed
their own versions of the mechanical television. In 1928, General Electric (Thomas Edison's old company) opened a station in upstate New
York and began broadcasting three times a week (ironically forcing Jenkins once again to compete against an Edison version of something he
had invented). While its technology was not as advanced as Jenkins's,
GE had a certain flair for flashy programming that grabbed headlines.
In the summer of 1928, for instance, using a portable unit, GE managed to broadcast Alfred Smith accepting the Democratic nomination
for president of the United States. A month later it would broadcast
what it heralded as the first television drama, a one-act performance
that left *The New York Times* spellbound. "For the first time in history,"
declared the front-page report, "a dramatic performance was broadcast
simultaneously by radio and television. Voice and action came together
through space in perfect synchronization. . . ."[8]

Meanwhile, in Britain, John Baird formed Baird Television Limited
in 1929 and secured the BBC's permission to broadcast one half-hour of
television, twice a week, at midnight. British television was far behind
its American counterpart in becoming a mass medium: asked, after his
very first BBC broadcast, to speculate on the size of the audience, he
answered, "Twenty-nine." Nevertheless, like Jenkins, he had proved
something—it was, as the *Daily Herald* wrote, "a British triumph."[9]

If their programming schedules and content were modest, Jenkins,
GE, and Baird had big plans. Jenkins in particular, with his two stations

and claims of substantial audiences, impressed would-be investors. In 1929 he made an initial public offering and raised $10 million in cash to fund his expansion. Baird, meanwhile, would be obliged to come to America to raise cash. Striking a deal with radio station WMCA for joint operation of an affiliated television station, he promised one million sets for the American market, at $25 apiece.[10]

Just what was the early technology like? Mechanical television was somewhat similar to photography. A disk is perforated with holes arranged in concentric rings, so that when an image from some lens is projected onto it, the disk as it spins successively "scans" the image, the light captured by each ring of holes representing a "slice" of the image that passes through to a photosensitive cell. At the other end of the wire, the process is repeated. There were inherent limitations: the image was always somewhat distorted, its resolution limited by the number of rings of holes. The first broadcasts managed between 30 and 60 lines of resolution, compared to 525 for standard television in the 1940s—as against the more than one thousand lines required now for "high definition." Inevitably, therefore the televised images would seem like novelty add-ons to the sound, still the primary element of the broadcast.[11]

Many historians assume that mechanical television was inherently too primitive to succeed as a consumer product. But while it doubtless would have been replaced by electronic television, to insist that mechanical television was doomed seems an overstatement. By the mid-1930s, Baird Television had developed to the point of over 400 lines of resolution, reasonably comparable to the standard resolution of later televisions. And it is hard, if not impossible, to predict what will be a technology's threshold of consumer acceptability. Looking at today's PlayStation, who would guess that Pong had once been a transfixing game? Or, for that matter, that in the age of hi-def, YouTube, with poorer resolution than television of the 1940s, could have caught on as it has? In fact, the primitive prototype is typical in the founding stage of a new industry, as are the "early adopters" prepared to take a chance on it. Recall that Bell's telephones, when they were first sold, hardly worked—it took Thomas Edison to create a truly functional device.

The reason such prototypes are sustainable, however briefly, and ultimately important is not their capacity to do what the technology is meant to do; rather, their value is in exposing a working model to more minds that might muse upon it and imagine a more evolved version. And so the clunky first telephones could inspire Theodore Vail and the Independents to envision a network reaching every citizen, just as the first personal computers, virtually useless, nevertheless set minds dreaming of future wonders. So it was with the television, too, the mechanical version stirring not frustration but imagination. "When perfection has been attained," wrote a reporter for the *Daily News* in 1926, television will transmit to London "the jostling crowds of Broadway, the millions of electric colored lights at night. . . ." So what if not every vision would come to pass, such as those of the otherwise successful inventor Archibald M. Low, who in 1926 predicted that there "may come a time when we shall have 'smellyvision' and 'tastyvision.' When we are able to broadcast so that all the senses are catered for . . ."?[12]

Yet even with the most imaginative early adapters, an industry must meet some basic requirements to take flight. The TV had to be transformed into, if not a perfect consumer product, at least a workable device for hobbyists, and to attract adequate capital, whether from public or private sources. Jenkins, like his colleagues a veteran of the radio pioneer days, understood the importance of a device targeting ordinary consumers, and in the beginning, he sold a kit for $7.50, before going on to sell the first ready-to-operate television set, which he called the "Jenkins Radiovisor."[13] Not knowing which name would catch on, he fudged it in his ads:

> It works! You can now enjoy radiovision programs. . . . Television is here! It is ready for the experimenter, service man, and dealer! Television programs are steadily improving. Now is the time to get into television. Experience the thrills of pioneer broadcast days all over again!

Forward-looking magazines like *Popular Mechanics* took the bait. A 1929 article pictures a smiling American family sitting in front of a Jenkins radiovisor. They are riveted by the device that is unrecognizable to us, with its small circular screen and its huge cabinet shaped like a table. The caption optimistically proclaims, "Already Television Has Reached the Stage Where the Whole Family Can Enjoy It."[14] It was just the toehold Jenkins needed in America, especially considering the backwardness of things in Great Britain. Unfortunately for him, others were already plotting against him.

SARNOFF ON MECHANICAL TELEVISION

"It would be easy to cry 'television is here,'" wrote David Sarnoff in the Sunday *New York Times* in 1928. "It would be easy, and it might be profitable, but it would not advance the day when sight is added to sound in an adequate service to the home in radio communication."[15]

The mechanical television, in Sarnoff's view, was a shoddy product that should not be sold to the public just so a few could make a quick buck. No less calculated is his implicit insistence that television's legitimate arrival could only be as an adequate *addition* to the service provided by nation's greatest (and only) radio network, NBC. Having sole proprietorship of the nation's ears, Sarnoff was laying his forbidding claim to their eyes as well, such as would follow adding "sight to sound." "We have created a nation-wide service for listeners-in," wrote Sarnoff, referring to the radio, "We must establish a similar service for lookers-on, once the road of transmission through space has been cleared."

David Sarnoff was a true visionary, but not of the progressive kind. He could foresee even in the 1920s, as he founded NBC, that television had the potential to destroy both the broadcast network and its parent company, RCA, increasingly dependent on broadcast revenues. And so as early as 1928 he was establishing a clear, consistent message: television is not ready for prime time. He was, of course, entitled to his opinion, just as any private citizen would be in expressing himself vis-à-vis a technology with which he was not enamored. (Many I know, for instance, quite disliked Twitter when it appeared, and discouraged friends from using it.) But Sarnoff was no ordinary citizen, having the ear of the Federal Communications Commission.* And his campaign of technological carping was disingenuous, his real aim being to persuade the FCC to freeze the television industry until RCA and the rest of the radio industry were ready to make it theirs.

From the late 1920s, Sarnoff, RCA, NBC, and later CBS repeatedly lobbied the FCC to adopt their view that television was simply an outgrowth of radio, and one that only the established radio industry could be entrusted to bring to proper fruition. An RCA submission to the

---

*The Federal Radio Commission was renamed the Federal Communications Commission in 1934. In this chapter, for simplicity, I refer to both as the FCC.

FCC from the 1930s, for instance, argues, "Only an experienced and responsible organization such as the Radio Corporation of America, should be granted licenses to broadcast material, for only such organizations can be depended upon to uphold high ideals of service."[16]

Unfortunately for Jenkins and the rest of the infant television industry, the commission proved receptive to such cajoling. While never coming out against mechanical television as Sarnoff had, the commission would tend to agree that in such form the technology was inadequate for marketing to the public. And so, in the name of progress and a brighter future, the FCC halted television in its tracks, just as it had done to FM radio to avoid unsettling AM.

The commission's motivations were perhaps not as conspiratorial as they might seem. True, it was staffed largely by men who had once worked for either Bell or the radio industry, in which contexts they would have absorbed the thinking that television was simply radio with pictures. But the FCC, at the time, was obsessed with the perceived benefits of "planning"—of setting out America's technological future in an orderly manner. Just as the Soviet Union was launching its own Five Year Plan, the FCC was busily and, one might argue, less efficiently working out the future of broadcasting. It therefore had its own ideological reasons for accepting RCA's self-serving claim that television, if it got off on the wrong foot too early, might "fix" on an inferior standard.

Consider for a moment the oddness of this phenomenon in a putatively free-market economy. The government was deciding, in effect, when a product that posed no hazard to public health would be "ready" for sale. Consider, too, how incongruous this was in a society under the First Amendment: a medium with great potential to further the exercise of free speech was being stalled until such time as the government could agree it had attained an acceptable technical standard. Rather than letting the market decide what a technology in its present state was worth, a federal agency—not even a democratically elected body—was to forbid its sale outright. One can see the logic of such oversight in the case of, say, an experimental cancer treatment—but a television set?

This is not to deny the utility of thinking ahead in general and in smaller contexts (as the Boy Scout motto advises), but central planning undertaken at a national level is quite another thing, its limitations evident in the experience of every controlled economy, even the universally dazzling example of China today. As Friedrich Hayek would later argue, how can the government possibly have enough information to know when something as unpredictable as technology is "ready"? What fate might have befallen the telephone, the radio, motion pictures—or, more recently, a strange new device like the iPod or a site like eBay, if going to market required one first to gain federal permission?

Acting in what it believed to be the public interest, the commission arrested the marketing of television from its invention in the late 1920s until the 1940s. It issued only a few licenses to men like Jenkins, and specified that they were for experimental purposes only—all forms of commercial television were banned. And this ban on commerce was strict enough that when Jenkins aired an announcement of his $7.50 television kits, the FCC sanctioned him.[17]

To be sure—unlike, say, the iPod—television technology depended on access to the radio frequency spectrum, which since the Communications Act of 1934 was subject to the management of the executive branch. In this sense, some measure of regulation by the government was, of course, to be expected. But even this fact cannot justify a total freeze on commercial television lasting nearly two decades. The contrast with early radio is instructive. When Hoover headed the agency, virtually anyone was welcome to run a primitive station, an environment that the Internet pioneeer Vint Cerf would later term "permissionless innovation." To run a television station, however, one had to apply to the FCC for an experimental license, subject to strict standards for obtaining it and for keeping it. A licensed broadcaster had to file regular reports, and show, among other things:

> That he intended to engage in *bona fide* experimental operations related to television; . . .

> That he had adequate financial responsibility, engineering personnel and sufficient equipment and facilities to carry out a research program.[18]

It was based on these standards that the FCC rejected John Logie Baird's plans for entering the U.S. market via a joint venture with WMCA in New York. The FCC's theory was that American radio stations could fulfill the experimental mandate just as effectively as Baird, and hence there was no reason to extend a license to a foreigner. And so the Baird Television Corporation never got its start in the United

States, a missed opportunity that both slowed television's penetration in America and helped doom an independent variant of the medium.

Apart from keeping out foreigners, the most stultifying effect of the FCC's television freeze was to inhibit potential investors, compounding the capital scarcity created by the Depression. As we have already seen, time and time again, it is investors as much as inventors who decide what our future will look like, and what we call genius might better be described as smarts coupled with capital. Unable to make money or attract funding, potential American manufacturers of mechanical television would all collapse or abandon their efforts within a few years. There was simply no business model or means of support after the initial novelty wore off. Bluster alone could not save even Jenkins, and so, by February of 1932, his reign as the master of television was over. With his bankrupt company in receivership, the radiovisor was no more. Jenkins himself died a quiet death two years later, having tried and failed twice to found his own information empire.

What was at stake here? What difference would it make that the FCC slowed down television to let RCA and Sarnoff get their ducks aligned, as opposed to giving the Independents their head? While not so obvious, the greater consequences may have been not for television per se but for the cause of innovation more broadly defined. Yes, Sarnoff would eventually bring television to market himself. Yet, as usually happens, the radio industry was led to television in the first place by the exertions of the independent strivers—it did not get there on its own. But if government makes clear that the game is rigged, that there is little room for the independent inventor to score, it removes the potent incentive for becoming a Jenkins, a Bell, or an Edison. As the Hush-A-Phone affair makes plain, the conditions facing entrepreneurs determine how much innovation happens.

There are yet more subtle social costs to a rigged game. Sarnoff's RCA was, like Zukor's Paramount studios, an integrated business. RCA both sold radios and owned the nation's main broadcast network, NBC. Naturally, then, as part of its plan to take over the emergent industry, Sarnoff wanted to see television, both as technology and as content, tailored to NBC's formula of "entertainment that sells." It was his intention (one he ultimately achieved) that, when the time came, all radio programming would migrate to television. Conversely, television would be nothing more than visible radio. And so we see the now familiar effect of a vertically integrated industry subsuming a new information medium, wresting control of it from a disorganized industry that in all likelihood would have nurtured more experimentation and more diverse ideas of what television should be. There can be little doubt that on the Procrustean bed of NBC's business model, television's growth as an expressive medium would be stunted.

Even if we accept the genuineness of its benign intentions, the FCC was oblivious or indifferent to the consequences of its planned future for freedom and variety of expression. It did not consider, or perhaps didn't recognize, that simply to hand over to the radio industry a medium with the potential power of television would be to determine who was to be heard (and seen) and who not. It did so, I would argue, out of a preoccupation with facilitating commerce, convinced that the NBC model represented the perfection of broadcasting. As with radio, the FCC, since its founding, had tended to accept NBC's view that independent TV stations were irresponsible organs of propaganda (in the value-neutral sense), and that only commercial networks like NBC could provide what the public needed. It is to our sensibilities an unlikely alignment of corporate and public interests, all the more in that NBC had never acceded to responsibilities of public trust as AT&T had in exchange for the blessings of authorized monopoly. While the telephone had been regulated with a full awareness of the power of the medium, television came into industrial being with little more than an extension of the philosophy underlying mass production as inspired by the automobile: Henry Ford's dictum, "the business of America is business."

In the end, it is perhaps incalculable whether the Depression, the insufficiencies of the mechanical version, or the hand of the FCC was most responsible for killing off America's first television industry. The Depression, although global, did not prevent other countries from launching their own television industries during this period, usually with government support. Since its inception, BBC television, for one, grew in quality and quantity, until by the mid-1930s it had begun to broadcast in what would become the standard resolution for electronic television. By 1935, Nazi Germany, too, would begin limited electronic broadcasts, and would be on the air all day with the 1936 Summer Olympics in Berlin.[19]

We fancy having in the United States the most open of markets for innovation, in contrast to the more controlled economies of other nations. In truth, however, the record is decidedly uneven, even given to excesses that would shame a socialist, with the federal government, at the behest of an entrenched industry, putting itself in charge of the future. Fortunately for the Free World, while the Nazis may have beaten us to television, we nicked them out for the Bomb.

David Sarnoff was silent during the years that saw the rise and fall of mechanical television, but he was not inactive. Having seen the future, he was discreetly planning to get there first. As early as the late 1920s, Sarnoff had secretly ordered RCA's laboratories to channel all efforts into developing a working *electronic* television. The FCC's suppression of the mechanical television industry (at his urging) ensured that neither Jenkins nor Baird would be in any position to challenge RCA once the next iteration came into the American market. But between Sarnoff and his goals stood one more obstacle: the last of the loners, Philo Farnsworth, the most formidable of television's inventors, and the holder of the seed patent.

The Third Independent

"S.F. Man's Invention to Revolutionize Television" was a headline in the *San Francisco Chronicle* on September 3, 1928, just a week before David Sarnoff's editorial denouncing early television. The article went on, "Two major advances in television were announced yesterday by a young inventor who has been quietly working away in his laboratory in San Francisco and has evolved a system of television basically different from any system yet placed in operation."[20]

Farnsworth's invention was a radical improvement on mechanical television. The old technology, with its spinning disk full of holes to scan and transmit an image, offered poor resolution and a picture that was always slightly warped. It is true that Baird Television had gone a long way toward improving the product, but in the end there was no overcoming the fundamental limitations of the technology. Enter Farnsworth's "image dissector," later simply called a "TV camera," which performed a line-by-line conversion of the photons making up an image into electrons, and line-by-line re-created the image on the receiving end. Compared to mechanical television, it offered a resolution that was, in principle, unlimited.[21]

Like most of the key inventors in this book, Farnsworth, who conceived electronic television at the age of twenty-two, was an outsider. Born in 1906, and raised a Mormon in Idaho, he made his first experiments as a teenager in a farm shed. He was also perhaps the first tech entrepreneur to base himself in the Bay Area. Farnsworth would patent his scanning technology in 1930 and have working models by the early 1930s. He was, then, in a perfect position to be the "first mover" in the electronic television market—if it were not for the subtle and not so subtle ways that Sarnoff, the radio industry, and the FCC set about neutralizing him.[22]

Sarnoff, to his credit (as an industrialist if not a moralist), realized that Farnsworth owned a technology that could be the foundation of a true television industry, and a dagger to the heart of radio and the RCA. He had a bit of help reaching this insight from his unwitting and naïve adversary. As Evan Schwartz, author of *The Last Lone Inventor,* documents, Sarnoff was tipped off by Farnsworth's demonstrations that revealed far too much about how his product worked. Even more foolishly, in 1929 Farnsworth gave a three-day tour of his laboratories to Sarnoff's top television scientist, Vladimir Zworykin, who gained admittance on the pretense of scientific curiosity and through playing on Farnsworth's hope that RCA might be considering an investment. Of course, Farnsworth needed publicity as well as investors, but he did not anticipate that Sarnoff would seek to displace him, not fund him. The most Sarnoff would ever offer him was a measly $100,000 for everything Farnsworth owned.

With word of Farnsworth's invention growing, Sarnoff started putting it about that in fact the young man had nothing of any interest to RCA or the market. He was bluffing: Schwartz and others have shown that from the late 1920s onward, RCA labs were feverishly trying to reverse-engineer Farnsworth's machine based on the information gleaned from Zworykin's visit. For while RCA had by now developed its own patented technology, Zworykin's kinoscope, the demonstrations of Farnsworth's image dissector showed his technology to be superior. Indeed, it would ultimately prove the starting point for modern television.

And so, as the 1930s began, Sarnoff and his allies were simultaneously trying to discredit the Farnsworth television and to reproduce it—such was the perverse genius of the Sarnoff plan. The talk campaign was similar to the one he'd lodged against FM and Edwin Armstrong: Farnsworth's invention, while of some scientific interest, did not work, and so his patents were invalid. The key audience for this disinformation was, of course, the investment community, and in 1935 RCA intensified the effort to keep them away from Farnsworth by means of what we now call a "vaporware" strategy. In 1935, Sarnoff announced that the company was commiting millions to build its own television, with its mighty industrial laboratories and its own resident genius, Zworykin. In the face of the imminent release of a "real" version of the television, covered by its own patents, only a fool would invest in Farnsworth.

Combined with the credit scarcity of those years and the FCC's ban on commercial television, the Sarnoff squeeze made survival quite a struggle for Farnsworth's "Television Laboratories." His machine, although it worked, still required powerful lighting, which problem he needed to address in development while he tried to build his own industry. He would find money to do neither, and one can well imagine the frustration of a man who in 1935 owns a working television camera and television sets of the highest quality anywhere, yet who is legally and financially constrained from bringing his surefire product to market. Little wonder that Farnsworth, despite his Mormon upbringing, turned to the bottle. As the years flew by, he would be stuck doing endless public demonstrations—one in 1934 a ten-day affair at the Franklin Institute in Philadelphia. He could still attract media attention, but without capital, he had nothing to build a company on.

Meanwhile, by the mid-1930s, Sarnoff's secret research and development effort had taken Farnsworth's invention beyond anything Farnsworth himself had managed. The RCA laboratories had overcome the need for strong artificial lighting and now had a TV camera that could work in mere daylight. In short, thanks in part to industrial espionage, RCA now had everything it needed to transform itself into a television company.

BAIRD AND FARNSWORTH UNITE

Given the FCC ban, it became clear to Farnsworth that his best chance to exploit his technology lay overseas. Britain was the world's leading television market in the 1930s, and in 1932 Farnsworth met with Baird to discuss a joint venture combining Farnsworth's technologies with Baird's organization and his contract with the BBC. Here was a partnership with much potential, for Baird Television Limited was at the time the most successful company of its kind anywhere, with more than two hundred employees and the world's only regular broadcasts via the BBC.

In 1936, Farnsworth visited Baird's operations in south London's gigantic Crystal Palace. Inside the massive structure, originally built for the Great Exhibition of 1851, Baird's company had constructed an astonishing complex combining a laboratory, television station, and production studio. Wedding some of Farnsworth's technologies with some of Baird's, the pair produced what were then some of the world's most advanced electronic television sets.[23]

Unfortunately, soon thereafter, disaster intervened. Farnsworth had arrived in April, and in late November of that year the entire Crystal Palace was consumed in a massive conflagration of mysterious cause—"the most spectacular night fire in living memory," as one newsreel put it.* In addition to the magnificent structure of cast iron and glass, also consumed by the flames was any hope for Farnsworth's technologies getting their start in England. All Baird's work, too, went up in smoke. Soon thereafter, the BBC would abandon his technologies for an electronic system from another company, EMI, whose technologies were based on RCA's. His business ruined, Baird returned to the life of a solo inventor. He would spend his last years creating a prototype of color television in high definition (about a thousand lines), television of an order that would not reach the public until the twenty-first century.

Now We Add Sight to Sound

It was at the 1939 World's Fair, held in the borough of Queens in New York City, that David Sarnoff would uncloak his scheme to establish RCA as the champion of American television. At the fair, RCA erected a nine-thousand-square-foot pavilion shaped like a giant vacuum tube and dedicated to the "Radio Living Room of Tomorrow." Its centerpiece, of

---

*Dramatic film footage of the fire can be seen on the Internet.

course, was television. Ten days before the fair opened, Sarnoff held a press conference, among the most effective presentations of its kind in the history of technology and communications. Alone at the podium, surrounded by rows of television sets, a curtain draped over them, Sarnoff would, to borrow a word from the future, reboot the history of television, proclaiming himself and RCA the founders of a new age.[24]

Baird had demonstrated the first working mechanical television in 1926; Charles Francis Jenkins had begun TV broadcasting in 1928; Farnsworth had patented the electronic television in 1930, starting experimental broadcasts in 1936; and the BBC had been creating television broadcasts of high quality since the mid-1930s. Nonetheless, in 1939, Sarnoff determined to hijack the narrative, rewriting the official story as the public would understand it. He made no mention of television's history or inventors. Instead, he said:

> It is with a feeling of humbleness that I come to this moment of announcing the birth in this country of a new art so important in its implications that it is bound to affect all society. Television is an art which shines like a torch of hope to a troubled world. It is a creative force which we must learn to utilize for the benefit of mankind. . . . Now, ladies and gentlemen, we add sight to sound!

At that moment, the veils lifted to reveal the rows of television sets, each of them tuned to the spectacle of Sarnoff standing at the podium. It was an image of such power as to overwhelm facts. The news media, having ostensibly forgotten every dazzling demonstration of the technology they had reported over the past thirteen years, fell lazily into line to inform an equally forgetful public that RCA and Sarnoff had invented television—and now were launching it in the United States. The "Talk of the Town" column in *The New Yorker* put it with laconic knowingness: "Last week, of course, witnessed the official birth of television." Decades later, that impression would stand unchanged. In 1999, *Time* magazine, celebrating Sarnoff as the "Father of Broadcasting," would harken to that "fateful day in 1939" when "Sarnoff gave the world a look into a new life."[25]

In a sense, of course, *Time* was right: Sarnoff had made American broadcasting in his image, though it was less an act of creation than of re-creation. The Independents had had their chance, if not quite a fair one. All the same, ten days after the Sight to Sound speech, Franklin Roosevelt made his first television appearance, and the game was over. From then on, American television, in practice and by common consent, did indeed belong to David Sarnoff, the Radio Corporation of America, and the networks NBC and CBS.

When TV reached consumers after the war, it was, as prophesied, a replica of radio in all respects. The programming was sponsored by advertisers, most of the shows simply adaptations of existing radio programs. Sarnoff had planned it all out in *The New York Times* in 1928 and, one step at a time, had made it all happen.

THE SEED PATENT

What ever happened to Farnsworth's patent? We might recall that in the 1870s, Bell had managed to force Western Union's retreat from telephony by means of a patent lawsuit. By such means, too, Sarnoff himself had pushed AT&T out of the radio industry in the 1920s. Farnsworth might have done likewise to RCA in the 1930s. But in Sarnoff he was facing ruthless genius not to be cowed by any man or law.

Reprising his strategy against FM radio, Sarnoff was willing to simply break the law, gird his lawyers, and force Farnsworth to seek redress. In fact, he went further, ordering his lawyers to challenge Farnsworth's patents and claiming audaciously that all the relevant ideas had been Zworykin's. Like the founders of the film industry, Sarnoff understood full well that obtaining justice can be expensive and time-consuming, and, with enough adroitness, one could found an industry on lawbreaking.

Forced into litigation, Farnsworth did eventually prove the validity of his patents in 1934, and late in 1939, after the World's Fair, he did force Sarnoff to license his technologies, nonexclusively, for approximately $1 million plus royalties.[26] In this sense Farnsworth's patents did pay off. But he was never able to accomplish the greatest power move of a patent holder and force RCA out of television altogether. He needed the cash, and in a deeper sense, was not ambitious enough perhaps, or simply lacked such aggressive lawyers as Bell had summoned in the 1870s. It's hard to know, exactly, what Farnsworth wanted; certainly he wished to be credited as an inventor, and to be paid his royalties, but did he, or anyone else in his company, have the dream of domination of the Schumpeterian "private kingdom"? Absent any evidence to that effect, one tends to see him as that variety of inventor ill equipped to be a founder, and one who furthermore failed to find a champion up to doing battle with the radio industry. There was no great warrior on Farnsworth's side, no Hubbard or Vail, and no one with the foresight to realize that the inventor had everything necessary to defeat Sarnoff.

Not just Sarnoff but luck, too, went against Farnsworth. The combination of the FCC's stultifying policies, the Depression, and World War II virtually ensured he would run out of time. Once his patent expired in 1947, and with it RCA's exclusive license, Farnsworth did market his own television sets, but by then, even as the acknowledged inventor of the modern television, he had no comparative advantage over RCA, General Electric, or any other firm. With his technological breakthrough, he'd had the potential, for a time, to be a disruptive founder, but now he was just a minor manufacturer. Mired in debt and short of parts, his firm soon went bankrupt.

Personally, too, it had all been too much. Farnsworth's drinking would grow more serious, and he would fall into a deep depression that nearly killed him before he abandoned the television industry altogether. In the end, he would be acknowledged on the medium he invented only once, in 1957, when a quiz show panel failed to come up with his name as the man who had invented electronic television.

By then, the medium was well on its way. The two networks, NBC and CBS, enjoyed a comfortable duopoly, aided in part by the FCC's policy of issuing just two licenses per community. Television sets were manufactured primarily by the old radio industry, dominated by RCA. And there was no debate over advertising or what the content of television should be. Unlike the telephone, radio, the Internet, and other technologies, electronic television in America simply skipped any amateur or noncommercial phase.*

---

*Right at the birth of television, there was one other chance for the industry to open up. As television gained popularity across the United States, a broad range of interests clamored to start their own stations. The film industry, notably Warner Bros., among others, applied to the FCC for television licenses. In some other version of history, Hollywood might have started a decidedly different brand of television in the 1940s, drawing on its vast stockpiles of content.

Who Lost Television?

Industry structure, as I have suggested, is what determines the freedom of expression in the underlying medium. Sarnoff did not aim to be a censor in the classic sense. And though he did, in 1938, propose a voluntary radio code modeled on the Hollywood Production Code ("It is the democratic way in a democratic country," he wrote), this was no doubt a typically calculated accommodation to the cultural climate—that is, a business decision—rather than any vision of the good.[27] Actually it is not clear that he had any particular opinions about programming. But that ultimately was the problem. Even more than Hollywood, Sarnoff and his cohort cared little about anything but profit and retaining control of their industry. With no need to fix what wasn't broken, there was only the slightest bit of experimentation in the early days of television, and very little reflection on what television could become. Rather, the unexamined vision of the commercial radio networks would continue to dominate, as it had from the start: light amusement produced by advertising firms, Huxley's soma for the masses.

The 1940s and '50s were mostly given over to awe at the new technology, and in the relatively untransgressive cultural climate in which television grew, and which it would help to sustain, it would be a long time before anyone began to question whether the medium had lived up to its potential for good. It was not until the late 1950s that industry figures such as Fred Friendly, head of CBS News, or media critics such as Walter Lippmann, began to meditate on what they saw as the great lost opportunity that was television. The tipping point of this accumulation of doubt came with the revelation, in 1957, that the networks' popular quiz shows were fixed; in response, Lippmann published a now-famous column entitled "The TV Problem":

"There is something radically wrong with the fundamental national policy under which television operates," he wrote. "The principle of that policy is that for all practical purposes television shall be operated wholly for private profit." By that logic, Lippmann reasoned, it was worth defrauding the public with the quiz shows if that was what it took to "capture the largest mass audience which can be made to look at and listen to the most profitable advertising." He concluded—exposing the transaction to which the generality were oblivious—"while television is supposed to be free, it has in fact become the creature, the servant, and indeed the prostitute, of merchandising."[28]

Such was the ignominious birth of the most influential medium of the postwar era. Under the suspiciously un-American banner of planning and progress, the indubitably American ideals of free exchange and merit rewarded were forgotten. Above all, the early history of the television shows us that the Cycle, at least respecting industrial destruction, is not, after all, inevitable. For the combined forces of a dominant industry and the federal government can arrest the Cycle's otherwise inexorable progress, intimating for the prevailing order something like Kronos's fantasy of perpetual rule.

The Rebels, the Challengers, and the Fall

IN THE SMALL CRACKS of the twentieth century's empires, challengers were slowly born over the decades of dominance. Interestingly, each of these would come to life as a tiny irrelevancy, a speck off the map. Small-town entrepreneurs invented the community antennas that would become cable television. A failing UHF broadcaster from Atlanta, Ted Turner, pioneered the idea of the cable network. Filmmakers until then excluded from all but the most obscure theaters would remake Hollywood, damaged by television and the antitrust division of the Justice Department. And an impractical, highly abstract academic project became, eventually, the first universal network: the Internet.

Part III tells the story of how information monopolies disintegrate. An industry is dominated by one ruler, an oligarchy or trust of some sort. What forces can break such hegemony?

CHAPTER 11

The Right Kind of Breakup

Toward the end of World War II, before the atomic bombs were dropped on Hiroshima and Nagasaki, the Sandia National Laboratories were founded in New Mexico. Situated near the better known Los Alamos labs, Sandia was to extend the basic work of the Manhattan Project into more sophisticated weapons development. The labs' ongoing mission was to serve as the "steward" of the United States' nuclear arsenal. What may seem surprising is that this top-secret effort should have been overseen not by the Department of Defense or some other government agency but, as late as 1992, by the telephone company. It all began when President Truman wrote a letter to AT&T subsidiary Western Electric. "In my opinion," Truman wrote in 1949, "you have here an opportunity to render an exceptional service in the national interest."[1]

Perhaps no other arrangement more clearly bespeaks the trust and intimacy that existed for decades between the U.S. government and the nation's great communications empires than the privilege enjoyed by the authorized telephone monopoly. Nor was Sandia Laboratories AT&T's only contribution to the Cold War. AT&T built a system of towers across the top of Canada and Alaska designed to warn of approaching ICBMs, a secret radio network to provide communications for Air Force One, and at least sixty hardened underground bunkers housing emergency equipment. Indeed, so essential to the common defense did AT&T seem that the Defense Department would intervene force-fully to prevent the company's breakup by antitrust suit in 1956, citing a "hazard to national security." Fittingly, during the 1950s, AT&T, for its part, adopted the notably Orwellian slogan "Communications Is the Foundation of Democracy."[2]

AT&T's relationship with the federal government may have been a uniquely intimate entanglement of interests. But, in fact, the blessing of the state, implicit or explicit, has been crucial to every twentieth-century information empire. We have seen how it influenced the course of radio (initially for military reasons) and later of television. In the case of Hol-lywood, and the government's decades-long acceptance of that indus-trial concentration, there may have been no national-security imperative beyond the morale of weary soldiers lifted by celluloid apparitions of Betty Grable. But as long as the studios had friends in Washington, their empire was secure. In every information industry, the government medi-ated what would have otherwise surely been a more tumultuous course of the Cycle.

Theorists of industrial evolution, Schumpeter foremost, have always understood the alternation of birth and destruction to be a natural inevi-tability of markets. Nothing, the theory goes, can stop an idea whose time has come. But what if in an otherwise free-market setting, an indus-trial entity enjoys a special forbearance or favor of the state. What if, as with AT&T, that favor amounts to its being a virtual organ of govern-ment? Can the natural ecology of the market still function, or is indus-trial creativity arrested? The power of state patronage or sufferance to any degree would seem to be more than any would-be competitor, even one armed with a technical breakthrough, can overcome. Herein lies the greatest complication to Schumpeter's idea of how capitalism works.

You cannot expect creative destruction to proceed normally in such circumstances: unseating such a monopolist thus becomes less a ques-tion of market dynamics than of politics. After the Second World War, the state would twice abandon its habit of tolerance and sponsorship, intervening in communications industries to break up dominant players. In 1984, it would have another run at Bell, this time finishing the job in had aborted in 1956. But even before the first attempt to break up Bell, in 1948, the government would take action against another information empire, forcing the Hollywood studios to sell their theaters and thus precipitating the collapse of the carefully designed studio system.

Both breakups—that of AT&T and that of the studios—would generate significant controversy, at the time and later. For in each case, there would be those who saw the dismemberment as a senseless summary execution of a robust, if restrictive, industry. With Bell particularly, a case in which the Justice Department had deferred action until a combination of the monopoly's arrogance and technological stagnation made it seem to many ludicrously overdue, there were nonetheless those who would regard—indeed, still do regard—the breakup as the crime of the century. In 1988 two Bell engineer-managers, Raymond Kraus and Al Duerig, would write a book called *The Rape of Ma Bell,* decrying how "the nation's largest and most socially minded corporation was defiled and destroyed." Barry Goldwater, the conservative icon and candidate for president, put it this way: "I fear that the breakup of AT&T is potentially the worst thing to happen to our national interests in telecommunications that will ever occur."[3]

The critics have a point: a federal breakup is an act of aggression and arguably punishes success. In the short term, the consequences of the state's interventions in both communications cases were ugly indeed. Each industry lapsed into an immediate period of chaos and experienced a drop in product quality. The decline of the film industry, which had been so grand and powerful in the 1930s and 1940s, would last into the 1970s. And in the immediate aftermath of the AT&T breakup, consumers saw a drop-off in service quality utterly unexampled since the formation of the Bell system. In fact, the "competitive" industries that replaced the imperial monopolies were often not as efficient or successful as their predecessors, failing to deliver even the fail-safe benefit of competition: lower prices.

Whether sanctioned by the state or not, monopolies represent a special kind of industrial concentration, with special consequences flowing from their dissolution. Often the useful results are delayed and unpredictable, while the negative outcomes are immediate and obvious. Deregulating air travel, for instance, implied a combination of greater choice, lower prices, and, alas, smaller seats, among other downgrades, as one might have more or less foreseen. The breakup of Paramount, by contrast, and the fall of the studio system ushered in something less expected: the collapse of the Production Code system of film censorship. While not the only factor transforming film in the 1960s and 1970s, the end of censorship certainly contributed to an astonishing period of experimentation and innovation. Likewise, the breakup of Bell laid the foundation for every important communications revolution since the 1980s onward. There was no way of knowing that thirty years on we would have an Internet, handheld computers, and social networking, but it is hard to imagine their coming when they did had the company that buried the answering machine remained intact.

The case for industry breakups comes from Thomas Jefferson's idea that occasional revolutions are important to the health of any system. As he wrote in 1787, "a little rebellion now and then is a good thing, and as necessary in the political world as storms in the physical. . . . It is a medicine necessary for the sound health of government."

Let us now evaluate the success of the government's first breakup of an information empire. It is not a tale to rival the epic of AT&T's breakup, which we take up in greater detail later. But it is the first crack in the ancien régime of state connivance with information industries and as such a fitting place to start.

THE STUDIOS

By the 1940s the Hollywood studio system had been perfected as a machine for producing, distributing, and exhibiting films at a guaranteed rate of return—if not on every film, on the product in the aggregate. Each of the five major studios had by then undergone full vertical integration, with its own production resources (including not just lots and cameras but actors, directors, and writers as human cogs as well), distribution system, and proprietary theaters. There was much to say about this setup in terms of efficiency, which was effectively an assembly line for film. Out of the factory came a steady supply of films of reliable quality; yet on the other hand, like any factory, the studios did not admit a lot of variety in their product. Henry Ford famously refused to issue his Model T car in any color but black, and while Hollywood didn't go that far, there was a certain sameness, a certain homogeneity to the films produced in the 1930s through the 1950s. That homogeneity was buttressed by the ongoing precensorship under the Production Code, which ensured that films would not stray too far from delivering the "right" messages: marriage was good, divorce bad; police good, gangsters bad—leaving no room for, say, *The Godfather,* let alone its sequels.

The cornerstone of the studio system was the victory Zukor won over the large first-run theater in a major cities and the ongoing block booking system. In America's ninety-two largest cities, the studios owned more than 70 percent of them. And though these first-run movie palaces comprised less than 20 percent of all the country's theaters, they accounted for most of the ticket revenue.[4] As the writer Ernest Borneman put it in 1951, "control of first run theaters meant, in effect, control of the screen."

The man inspired to challenge this system was Thurman Arnold, a Yale law professor turned trustbuster with some rather striking ideas about industrial concentration. Arnold, whose name continues to grace one of Washington, D.C.'s most prestigious law firms (Arnold & Porter), was by today's standards an antitrust radical, a fundamentalist who believed the law should be enforced as written. In *The Folklore of Capitalism* (1937), Arnold compared the role of U.S. antitrust law to statutes concerning prostitution: he deemed that both existed more to flatter American moral vanity than to be enforced.[5]

His language may have been strong, but Arnold had a point. By the time he took over the antitrust department in the 1930s, the United States, once a nation of small businesses and farms, was dominated by monopolies and cartels in nearly every industry. As the economist Alfred Chandler famously described it, the American economy was now dominated by the "visible hand" of managerial capitalism.[6] This despite the fact that the text of the Sherman Act, the main antitrust law, wasn't (and isn't) all that ambiguous. The law explicitly made monopolization and deals in restraint of trade illegal. A nonlawyer can understand this from reading sections one and two of the Act:*

> Every contract, combination in the form of trust or otherwise, or conspiracy, in restraint of trade or commerce among the several States, or with foreign nations, is declared to be illegal.

> Every person who shall monopolize, or attempt to monopolize . . . any part of the trade or commerce among the several States, or with foreign nations, shall be deemed guilty of a felony.

---

*The argument that the text *is* ambiguous comes from the idea that the law would make so much illegal that it couldn't possibly mean what it literally says.

Arnold, as soon as he gained Senate confirmation, acted quickly to implement his literalist view of the antitrust laws. His aim was to bring quick, high-visibility lawsuits breaking up cartels in whose evils American citizens could easily understand. His first lawsuits were brought against the car industry (GM, Ford, and Chrysler, the "Big Three"); the American Medical Association, which he charged with preventing competition among health plans; and most relevant for us, the film industry. Arnold's 1938 lawsuit against Hollywood charged twenty-eight separate violations of the Sherman Act and demanded that the film studios "divorce" their first-run theaters. And he repeatedly denounced the film industry as "distinctly un-American," and characterized its structure as a "vertical cartel like the vertical cartels of Hitler's Germany, Stalin's Russia."[7]

A decade would intervene, with various near-settlements and consent decrees, but the Antitrust Division finally achieved what Arnold wanted. In 1948, the United States Supreme Court agreed with the Justice Department's petition that Hollywood was an illegal conspiracy in restraint of trade, whose proper remedy lay in uncoupling the studios from the theaters. The Court's ruling by Justice William O. Douglas readily accepted Arnold's contention that the first-run theaters were the key to the matter, and with that acceptance disappeared any hope the studios might prevail. The Court ruled that they had undeniably fixed prices and, beginning in 1919 with Zuckor's Paramount, unfairly discriminated against independent theaters by selling films in block. There were various other offenses, but that was enough. Over the next several years, every studio would be forced to sell off its theaters.[8]

For the new information industries of the twentieth century, the Paramount decision was the first experience they would have of the awesome power of the state. The government had induced a paroxysm of creative destruction, seizing an industry by the throat. The infractions were indisputable, but there was nevertheless a degree of arbitrariness in the exercise of state power. Was this, after all, not the same government that had encouraged and supported the broadcast networks and the Bell system in their hegemonic forms? It was indeed, but Thurman Arnold was a different head of the hydra. Stripped of their control over exhibition, the Hollywood studios lost their guaranteed audiences. The business as they knew it would have to be entirely rethought.

In the short term came the chaos of breakup without the economic efficiencies. Robert Crandall, an economist at the Brookings Institution and a critic of the antitrust laws, has argued that the Paramount decree, as it was known, failed to lower the price of theater tickets.* And while there may never be a good time to sustain such a body blow, the action came at an especially bad moment for the studios; the arrival of television and the rise of suburbs after the war cut sharply into film viewership and revenues from the key urban markets. Still, in some sense the Paramount decree may have been just the bitter pill that the already listless studios needed: losing the first-run advantage would force them to reorganize and change the way films were made sooner rather than later. Institutional inertia being what it is, systems are rarely fixed unless they are broken, and this one, against its will, was broken utterly.[9]

Whatever its immediate consequences, the Paramount decision launched a transformation of American film as cultural institution, throwing the industry back into the open state it had emerged from in the 1920s. As Arnold had hoped, the independence of theaters cleared the way for independent producers, and even for foreign filmmakers, long excluded, to now sell to theaters directly. But the most profound effects of the decree would not emerge for decades. The industry would remain in an odd stasis through the 1950s and into the early 1960s. Eventually, though, as the mode of film production changed, returning to a decentralized style not seen since the 1910s and 1920s, so, too, did the product. After the decree, films were increasingly made one at a time rather than from a mold, according to the vision of a director or producer. "What replaced film production in the dismantled studios was a transformed system," writes the economist Richard Caves, "with most inputs required to make a film coming together only in a one-shot deal. . . . the same ideal list of idiosyncratic talents rarely turns up for two different films."[10]

With the fall of the studios, perhaps even more decisive than the transformation of production structure was the end of the censorship system. The power of the old Production Code written by Daniel Lord and enforced by Joseph Breen was effectively abrogated when the studios lost control over what the theaters showed.[11]

---

* Of course, there is no knowing whether prices would have risen even higher were the industry still intact but operating under new market pressures.

A very different type of production was feasible once theaters were free to buy unapproved films and ignore the regime that the studios had enforced in exchange for Breen's blessing. Producers took the cue, creating darker, countercultural, and controversial works—everything the Code prohibited. The Code itself was still around, but it had lost its bite. In 1966, Jack Valenti, the new, forty-five-year-old head of the MPAA, decided he wanted to "junk it at the first opportune moment." He noticed something obvious in retrospect: "There was about this stern, forbidding catalogue of 'Do's' and 'Don'ts' the odious smell of censorship."[12]

Valenti instituted the familiar ratings system (G, PG, R, X) in 1968, and far from marking a return to restraint, it was a license to make films patently unsuitable for children—even to the point of being what is euphemistically called "adult." At the same time, the freedom to import European films had its own influence on American production. Seeing the popularity of foreign offerings—typically moodier, more cerebral, and erotically explicit—the desperate studios were forced to invest in a new kind of American film. The result is known by film historians as the New Hollywood era, among its emblematic products *Bonnie and Clyde, Easy Rider,* and *Midnight Cowboy,* all edgy, defiant affairs announcing a new day for the industry and the culture.*

So great was the range of experimentation in film in the 1970s that for a time, as surprising as it sounds now, X-rated films—that is, pornography—went through "normal" theatrical releases. The most famous example is 1972's *Deep Throat,* which played in basically the same kind of theaters and made the same kind of money that a Hollywood blockbuster might today. Here was the medium as far as it could get from the days when the Production Code required preapproval of all films and obliged filmmakers, as a matter of course, to give audiences the "right" answers to all social questions.

Of course not every production of the period, which lasted until the early 1980s, would prove well made or enduring. Nevertheless the free-

---

*It is perhaps difficult to imagine that even without the antitrust action of the Roosevelt administration, Hollywood would not have evolved with the national mood in the 1960s. Changing sensibilities might well have upended the Code. But one shouldn't underestimate the capacity of an entrenched industry to avoid the risk of innovation, the initial resistance to features providing perhaps the most stunning example in the history of film.

dom to fail and sometimes to offend was extremely salutary for the medium in the era following the age of guaranteed success. What greatness did result came because directors and producers were allowed to experiment and probe the limits of what film could be. Whatever the merits of the individual outcome, the variety of ideas, in style and substance, was the widest it had been since before the 1934 Code.[13]

Antitrust action rarely takes the promotion of such variety and cultural innovation as one of its goals. The purpose of the statutes is to facilitate competition, not cultural or technological advancement (they were, after all, enacted under Congress's constitutional authority over interstate commerce). Innovation in an expressive form isn't ordinarily something one can patent, nor can creativity be satisfactorily quantified. But in considering whether government action was worthwhile, let us not, particularly where information and culture industries are concerned, fall into the trap of looking to results that only econometrics can reveal.

Films are not screwdrivers. As with all information industries, the merits of a breakup cannot be reduced to its effect on consumer prices, which may be slow to decline amid the inefficiencies and chaos of the immediate aftermath. But who would deny there are intangible costs to censorship? It is useful to consider whether Hollywood would be the peerless cultural export that it is were the industry not open to the full variety of sensibilities and ideas a pluralistic society has to offer.

CHAPTER 12

The Radicalism of the Internet Revolution

In late April 1963, in the D Ring of the massive new building called the Pentagon, J.C.R. Licklider sat before a typewriter in his office, working on a memo. A member of the Defense Department's Advanced Research Projects Agency (ARPA)—he wore the thick-rimmed black glasses popular among engineers of the era to prove it—Licklider addressed his memo to "Members and Affiliates of the Intergalactic Computer Network," as a sort of joke. But in this message he sent around to the nation's top computer networking scientists, Licklider argued very much in earnest that the time had come for a universal, or intergalactic, computer network: "It will possibly turn out, I realize, that only on rare occasions do most or all of the computers in the over-all system operate together in an integrated network. It seems to me to be interesting and important, nevertheless, to develop a capability for integrated network operation."[1]

That may not sound terribly exciting. "We would have at least four large computers," he continued, "and a great assortment of disc files and magnetic tape units—not to mention the remote consoles and teletype stations—all churning away." A collection of hardware churning away; but to what end? Actually, the "intergalactic memo" was the seed of what we now call the Internet.

We may exaggerate the degree to which an invention can tend to resemble the inventor, just as a pet can resemble its master, but "scattered" and "quirky" are terms equally befitting both the Internet and Licklider, one of the first to envision it. He carried with him everywhere a can of Coca-Cola, his trademark. "He was the sort of man who would chuckle," said his colleague, Leo Beranek, "even though he had said something quite ordinary." We met both men, readers will recall, during the Hush-A-Phone affair.

Born in St. Louis in 1915, Licklider undertook a protracted curriculum at Washington University, emerging with undergraduate degrees in psychology, mathematics, and physics. He made his name in acoustics, which he studied as a psychologist, aiming to understand how sound reached the human mind. After earning his Ph.D. at the University of Rochester, he taught at MIT in the late 1950s, where, at Lincoln Laboratories, he first used a computer and got hooked.[2]

Licklider came to believe that the computer would realize its deepest potential in linking man and machine. He was interested in all forms of technologically augmented human life—what science fiction writers call cyborgs, and what Sigmund Freud meant when he described man as a "prosthetic god."* In the 1960s, Licklider imagined a great universal network by which the minds of all humanity might be linked via computers. This strange idea was the basis of what we now call the Internet.

The basic story of the Internet's early development has been told many times; but our specific concern is to understand what was the same and what was different about this network as compared with radio, television, and the telephone system. Licklider and other early Internet founders believed that they were building an information network like none other. Some of its innovations, like packet switching, were obviously radical even in their day. Yet as we have seen time and time again, one generation's radical innovation is the next generation's unyielding dinosaur.

In this chapter, we begin the pursuit of a central question: Was the Internet truly different, a real revolution? We don't yet know the answer. But here, at its origins, we can gain the first inklings of what might account for that sense. The evidence boils down to the idea that of its singularity, the computer and the Internet attempted to give individuals a degree of control, of decision-making power unprecedented in a com-

---

*Freud wrote, "Man has, as it were, become a kind of prosthetic god. When he puts on all his auxiliary organs he is truly magnificent; but those organs have not grown on to him and they still give him much trouble at times."

munications system. These were systems whose priority was human augmentation rather than the system itself. The aim was therefore an effort to create a decentralized network, and one that would stay that way.

THE NETWORK AND THE COMPUTER

To understand how far any notion of the Internet in the 1960s might be from our present experience, consider how far were the machines it meant to link from any we would call by the same name today. Computers were fearsome creatures, the size of rooms, jealously guarded by companies and government agencies. Their main function was mass-produced arithmetic—"data processing." The archetype was the IBM AN/FSQ-7, the largest computer in human history, an electronic version of the Flying Fortress. As the scholar of media Howard Rheingold describes it, "the computers weighed three hundred tons, took up twenty thousand feet of floor space, and were delivered in eighteen large vans apiece. Ultimately, the air force bought fifty-six of them."[3]

There could be no Internet as we know it without a concept of the computer as something beyond an adding machine—this had to come first. The philosophy of the Internet and the computer are so intertwined that is difficult to discuss just one of the two. They are in the same relationship as the telephone and its wires, or the film industry and its theaters: one could not advance without the other.

In 1960 Licklider wrote his famous paper, "Man-Computer Symbiosis." Until then, if the dominant real-life vision of computing was IBM's giant abacus, the prevailing imaginative alternative was the cliché of 1950s science fiction and some of the day's more outlandish computer science speculation. It was an autonomous machine whose spirit would endure in the robot character of *Lost in Space* and the droids of *Star Wars*. Theorists of Artificial Intelligence foresaw computers that were intelligent machines, that could walk, talk, and help us with household tasks like washing dishes or greeting guests. This vision didn't suffer the problem we've identified as technological shortsightedness. On the contrary, it was just too far out.

Licklider's idea was different. "The hope," he wrote, "is that, in not too many years, human brains and computing machines will be coupled together very tightly, and that the resulting partnership will think as no human brain has ever thought. . . ." The idea is one we take for granted now: computers would be used by humans in the process of thinking, as analytic aids rather than as calculators (the status quo) or as surrogates (the stuff of fantasy).[4]

The idea wasn't Licklider's alone. As with other conceptual leaps we've described, several individuals also made it at about the same time. Ten years before Licklider wrote his paper, for instance, a young engineer named Douglas Engelbart was pondering what he might do with his life. He was recently married yet felt himself lost, an idealist in search of a meaningful contribution. One evening in 1950 he was struck with a powerful vision: a general purpose machine that might augment human intelligence and help humans negotiate life's complexities. John Markoff, who has documented Engelbart's life carefully, describes the vision in some detail: Engelbart "saw himself sitting in front of a large computer screen full of different symbols. He would create a workstation for organizing all of the information and communications needed for any given project."[5]

Engelbart's ideas were similar to Licklider's, if a bit further along in their development. But neither was as yet close to describing how one might practically wed human and computer capacities. Eventually Engelbart's work caught Licklider's attention, and with that, ARPA funding flowed to Engelbart to create the "Augmentation Reseach Center" at the Stanford Research Institute in Menlo Park, California. His immediate objective was finding better ways to connect the human brain to the power of a computer—what we now call "interfaces."

It's easy to forget that computers once took all of their questions and delivered all of their answers in numerical form. The basic ideas of a screen, a keyboard, and, most famously, a mouse are owed to Engelbart, who was the first to model those concepts, however crudely. He invented what would be called the "personal computer paradigm," and even if there is much more to the history of the PC than what he did, the degree to which our present matches his drawing-board vision of 1950 is a little unnerving—every day billions at home or at work sit down in front of something that is essentially what he imagined that evening.*

---

*Any who doubt the prescience of his vision are invited to watch a video of his model PC that Engelbart made in 1968. Some of the components are a bit off—the keyboard, for instance, is not QWERTY—but there is no mistaking the basic form of the computer that would not become commonplace before the arrival of the Apple Macintosh and the first browsers in the mid-eighties and early nineties.

Today, not only that interface but also that notion of what a computer is for—the vision Engelbart shared with Licklider—reign supreme. Nearly every program we use is a type of thinking aid—whether the task is to remember things (an address book), to organize prose (a word processor), or to keep track of friends (social networking software). This purpose of *personal* computing would go hand in glove with the idea of computer network communication. Both were radical technology; and fittingly, both grew out of a kind of counterculture.*

AT&T AND THE INTERNET

If, in the 1960s, computing was dominated by the giant data processing mainframe, the first (and last) word in communications and networking was still AT&T. AT&T owned the physical long distance lines that connected cities. If you wanted to send information from place to place, it was to AT&T you turned. Thus, improving communications meant improving the Bell system. Toward this end, a man named Paul Baran would spend years of his life trying to persuade AT&T to adopt the networking technologies that would ultimately underlie the Internet.[6]

In the early 1960s, Baran, a researcher at the RAND Institute, was thinking about how America could survive a nuclear attack. His goal, as he wrote at the time, was to "do all those things necessary to permit the survivors of the holocaust to shuck their ashes and reconstruct the economy." Chief among his concerns were communications systems. Having concluded that AT&T's long distance system was vulnerable to a Soviet strike, Baran came up with an ingenious means to harden the system. The idea was to try to turn the telephone infrastructure, a point-to-point system, into a highly redundant network—that is, one with various paths between any two points, so that if one route were taken out, the others would survive.[†] Baran's inspiration was the human brain, which can sometimes recover from damage by reassigning lost functions to neural paths still intact. In order for his approach

---

* Concurrent with Engelbart's design efforts was his participation as a subject in trials to evaluate the effects of LSD on human creativity.

[†] This, by the way, is the source of the commonplace that the Internet was designed to survive a nuclear attack.

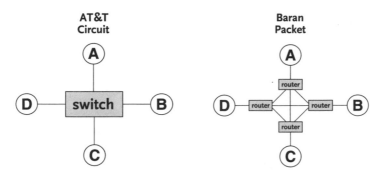

Circuit switching and packet routing

to work, Baran envisioned breaking up every message into tiny pieces, which would be sent over the network by any path available at a given moment. Today we call Baran's concept "packet networking," and it is the basis of almost every information network in the world.

These diagrams distinguish Baran's idea from AT&T's. On the AT&T network, a centralized switch picks a single route (a "circuit") between two points, A, B, or C. On the Baran network, the packets of information can travel between any two points in multiple ways. There are, as pictured, three different ways between A and B, for instance.

The key, however, is understanding how these different types of networks embody different systems of decision making. The AT&T system on the left is centralized, or hierarchical. The switch at the center decides how A will reach B. But, Baran's system features multiple decision makers of equal weight. Each "router" must help decide how information should get from A to D, and as you can see, there are three different paths. Hence, in the same way that Licklider's computer was meant to empower the individual, Baran's packet networks contemplated a network of equals.

Perhaps it is a philosophical impulse that helps explain AT&T's lack of enthusiasm for Baran's ideas. As Katie Hafner and Matthew Lyon write in *Where Wizards Stay Up Late,* packet networking struck AT&T officials as "preposterous." "Their attitude," Baran said, "was that they knew everything and nobody outside the Bell System knew anything. So here some idiot comes along and talks about something being very simple, who obviously doesn't know how the system works." AT&T

even went to the trouble of hosting a series of seminars to explain to Baran and others how the Bell system operated, and why a packet network was impossible, which is to suggest that there is more to their demurral than just the usual myopia. Ideologically, AT&T was committed to a network of defined circuits, or reserved paths, controlled by a single entity. Based on the principle that any available path was a good path, the packet concept admitted, however theoretically, the possibility of a network with multiple owners—an open network. And such a notion was anathema to AT&T's "ONE COMPANY, ONE SYSTEM, UNIVERSAL SERVICE."

Baran would spend four years at RAND trying to persuade AT&T to build the world's first packet network, which he saw as simply an advance, not a threat. Yet even with the Air Force offering to pay for an experimental network, AT&T would not be budged. Baran would have to look elsewhere to try out his ideas.

COMMUNICATIONS

"In a few years," Licklider wrote in 1968, "men will be able to communicate more effectively through a machine than face to face." If we owe the computer's interface more to Engelbart's vision, we owe its status as communications instrument par excellence more to Licklider's. It was his conviction that one day, the computer would displace the telephone as the dominant tool for human interaction. He was first to see the great coming rivalry between the telephone and the computer.

In his 1968 paper "The Computer as a Communication Device," Licklider and a fellow scientist, Robert Taylor, made the following prediction: "We believe that we are entering a technological age in which we will be able to interact with the richness of living information—not merely in the passive way that we have become accustomed to using books and libraries, but as active participants in an ongoing process, bringing something to it through our interaction with it, and not simply receiving something from it by our connection to it."

It is an astonishingly prescient comment, though it might have amounted to far less had Licklider not been appointed, by the Kennedy administration, to direct ARPA funding at the Pentagon in 1962. That position allowed him to direct capital toward individuals whose work he believed could make his great multiaccess network a reality—Engelbart, as we have seen, and also most of the other fabled fathers of the Internet.

Here, then, in the story of its origins, is the case that the Internet was different, that fundamentally different and indeed radical ideas were in play. But if computers had the potential to revolutionize communications, the challenges remained vast to put those ideas into effect. Computer communications required the development of a common language, which effort we shall follow in chapter 15, and some way to reach the masses. Both of those problems would take decades to solve— and so not until the 1990s would the seed bear fruit.

CHAPTER 13

Nixon's Cable

In the late 1960s, Ralph Lee Smith was at home one afternoon in New York's Greenwich Village when the telephone rang. It was an editor at *The New York Times Magazine,* well known to Smith, a freelance writer and frequent contributor to all the leading magazines in town. Familiar with Smith's progressive social criticism, including his ably researched books *The Health Hucksters* (an exposé of food and drug advertising) and *At Your Own Risk: The Case Against Chiropractic,* the editor wanted to suggest a subject: "cable television." Smith had never heard of it—in fact, he didn't even own a TV. But he thanked the editor for thinking of him.

Deciding the subject was worth a sniff, Smith began to talk to engineers, futurists, and government officials, and he became tremendously excited. All who spoke to him described the coming technology as having near-utopian promise for social liberation. Cable, they believed, might well prove more revolutionary than the printing press. With the capacity in theory to bring an unlimited number of channels of information into the home, it had the potential to heal American politics, revive local communities, and offer every American direct access to the world's knowledge and wisdom: "a communications center of a breadth and flexibility to influence every aspect of private and community life."[1]

Smith became a believer. The idea of a technology that might democratize information resonated with the values of late-1960s New York, a folk music hotspot where only recently the city government had managed to vote down such imperious designs as Robert Moses's Lower Manhattan Expressway. Smith wrote a manifesto called *The Wired Nation,* which won awards as a magazine article and later a book. Smith thus suddenly found himself at the vanguard of a visionary—and today mostly forgotten—movement to promote cable television as a technological savior of liberal values.[2]

By the 1940s the major media industries had all assumed their stable, apparently invincible forms; they seemed to be permanent fixtures of the American landscape, like the Democratic Party or Mount Rushmore. NBC and CBS ruled broadcasting. AT&T ran the telephone system. The Hollywood studios controlled film. Each monopoly or oligopoly had been blessed by the government in one way or another. And within two decades each would lie in the ruins of its former self.

The empires of AT&T and the Hollywood studios would be broken up by court orders. But broadcasting's fate would be different. The stations, and ultimately the networks would be natural victims of the Cycle. Cable was the first disruptive innovation since the war, and one that would shred the prevailing power structure of television. Ralph Lee Smith was thus the 1970s avatar of what to us is a familiar figure: the idealist who helps to usher a closed established industry into a wide-open, expansive phase.

What few people know is that Ralph Smith's arguably most important ally in this power-to-the-people crusade was President Richard Nixon. In the 1960s, cable was a technology serving small towns and remote localities, barred by federal law from expansion. It seemed doomed to being but the handmaid of broadcasting. Indeed, in another version of history, the cable networks would have emerged only as offshoots of NBC, CBS, and ABC, as has been the fate of cable in other major economies, among them Japan and Germany. But the Nixon administration had a different vision for cable. Nixon's young head of communications policy, Clay Whitehead, ran the Cabinet Committee on Cable, which foresaw a life for the medium as a highly deregulated common carrier. And it was Nixon's FCC that would launch the reforms to set cable free, for reasons somewhat more complicated than the general advancement of freedom.

CABLE IMPRISONED

In the late 1960s, cable had a distinctive identity. It was a scrappy industry of small-town entrepreneurs in perpetual trouble with the law, something akin to the file sharing sites of the early twenty-first century—a
band of outsiders, certainly; outlaws, maybe.[3]

In the gleaming media metropolis, cable was the dive bar. It attracted
shady characters, and its function was, one might say, parasitical. Cable
founders were offering something that was hardly new or bold. The
concept was called Community Antenna Television, a system to capture and retransmit TV to places that the broadcast signal didn't reach.
As with broadcast radio in the 1910s, the origins of cable television are
obscure, because it was the work of amateurs.

In the late 1940s or so, men like John Walson, the owner of an appliance store in the Pennsylvania mountains, began erecting giant antennas to "catch" the weak signals and then transmit them over wires to
paying customers. As with the farmer's telephone in the first years of
the twentieth century, cable was a do-it-yourself business for anyone
with will and wires. It was a genesis and a business model that would
ever after be stamped as pugnacious and cut-rate, a sharp contrast to the
affected regal bearing of NBC (the Peacock Network) and CBS (the Tif-
fany Network), to the ultra-establishmentarian self-importance of the
Bell system or the glamour of classic Hollywood.

At first, broadcasters could ignore cable as an irrelevancy at worst and,
in extending the reach of broadcast at the margins (a bonus audience for
advertisers), a minor help at best. The friction started when enterprising
cable operators began to set up shop in larger towns and offer "imports"
of channels from other vicinities. The signal-obstructing hills of Pennsylvania, for instance, made it a prime market for cable TV (as they had,
interestingly, also for early radio). A small-town operator might offer his
customers not just the local broadcasts but Pittsburgh stations as well.

The fat did not really hit the flame until the late 1950s, however,
when the cable operators began to lease time on microwave towers,
allowing them to import stations from even farther away. Now mostly
supplanted by fiber optic technology, these structures were then cropping up across the country, originally erected to provide a cheap way
for radio and television networks to move their signals from one city

to another in relay fashion via high frequency microwaves. The real significance of these towers to our larger narrative, however, will be as the first alternative to AT&T's long lines, a new channel for sending information across the country instantaneously. But so far as broadcast/cable relations were concerned, microwaves were the last straw. The two sides would become the implacable foes they remain to this day.

The broadcasters banded together to squash cable, or at least beat it back into the boonies. Their campaign would mark another instance of the Kronos effect: an effort by an existing media power to devour a suspected challenger in its infancy.

The broadcasters were not paranoid: cable was indeed an idea with the power ultimately to destroy broadcasting by stripping it of its audience. By freely importing and exporting stations between cities, the cable operators threatened to fragment what were once fixed, guaranteed audiences for local broadcasters, who enjoyed an effective monopoly and could charge advertisers accordingly. (If there was a chance someone in town was getting their television from a nearby city, the local broadcaster's claim to audience and thereby revenues was diminished.) The campaign against cable, however, was waged in terms of loftier principles than simple commerce. The broadcasters framed "the cable problem" as a crisis of infringed rights of intellectual property, an attack on free television (cable, unlike broadcast, was from the start by subscription), and even a threat to social mores.

As one local broadcaster testified in 1958, "We believe that when a community antenna system takes our programs out of the air, without our permission, and sells that program material at a profit—and in many cases, a fantastic profit indeed—this is a violation of our property rights."[4] The copyright complaint was also brought before the U.S. Copyright Office, summarized as follows: "The activities of the CATV operators constitute a 'clear moral wrong' comparable to the old practice of 'bicycling' movies from one theatre to another in order to get two performances out of a single license."[5] It was the broadcasters' contention that cable would destroy local media and with it local communities, for the importation of big-city stations would bring with it big-city values. Jack Valenti, Hollywood's lobbyist, lent moral support from a sister industry, calling cable "a huge parasite in the marketplace, feeding and fattening itself off of local television stations."[6]

In a sense, some of what the broadcasters charged was true, especially in the long run. Cable did in fact virtually destroy the world of "free" television: by 2010, the vast majority of American households would be paying for access to television, either through cable, satellite, or fiber optics. It is also true that the importation of channels made the local broadcaster less important or viable. On the other hand, these local operators were typically affiliates of one of the national networks, making their claim to local legitimacy more a matter of form than of substance.

In addition to this rhetoric, the broadcasters brought a full-scale legal attack to bear against cable before the federal courts and the FCC. They began by accusing cable of unfair competition and copyright infringement. It is one thing for an individual's antenna to capture a transmission out of the air, argued the broadcasters, but when a firm retransmits it to thousands of its subscribers, it is "performing" the program without permission, therefore illegally. The broadcasters' copyright infringement suit went all the way to the Supreme Court, where in *Fortnightly Corp. v. United Artists*[7] in 1968 the Court ruled that the cable operators had done no wrong. The Court's reasoning was simple, if a bit circular: cable operators were part of the free audience for the work, albeit possessed of an unusually powerful antenna, and their retransmission no more constituted a "performance" under statute than if an apartment building owner had set up an antenna for the benefit of all his tenants. The majority opinion by Potter Stewart (whose famous common sense allowed of pornography, "I know it when I see it") represented a fairly clear effort on the part of the Warren Court to throw the cable industry a lifeline, despite the anchor of the copyright law.

But if the courts were hesitant to throw the book at a new industry on behalf of the broadcasters, the FCC would prove much more receptive to the plea. As in the 1930s, the commission was gripped by fear of new technology, and when asked to consider cable, they acted like the farmer who is dismayed by a tractor's lack of horses. The merits of the innovation—access to dozens of channels and a high quality picture—did not fit their mandate as they saw it, which was to bring free television to the people, improve the quality of what was broadcast, and encourage the rise of as many local stations as possible.

Cable television just didn't fit the mission. The technology, moreover, such as it was, came not from Bell Labs, MIT, or some other pedigreed institution, but from a collection of small-town wheeler-dealers—owners of appliance stores, for example, who saw cable as a way to supplement their income. Besides, by the 1960s the FCC had found another idea of what the future of television would be: UHF or Ultra High Frequency broadcasting (also known as the "bottom dial" on old TV sets, before electronic tuners). UHF was similar to VHF, the existing broadcast television, but tended to propagate more weakly.

Siding with the broadcasters, the FCC began to use its regulatory powers to throttle the cable industry. Its most aggressive move came in 1966, when, having decided that cable posed a threat to the common good, they issued an order barring it from America's hundred largest cities or towns by population. It was, effectively, a cease and desist order; bringing TV to remote towns was fine, but for the sake of the public interest, cable would not become a major means of distributing television.

With such operational constraints, investment in cable dried up. And so it was that what had seemed as late as 1968 a classic case of creative destruction in the making, the emergence of a clearly disruptive technology in which even some broadcasters, seeing its inevitability, had already begun to invest, ended with deepening the entrenchment of the established broadcast industry. As the economic historians Stanley Besen and Robert Crandall would later write, "Cable entered the 1970s as a small business relegated principally to rural areas and small communities and held hostage by television broadcasters to the Commission's hope for the development of UHF."[8] Things could very easily have stayed that way. In many countries, cable television was effectively blocked by regulation, even today reaching just a small percentage of homes. As recently as the 1990s, most Britons received just four broadcast television channels. But American cable's redeemer was coming, and in a most unlikely guise.

CABLE'S SAVIORS

In the annals of television, Fred Friendly is known for having, among other things, collaborated with his close friend Edward R. Murrow to create *See It Now* on CBS, a totally new type of program that aimed to use the power of network television as a counterweight to political authority. The content varied, but the fundamental idea was to offer a forum for otherwise unheard voices, perhaps the most famous of these being that of Milo Radulovich, a U.S. military officer victimized by Joe McCarthy's Communist witch hunt. Conceived in a crusading spirit going back at least as far as Upton Sinclair, the show certainly didn't represent a revolutionary mission for journalists, or for the media for that matter. Yet it was a novelty for network television, and one that would change the face of the medium completely. Friendly would eventually abandon the networks to become a founding advocate of public broadcasting, and by the late 1960s, he was in the vanguard with Smith and others evangelizing for cable.[9]

Smith and Friendly were both residents of an American city critically in need of cable TV: New York, or more precisely, Manhattan. Like those who inhabited the mountain towns of Pennsylvania or the West, Manhattanites, caught in the canyons formed by skyscrapers, had difficulty receiving TV broadcasts, and so the cable companies found a ready market. Unsure of where he should stand on the matter, Mayor John Lindsay in 1967 put Fred Friendly in charge of a commission to study cable television in New York.[10]

Friendly had slightly less utopian ideas than Smith about the promise of cable. He had by then spent a decade trying to build up public TV as an alternative to the networks, and in cable he saw another means to the same end. His vision was less Smith's brave new world of radically democratic access to a limitless variety of content, and more a pragmatic way to alleviate the relative scarcity of options. As he described the problem in *The Saturday Review:*

> What ails us is not too many Brinkleys and Cronkites, not the broadcast executives who favor Nixon . . . not a conspiracy of white supremacist station owners who will not give minority groups the prime time of day (although there are a few of these). Rather the major restrictive and malevolent force is the absurd shortage of air time.[11]

With just three TV channels, producers had to make hard choices, as in 1964, when faced with broadcasting the cash cow *I Love Lucy* or the congressional hearings on the Vietnam War. Friendly's vision of public television was of a channel free of commercial considerations and thus always available to serve up "alternative" content. In cable, Friendly saw the same, only more so: a giant, wide-open medium where the public interest could be given its due respecting all sorts of issues, and the people thereby empowered.

Friendly had identified a new reality of the age of mass information: the power of concentrated media to narrow the national conversation. It may seem paradoxical to suggest that new means of facilitating communication could result in less, not more, freedom of expression. But a medium, after all, is literally something that comes between the speaker and the potential listeners. It can facilitate speech only if it is freely accessible. And if it becomes the means by which most people inform themselves, it can decisively reduce free speech by becoming, whether by malign intent or merely benign effect, the arbiter of who gets heard. It was by such means, Friendly believed, that the shortage of TV stations had given exclusive custody of a "master switch" over speech, creating "an autocracy where a very few citizens are more equal than all the others."

Based on this logic, Friendly developed a notion of how cable might cure what was ailing the nation, including electoral politics, an arena in which he thought television had "made the high cost of campaigning an aberration of democracy." Ralph Smith had a similar hope, prophesying that "CATV could arrest and reverse some ominous developments in American electoral politics." The two shared the view that the cable system could simply open an extra channel exclusively for politicians to speak to the public. Without the need to spend money on broadcast time just to be heard, all politicians, presumably, could compete on equal footing in the marketplace of ideas.

The Alfred P. Sloan Foundation, a philanthropic organization founded in 1934 by a former CEO of General Motors, had also become a voice in the debate over cable and went further than Friendly or Smith, proposing a future including both open and partisan cable channels. It would be an expensive arrangement, conceded the Sloan Foundation, but the latter sort of outlet, they theorized, could be "an extremely valuable fund raising instrument, and might well pay its own way."[12] In suggesting the concept of an all-news channel that could also be of use to campaigns, the foundation had perhaps a premonition of the stroke of genius that would give birth to Fox News. But it remains a matter of heated debate whether the existence of such a commercial outlet relieves or exacerbates what Friendly termed "an aberration of democracy."

These were some of the great hopes for cable. But the cable dreamers were also mindful of dangers. Should cable come out from under the heel of broadcast, it must not then become a monster in turn. Above all, its champions pressed for some form of common carriage regulation. What this meant exactly wasn't clear, but there was a general concern that cable should carry content without discrimination, and should be impressed with duties of public service. Friendly, of all the cable evangelists the most cynical about what might go wrong, wrote, rather presciently, "If not regulated, the current Monopoly could give way to a new Tower of Babel, in which a half-hundred voices scream in a cacophonous attempt to attract the largest audience." It was clear to him that cable could equally be a force for the worse as for the better—and if the former, he predicted, then "a debilitating and decaying force that could one day make us look back at the Sixties as the Golden Age."

As Friendly, Smith, and the others championed cable, its true white knight lurked unsuspected in the Nixon White House in the person of Clay Whitehead. At the age of thirty-two, Whitehead had been asked to lead the newly created Office of Telecommunications Policy, from which, working with a now less hostile FCC, he proceeded to launch the initiatives that would untether cable.

First among these was the creation of the Cabinet Committee on Cable Communications—which, as the name suggests, was a cabinet-level body appointed to decide the future of the cable industry. What's less predictable, perhaps, is that the Nixon administration's vision for cable was in some ways almost as idealistic as Friendly's and Ralph Lee Smith's. The administration, first of all, wanted to repeal the regulatory blocks imposed on the industry. Yet it also proposed a strict division between ownership of the cable lines and power over programming. The cable operator was to be granted discretion over the content of only one or two channels; the rest would be reserved for public interest programming, or freely available for lease by anyone. This arrangement the Nixon administration called its "separations policy."[13]

One cannot fail to be impressed by the radicalism of the Nixon policy.

Both in freeing the cable industry from geographical restrictions on its business and in denying both the federal government and the operators themselves any power over programming, the administration was evincing a hard-core libertarian streak not always associated with a White House that also spied on its enemies. It can't be accounted for precisely. Possibly the Cabinet Committee's views were really just those of Clay Whitehead, who, while believing in deregulation, tended to view the cable operators as no less a potential threat to diversity of speech than the government. This wariness of corporate power was also at the heart of his most famous initiative, the so-called Open Skies policy, which permitted any qualified company to launch a satellite, a technological shift that would liberate not only cable but long distance calling, too, as we shall see.*

Still, one cannot overlook President Nixon's immediate and personal motivation to help out the cable industry. In an increasingly pernicious ecosystem of information created by his perceived enemies, the networks reporting on the war in Vietnam and Watergate, he had identified their natural predator. The president had already channeled considerable thought and emotion toward the goal of bringing down the networks and their news departments. The logic of giving new freedoms to the cable industry cannot conceivably have been lost on his ceaselessly strategizing mind. And so, unlikely as it may seem, the president better known for threatening the media—his attorney general would infamously warn *The Washington Post* that its publisher would "get her tit caught in the big fat wringer" if Robert Woodward and Carl Bernstein continued their explosive investigation—must be credited, in part, for one of the greatest liberalizations of media in postwar history, and for the launch of the cable industry. As a further irony, the Cabinet Committee's so-called Whitehead Report would come out seven months before Nixon resigned his office.[14]

The administration changed, but not the drift of reform. Through the later 1970s, Ford's and then Carter's administrations would pick up

---

*Whitehead was a character of fascinating contradictions. On the one hand, he would be a point man in Nixon's increasingly fierce battles with the news media over Vietnam and Watergate, imputing to them liberal bias by describing their content with the deathless phrases "elitist gossip" and "ideological plugola." On the other hand, he was one of five federal officials working behind the scenes to lay the groundwork for Gerald Ford's presidency as the cauldron of Watergate was boiling over.

where Nixon's left off. The rule requiring the creation of local content was turned into the much less onerous condition of providing "public access" channels. Eventually, as a kind of sop to the networks, the White House would also broker a deal whereby cable television was brought under copyright. Thus by the end of the seventies the cable experiment was in full swing. It remained to be seen, however, what the medium would become.

CHAPTER 14

Broken Bell

Clay Whitehead, Nixon's telecommunications czar, became in 1974 the first government official to call openly for an end to the Bell monopoly. Just one month before Nixon's resignation, he declared, "Unless the would-be monopolist [AT&T] or the public can demonstrate special public policy considerations that justify monopoly, it should not be permitted." The "anti-trust laws," he said, "should be enforced to ensure that regulatory mechanisms cannot become a haven for escape from competition."[1]

That kind of language from the White House was a shock to AT&T, the longtime friend of the federal government; and yet by the late 1960s, with everything in flux, the Nixon White House and FCC began to question the continuing benefits of the Bell monopoly. In fact, the FCC had begun to entertain the radical belief that a bit of competition was not only feasible (technically and politically) but indeed might just help the cause of efficiency in the telephone system (a line of reasoning paralleling that about the cable industry). It was a paradigm shift and an ideological reordering in process: bit by bit, Theodore Vail's faith in centralized monopoly was giving way to a new belief in the value of decentralization. And by the 1970s, both the White House and the FCC's official rhetoric began to use terms like "competition" and "deregulation" instead of "regulated monopoly."

To say AT&T wasn't receptive to the new paradigm of the 1970s would be an understatement. AT&T's new chairman, John deButts, responded with a speech that would not have been out of place in 1916. "There may be sectors of the economy," he said, "where the nation is better served by modes of cooperation than by modes of competition . . . the time has come, then, for a moratorium on further experiments in economics." He would spend the 1970s doing everything within AT&T's power to try to turn the river around, toward more monopoly and less competition, against all odds. He was a true Bell man, almost stereotypically so. *People* magazine called him "a one-company man and proud of it" and a man who "remembers the days when saying Ma Bell was a monopoly was an expression of pride." "The sacred public mission of the Bell System not only had been drilled into his mind," writes Steve Coll, "but had seeped into his soul." The collision between AT&T and the tides of history are what led to the second great communications breakup of the twentieth century.[2]

THE FCC TURNS ON AT&T

The trouble began when the FCC, increasingly convinced that competition had some place in the telephone system, began to think it would be a good idea to create a few little pools, as it were, of competition, here and there, while leaving the main Bell monopoly intact. That might sound moderate, but the reaction of AT&T at the time was resistant in the extreme. It had the mentality that to give an inch was to give the whole yardstick, and so it stuck stubbornly to the conviction that it, and it alone, needed to retain total control of every element of telephony. Echoing Theodore Vail's old condemnation of "wasteful competition," Bell argued that any degree of it, however minor, would create chaos. Judge Richard Posner, a consultant for AT&T in the 1970s, describes their rigid belief "that nobody should be permitted to interconnect with the network. . . ." In particular there must be no interconnection by MCI, AT&T's bête noire, and no attachment of terminal equipment by customers (that is, no "foreign attachments"). That was AT&T's absolute line of defense: "not one step backwards," as Stalin said when the Germans were approaching Stalingrad. In a 1968 submission to the FCC, Bell made the same point, if less colorfully: "Since the telephone companies have the responsibility to establish, operate and improve the telephone system," argued Bell, "they must have absolute control over the quality, installation, and maintenance of all parts of the system in order effectively to carry out that responsibility."[3]

The FCC went ahead with creating three main "pools" of competition with the Bell system: long distance services, attachments (or "consumer premises equipment"), and "data processing" services. Let us look at each in turn.

The company known as AT&T was born in 1885 as a long distance company, only later to become a holding company for the whole Bell Empire. AT&T had long regarded its long lines as the crown jewels of its kingdom. Indeed, long distance was the key to its power, and also central to Bell's whole idea of "universal service": the firm used inflated prices on long distance calls to subsidize phone service to rural communities, making good on its pledge of universality without suffering a financial loss.

So important were the long lines, you may remember, that in the early 1900s J. P. Morgan had used his financial muscle to prevent financing of any would-be alternative network. Denying access to the long lines, meanwhile, proved an instrument of terror in the campaign against the Independents in the 1910s. By the late 1960s there had still only ever been one national long distance network, and it was the hallmark of Bell's identity.

Suddenly, and unexpectedly, the FCC, AT&T's old ally in the suppression of competition, was a friend to competitors. An upstart known as Microwave Communications Inc. (MCI), founded in 1963, proposed to use microwave towers to sell cheaper private line service for businesses between Chicago and St. Louis. AT&T, of course, regarded MCI's service as inferior, redundant, and a challenge to the universal service system. In the Bell company vernacular, MCI was engaged in mere "cream skimming" of the most profitable services (long distance for business customers) without taking any responsibility for basic services. Remember, faith in its public duty was as much encoded in AT&T's DNA as its right to monopoly. Nonetheless, much to AT&T's outrage, the FCC let MCI go ahead, making the microwave company for a generation thereafter the chief thorn in AT&T's side.[4]

The second pool of competition was in devices that attached to the phone lines ("foreign attachments" or "consumer premises equipment").

In functionality, these devices made the Hush-A-Phone that Bell had fended off in the 1950s seem a mere doodad. In a seminal case in 1968, the FCC ordered Bell to allow the connection of the "Carterfone," a device designed to connect a mobile radio to a Bell telephone.* Based on Carterfone advance, the FCC went further and specified something simple but absolutely essential: the familiar RJ-45 telephone jack. You have probably used the phone jack hundreds of times without realizing the hard-fought battle behind it. The modular phone jack made it unnecessary for a Bell technician to come and attach one's phone to the phone line. More crucial, with the phone jack in place, any innovator—any person at all—was suddenly free to invent things that could be usefully attached to the phone line.[5]

That phone jack and the Carterfone decision made it possible to sell to the public devices like fax machines and competitively priced (non-Bell) telephones. They also made possible the career of Dennis Hayes, a computer hobbyist ("geek" is the term of art) who, in 1977, built the first modulator/demodulator (modem) designed and priced for consumers, the so-called Hayes Modem. He built, that is, the first consumer device that allowed personal computers to talk to each other, and with that you can spy the first causal relation between the federal deregulation of the 1970s and the birth of a mass Internet.[6]

The third pool of competition that the FCC created is the most obscure, but no less important. In 1971, the FCC issued a rule banning AT&T from directly entering the market for "data processing" or "online services." These were the earliest precursors of what we now call Internet services, though in those days it usually meant accessing a more powerful remote computer to help with number crunching. The FCC decided it would reserve this market for companies other than AT&T, though it did allow AT&T to participate at arm's length via a subsidiary. The reasoning was that if AT&T were allowed direct access to the market, it would immediately destroy any competitors and colonize the market for itself. And so, just as predatory fish are sometimes kept in separate tanks, AT&T was specifically banned from the burgeoning online services or data processing industry.[7]

---

* Today, in telecom jargon, a "Carterfone rule" is one that allows consumers to attach whatever they want to a network, physical or wireless.

In short, with strange and unprecedented foresight, the FCC watered, fertilized, and cultivated online computer services as a special, protected industry, and, over the years, ordained a set of rules called the *Computer Inquiries,* a complex regime designed both to prevent AT&T from destroying any budding firm and also to ensure that online computer services flourished unregulated. What matters so much for the fate of telecommunications and our narrative is that the infant industry the FCC protected in the 1970s would grow to be constituted of firms like America Online, Compuserve, and other online network companies (ISPs). While those names may no longer possess the luster they once had, in the 1990s they were the very firms that brought networking and the Internet to the American masses. In short, in these obscure and largely forgotten regimes, the new FCC played surrogate parent to the Internet firms that would later tear apart the traditional media industries and information empires, transforming the nation.

To the Breakup

Whereas in the 1970s the FCC saw itself as cultivating a garden of new and promising firms, AT&T saw a pestilent swamp in need of draining. AT&T was more than displeased with the government's promotion of competition; it was enraged and, as before in its history, willing to go to nearly any length to stop it. Dipping into a vast war chest, Chairman deButts in 1976 sent his lobbyists to Congress with a bill that would reverse all that the FCC was trying to do—with provisions, for instance, that simply outlawed MCI as a threat to universal service, and reversed *Carterfone* and even *Hush-A-Phone.* When that failed, AT&T returned to its most tried-and-true modus operandi: a campaign of industrial warfare designed to exterminate its competitors.[8]

Substantively, AT&T's campaign in the late 1970s might be described with the euphemism of "civil disobedience." Even as the FCC was encouraging firms to enter the telecommunications market, AT&T was laying traps for them that would make them regret that decision forever. But it would be Bell's course of retaliation, not the existence of their monopoly, that ultimately would put the Justice Department to the test and move them to seek a judgment against the firm for abusing its privileged power. In this sense, Bell's competitors served the FCC as a cape serves the bullfighter, goading the angry beast to action that led ultimately to the sword.

The full legal story of the breakup is complex, a major historical event to which no chronicler could do justice in brief. For our purposes, let the following lineaments suffice: Through the 1970s, AT&T came up with one scheme after another to nullify the effect of the FCC's orders and destroy the companies battening on them. For example, in the case of any foreign attachment (say, a fax machine), Bell would file a tariff requiring the competitor to establish something called a "protective connective arrangement." Supposedly a means to "protect" the network, the scheme was a thinly veiled way of imposing additional costs and regulatory burdens. The economist and Bell veteran Gerald R. Faulhaber contends that the scheme effectively bought AT&T eight more years of monopoly in this area.[9]

Meanwhile, AT&T's greatest wrath was saved for MCI, an upstart deemed beneath contempt, deserving of the nastiest tactics from the early Bell company's repertoire. The campaign against MCI was complex, subtle, and hard to summarize. In areas of MCI strength, AT&T took up the trusty lance of predatory pricing, trying to bleed MCI and any similar firms out of business. When enjoined by court order to allow MCI to connect their switches to its local circuits, Bell occasionally resorted to sabotage. According to one account from the other side: "on a Friday afternoon, AT&T pulled the plug on MCI's circuits between New York and Washington. Without warning, the data line at a major Washington department store went dead, and other corporate customers were similarly disconnected." Such conduct discouraged MCI's customers, but unfortunately for Bell, they also gave the antitrust lawyers at Justice plenty of proof to support a claim that Bell was abusing its monopoly.[10]

Although the business would be concluded under President Reagan, the will to break up Bell originated in the Nixon administration, as we've seen, in both the Justice Department and the White House Office of Technology Policy, headed by Clay Whitehead. Whitehead and others believed that the telephone system no longer needed to be a monopoly and that the nation would benefit from dividing AT&T into smaller parts. We do not know what President Nixon thought about breaking up AT&T—though perhaps, as the tragic outsider, he considered it yet another part of the establishment that he hated.

Like all antitrust lawsuits, the Bell case had gone on seemingly without end by the time Judge Harold H. Greene inherited it in 1978. At first, most knowledgeable observers considered it unthinkable the proceedings could end with AT&T broken up. It might, the smart money reckoned, be more aggressively forced into compliance with FCC regulations, or forced to sell Bell Labs. But an outright breakup of the nation's telephone company, which since 1921 had provided the world's best service? AT&T chairman deButts in 1974 made it clear that a breakup was unthinkable; he expressed disbelief that "Justice would take an action that could lead to dismemberment of the Bell System, with the inevitable result that costs would go up and service would suffer."

But unfortunately for deButts and AT&T, the enthusiasm in the age of Reagan for "competition" and "deregulation" exceeded even that of the Nixon era, and Bell's continuing opposition to anything other than monopoly was, in that environment, a blasphemy. As the years carried America farther into the new decade, and as more and more evidence was presented, the arguments of AT&T's defense counsel, virtually reconstituting Vail's speech of 1916—competition is bad, monopoly good; rivals are just "cream skimmers"; AT&T needs total control over telephony for the good of the nation—steadily lost their cogency.

True, the FCC had once agreed with all of these ideas. But times change, and so do regulatory regimes. As Harold Greene, the district judge who presided over the breakup, put it, "AT&T had an obligation to follow the more recent FCC policy rather than the Commission's previous policies which may have suited it better."[11]

At some point, and it's hard to say exactly when, AT&T finally began to recognize the existential peril it was in. Some say it happened after Judge Greene's incisive rejection of the firm's 500-page motion for dismissal.* It might have been earlier: John deButts, the most Bellheaded chairman in the firm's history, resigned in 1979, by some accounts because he realized that a breakup was inevitable and he wanted no part of a Bell divided. DeButts died two years after the breakup, which he never stopped referring to as a great tragedy. "Up until his death," read his obituary in *The New York Times,* he "fervently believed in the Bell telephone system."

---

* In 2000 Judge Greene's obituary in *The New York Times* would quote his son as saying the judge had always denied having made up his mind how he would have ruled had he not been presented with the compromise.

Whatever the trigger, what remains so fascinating is how much Bell, under its new chairman, Charlie Brown, conceded, and how quickly, once it took on board the likelihood of defeat.[12] In a compromise worked out with Justice, the greatest firm in communications history agreed to be divided into eight pieces—not slapped on the wrist, or even chipped away at, but virtually sliced and diced. The firm held on to its long distance services, Bell Labs, and Western Electric, its equipment manufacturer. But seven separate regional operating companies would be carved from the corporate carcass, the local monopolists now released as independent companies. Since each of the so-called Baby Bells would continue to have an effective monopoly over local services, however, each was placed in a newly designed regulatory cage of reinforced and toughened FCC rules. Each would be obliged to accept connections from any long distance company (not just their former parent), and all were explicitly shut out of new markets such as online service and cable.

AT&T did make a final attempt at challenging its own agreement in court. Yet Judge Greene was by now wholly unsympathetic; indeed, his decision affirming the breakup could easily have been written by Thurman Arnold, the antitrust fundamentalist. "It is antithetical to our political and economic system," wrote Greene, "for this key industry to be within the control of one company." The United States had tolerated, even encouraged, a monopoly of its most important industry for nearly seventy years. But it would no longer.

It fell to Charlie Brown, the last chairman of a united AT&T, to announce to the world in 1984 the breakup of the world's greatest communications monopoly. Trying to put a hopeful face on what was in effect a corporate funeral, Brown chose to introduce on the same day a new AT&T logo, a globe graphically girded by lines, replacing the old bell that had long signified corporate liberty virtually without limits

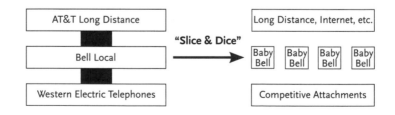

The Empire, divided

as well as the founder's name. "Today signals the end of an institution . . . the 107-year-old Bell System," announced Brown, "and the start of a new era in telecommunications in this nation." With that assertion no Bell antagonist could argue.[13]

DeButts and the old Bellheads were right about the immediate effects of the breakup. An American public wearied and bewildered by the years its government had spent hounding the nation's most reliable corporation* would very shortly face a rude awakening to inflated and complicated phone bills, including all sorts of mystifying connection fees and surcharges. (The degree to which the long lines had been subsidizing loss-leading markets throughout the country was greater than even Bell itself knew!) It would be some years before these inconveniences were offset by the fruits of innovation. On the other hand, when the innovation pent up by the Bell system came out, it was not a trickle but a tidal wave, in computing, telephony, networking, and everything else that has defined the information economy of the last thirty years.

It is always to be preferred that the Cycle proceed of its own accord. The examples of the Paramount decree and the Bell divestiture are both tales of much that is bad and ugly arising before the eventual good of an open industry. There is an undeniable efficiency that attends a monopoly's doing what it has been perfected to do, whether that be to turn out a certain kind of film or provide a universal phone service. What such well-oiled machines do not do so well, however, is initiate the sort of creative destruction that revolutionizes industries and ultimately multiplies productivity and value. And where information is the ultimate commodity, the multiplier effect is incalculably great. It is too much to ask of any corporate entity—pace Theodore Vail—to be the guardian of the general economic good. That interest will always be served by disruptive innovators, however much inconvenience they may visit upon us.

---

* During these same years, for instance, Japan overtook America in manufacturing quality. Few savvy shoppers would dream of buying, say, an American television set, firms like Zenith being at the nadir of their reputation.

CHAPTER 15

Esperanto for Machines

As a high school student in Bialystok, Russia, Ludwik Łazarz Zamenhof spent his spare time devising a language. He would work on it diligently for years, and in 1887, at the age of twenty-six, he published a booklet entitled *Lingvo internacia. Antaŭparolo kaj plena lernolibro* (*International Language: Foreword and Complete Textbook*). He signed it "Doctor Hopeful," or, in the language he had invented, "Doktoro Esperanto."[1]

Zamenhof's idea for a standardized international language was an ingenious idea poorly implemented. Consequently, his noble ambition is often forgotten: to dissolve what he considered the curse of nationalism. If everyone in the world shared a second language, "the impassable wall that separates literatures and peoples," he wrote, "would at once collapse into dust, and . . . the whole world would be as one family."[2] While there have been moments when it seemed that Esperanto might really take off—in 1911, for instance, the Republic of China considered adopting it as the country's official language—Zamenhof's invention has not remotely become a universal language.[3] Nonetheless, we live in a world in which his dream of a common tongue has been achieved—though not among humans, as he would have hoped, but among machines. It goes by a less hopeful sounding name, "the Internet Protocol," or TCP/IP, but for computer users it has succeeded where Esperanto failed.

Between Licklider's first enunciation of an intergalactic network and the mid-1970s, the idea of computers as communications devices had actually given birth to a primitive network, known as the ARPANET. The ARPANET was an experimental network that connected university and government computers over lines leased from AT&T. But it wasn't quite the universal network Licklider envisioned, one that could connect any network to any other. To achieve that goal of a true, universal computer network, one would need a universal language. One would need an Esperanto for computers. In 1973, this was the problem facing two young computer science graduate students named Vint Cerf and Robert Kahn.

One memorable afternoon in 2008 in a small Google conference room equipped with a whiteboard, I asked Vint Cerf what exactly was the problem he had been trying to solve when he designed the Internet protocol.[4] The answer surprised me. As Cerf explained it, he and Kahn were focused on developing not some grand design but rather a very much ad hoc accommodation. Running on a collection of government lines and lines leased from AT&T, the ARPANET was at the time just one of three packet networks in development. The others were a packet radio network and a packet satellite network, both privately run. Cerf and Kahn were trying to think of some way to make these networks talk to one another. That was the immediate necessity for an "internetwork," or a network of networks.

The Internet's design, then, wasn't the result of some grand theory or vision that emerged fully formed like Athena from the head of Zeus. Rather, these engineers were looking for a specific technical fix. Their solution was indeed ingenious, but only later would it become clear just how important it was. Cerf describes the open design of the Internet as necessitated by the particularities of the specific engineering challenge he faced. "A lot of the design," Cerf said, "was forced on us."

The Internet's creators, mainly academics operating within and outside the government, lacked the power or ambition to create an information empire. They faced a world in which the wires were owned by AT&T and computing was a patchwork of fiefdoms centered on the gigantic mainframe computers, each with idiosyncratic protocols and systems. Now as then, the salient reality—and one that too many observers don't grasp, or overlook—is that the Internet works over an infrastructure that doesn't belong to those using it. The owner is always someone else, and in the 1970s, that someone was generally AT&T.[5]

The Internet founders built their unifying network around this fundamental constraint. There was no other choice: even with government funding they did not have the resources to create an alternative infrastructure, to wire the world as Bell had spent generations and untold billions doing. Consequently, their network was from its beginning beholden to the power and autonomy of its owners. It was designed to link human brains, but it had no more control over their activities than that, an egalitarianism born of necessity, and one which would persist as the network grew over decades to include everyone.

The stroke of genius underlying a network that could interconnect other networks was the concept of "encapsulation." As Cerf said, "we thought of it as envelopes." Encapsulation means wrapping information from local networks in an envelope that the internetwork could recognize and direct it. It is akin to the world's post offices agreeing to use names of countries in English, even if the local addresses are in Japanese or Hindi. In what would come to be known as TCP (or Transmission Control Protocol), Cerf and Kahn created a standard for the size and flow rate of data packets, thereby furnishing computer users with a lingua franca that could work among all networks.[6]

As a practical matter, this innovation would allow the Internet to run on any infrastructure, and carry any application, its packets traveling any type of wire or radio broadcast band, even those owned by an entity as given to strict controls as AT&T. It was truly a first in human history: an electronic information network independent of the physical infrastructure over which it ran. The invention of encapsulation also permitted the famous "layered" structure of the Internet, whereby communications functions are segregated, allowing the network to negotiate the differing technical standards of various devices, media, and applications. But, again, this was an idea born not of design but of the practical necessity to link different types of networks.

To ponder the design of the Internet is to be struck by its resemblance to other decentralized systems, such as the federal system of the United States. The Founding Fathers had no choice but to deal with the fact of individual states already too powerful and mature to give up most of their authority to a central government. The designs of the first two constitutions were therefore constrained—indeed, overwhelmingly informed—by the imperative of preserving states' rights, in order to have any hope of ratification. Similarly, the Internet's founders were forced, however fortunate the effect may now seem, to invent a protocol that took account of the existence of many networks, over which they had limited power.

Cerf and Kahn pursued a principle for Internet protocols that was the exact opposite of Vail's mantra of "One System, One Policy, Universal Service." Where AT&T had unified American communications in the 1910s by forcing the same telephone on every single user, Cerf and Kahn and the other Internet founders embraced a system of tolerated difference—a system that recognized and accepted the autonomy of the network's members. Indeed, to do what Bell had done fifty years earlier might in fact have been impossible, even for an entity as powerful as Bell itself. For by the sixties, the charms of centrally planned systems generally were beginning to wear thin, soon to go the way of short-sleeved dress shirts.

DECENTRALIZATION

The economist John Maynard Keynes once said, "When the facts change, I change my mind. What do you do, sir?"[7] No apostle of central planning could live through Europe's Fascist and Soviet experiments without admitting that directed economies had their limitations and liabilities. The same ideas that had inspired Henry Ford and Theodore Vail had, in the realm of politics, led to Hitler and Stalin. And so a general repudiation of the whole logic of centralization was a natural fact of the Cold War era.

It was an Austrian economist who would provide the most powerful critique not just of central planning but of the Taylorist fallacies underlying it. Friedrich Hayek, author of *The Road to Serfdom,* is a patron saint of libertarians for having assailed not only big government, in the form of socialism, but also central planning in general.[8] For what he found dangerous about the centralizing tendencies of socialism applies equally well to the overbearing powers of the corporate monopolist.

Hayek would have agreed with Vail's claim, as with the Soviets', up to a point: ideally, planning should eliminate the senseless duplication that flows from decentralized decision making. There is a certain waste implied in having, say, two gas stations on a single street corner, and in this sense, as Vail insisted, monopolies are more efficient than competition.*

But what prevented monopoly and all centralized systems from realizing these efficiencies, in Hayek's view, was a fundamental failure to appreciate human limitations. With perfect information, a central planner could effect the best of all possible arrangements, but no such planner could ever hope to have all the relevant facts of local, regional, and national conditions to arrive at an adequately informed, or right, decision. As he wrote:

> *If* we possess all the relevant information, *if* we can start out from a given system of preferences and *if* we command complete knowledge of available means, the problem which remains is purely one of logic. . . . This, however, is emphatically *not* the economic problem which society faces. . . . [T]he "data" from which the economic calculus starts are never for the whole society "given" to a single mind which could work out the implications, and can never be so given.[9]

Such a rejection of central planning beginning in the sixties was hardly limited to those with conservative sensibilities. Indeed, the era's emblematic liberal thinkers, too, were rediscovering a love for organic, disorganized systems. Another Austrian, the political scientist Leopold Kohr in the 1950s, began a lifetime campaign against empires, large nations, and bigness in general. As he wrote, "there seems to be only one cause behind all forms of social misery: bigness. Oversimplified as this may seem, we shall find the idea more easily acceptable if we consider that bigness, or oversize, is really much more than just a social problem. . . . Whenever something is wrong, something is too big."[10]

Kohr's student, the economist E. F. Schumacher, in 1973 wrote *Small Is Beautiful: Economics As If People Mattered*, developing the concept of "enoughness" and sustainable development.[11] Jane Jacobs, the great theorist of urban planning, expresses a no less incendiary disdain for centralization, and as in Hayek, the indictment is based on an inherent

---

*In the parlance of economists, many market failures—externalities, collective action problems, and so on—can be eliminated by a central planner.

neglect of humanity. In her classic *The Death and Life of Great American Cities,* she relies on careful firsthand observations made while walking around cities and new developments to determine how Olympian planners like Robert Moses were going wrong.[12] There was no understanding, let alone regard, for the organic logic of the city's neighborhoods, a logic discernible only on foot.

All of these thinkers opposed bigness and prescribed a greater humility about one's unavoidable ignorance. No one could fully understand all the facts of the dynamic market any more than one could weigh the true costs of introducing a vast new flow of traffic through neighborhoods like New York's SoHo and West Village, which had developed organically for centuries. These thinkers were speaking up against a moribund belief in human perfectibility, or scientific management theorist Frederick Taylor's the "one right way."[13] Cities, like markets, had an inscrutable, idiosyncratic logic not easily grasped by the human mind but deserving of respect.

It was beginning to seem that the same might be true of information systems.

While its design had been born of necessity, through the 1970s and early 1980s the Internet's developers began to see a virtue in it. And their awareness grew with their understanding of what a universal network needed if it was to operate, evolve, and advance in organic fashion. In the final draft of the TCP protocol, Jon Postel,* another Internet founder, inserted the following dictum:

Be conservative in what you do. Be liberal in what you accept from others.[14]

It may seem strange that such a philosophical, perhaps even spiritual, principle should be embedded in the articulation of the Internet, but then network design, like all design, can be understood as ideology embodied, and the Internet clearly bore the stamp of the opposition to bigness characteristic of the era. Not long thereafter, three professors of computer science, David Reed, David Clark, and Jerome Saltzer, would

---

*Postel is profiled in my first book (coauthored with Jack Goldsmith), *Who Controls the Internet?*

try to explain what made the Internet so distinctive and powerful. In a seminal paper of 1984, "End-to-End Arguments in System Design," they argued for the enormous potential inherent in decentralizing decisional authority—giving it to the network users (the "ends").[15] The network itself (the "middle") should, they insisted, be as nonspecialized as possible, so as to serve the "ends" in any ways they could imagine.*

What were such notions if not the computer science version of what Hayek and Jacobs, Kohr and Schumacher, had been arguing. While we cannot say exactly that the network pioneers of the 1970s were disciples of these or any particular thinker, there is no denying the general climate of thought in which computer scientists were living, along with everybody else. Coming of age concurrently with an ideological backlash against centralized planning and authority, the Internet became a creature of its times.

In 1982 Vint Cerf and his colleagues issued a rare command, drawing on the limited power they did have over their creation. "If you don't implement TCP/IP, you're off the Net."[16] It was with that ultimatum that the Internet truly got started, as computer systems around the world came online. As with many new things, what was there at first was more impressive in a conceptual sense than in terms of bells and whistles, but as usual, it was the human factor that made the difference, as those who joined could suddenly email or discuss matters with fellow computer scientists—the first "netizens." The Internet of the 1980s was a mysterious, magical thing, like a secret club for those who could understand it.

What was the Internet in 1982? Certainly, it was nothing like what we think of as the Internet today. There was no World Wide Web, no Yahoo!, no Facebook. It was a text-only network, good for transmitting verbal messages alone. More important, it was not the mass medium of our experience. It still reached only large computers at universities and government agencies, and mostly ran over lines leased from AT&T (as

---

* Much later, in the early years of the twenty-first century, the phrase "net neutrality" would become a kind of shorthand for these founding principles of the Internet. The ideal of neutrality bespeaks a network that treats all it carries equally, indifferent to the nature of the content or the identity of the user. In the same spirit as the end-to-end principle, the neutrality principle holds that the big decisions concerning how to use the medium are best left to the "ends" of the network, not the carriers of information.

a complement, the federal government would begin to build its own physical network in 1986, the NFSNET). As far as the internetwork had come conceptually, a different kind of revolution would be needed to bring it to the people. And that transformation, less technological than industrial, would take another decade; first, the computer would have to become personal.

PART IV

Reborn Without a Soul

THIS PART IS ABOUT CORPORATE REINCARNATION, how the once mighty then fallen picked themselves up in the last two decades of the twentieth century. The film industries and broadcast networks recombined with the new cable industries and networks in a novel form, giant conglomerates on the model of Time Warner that spanned industries in new ways. Meanwhile, AT&T, broken up in the 1980s, by the first years of the twenty-first century managed to re-create itself, reestablishing the essential lineaments of the Bell system.

In each case the powerful, almost magnetic attractions of size and scale slowly put the pieces back together. Those attractions, as we shall see, are slightly different for the entertainment and communications sides, yet they lead to the same place. For the entertainment industries, size is a means of trying to flatten out the huge risk inherent in creating big, expensive products. In communications, the temptations of size and monopoly arise from an interest in running a fully integrated system and controlling every possible source of revenue.

Yet in the rebuilt industries something is missing, like a part overlooked, and this is the sense of civic responsibility. The old empires were suppressive and controlling in their own ways, yet each had some sense of public duty, informal or regulated, that they bore with their power. At their best, they were enlightened despots. But the new industries' ethos held that profit and shareholder value were the principal duty of an information company. What reemerged was similar in body but different in its soul.

The information empires created in the 1980s shared many of the worst aspects of both open and closed forms. The new giants had much of the power of the old, without the noblesse oblige. And by the late 1990s it began to seem inevitable that just two industries, the Bell companies and the media conglomerates, would soon control everything in the world of information.

CHAPTER 16

Turner Does Television

In the summer of 2008 I asked Ralph Lee Smith, folk musician, dulcimer virtuoso, and evangelist of cable television, whether the medium had lived up to his expectations. He paused before allowing it had certainly "helped loosen things up." He paused again. "Looks to me," he said at last, "as if the people with the money were way ahead of me."

Cable as it finally emerged in the late 1970s and early 1980s was an open and disruptive new industry, yet not quite what Ralph Lee Smith had envisioned. True, it hugely altered the nation's media environment. But it did so less by actualizing the utopian promise described by Smith and others, and more by becoming a medium with unprecedented reach into every marketing niche of American society.

In the 1970s, the cultural visionaries helped by elements of the Nixon administration gave impetus to the release of cable from its regulatory prison. But the firm government management that both the cultural idealists and the bureaucrats hoped for never came to pass.

Those who would shape the future of cable were cut from a rather different cloth than Smith, or Fred Friendly, for that matter. At about the same time that Smith, the Sloan Foundation, and the FCC were drafting the future of cable as they saw it, a businessman named Ted Turner was purchasing WJRJ, a small UHF station in Atlanta, still broadcasting in black-and-white. No sooner had Turner gotten his hands on his first station than he began to develop his grandiose ambitions for the conquest of television, a master plan founded on the idea of the cable network. "Television," announced Turner with prophetic zeal, "has led us, in the last twenty-five years, down the path of destruction. I intend to turn it around before it is too late."[1]

Ted Turner hardly needs introduction. Yet while he is known to the public mainly as the larger than life, bipolar enfant terrible who founded CNN, his greatest claim to immortal fame may actually be his role in opening up television and founding an entirely new industrial model. While it may seem unnecessary to add to the man's image, Turner is actually undercredited as an industrial innovator. For Turner made a critical imaginative leap respecting what cable could be, one that finally brought the medium invented in 1926 into an era of wide-open entrepreneurism and experimentation. He did so by making practical the use of cable lines not just to carry individual broadcast stations but as a platform for a national TV network. Seeing cable's potential as far more than an adjunct to broadcast television, he became the essential pioneer of the cable *network*.

In personality, Turner is undoubtedly a member of that rare breed described by Schumpeter—a man in the mold of Theodore Vail or Adolph Zukor, albeit with a much more public private life and wilder mood swings. "You only have one life," he once said; "you might as well make it a great one." Like those others in this cohort, he was seemingly incapable of thinking small, a trait that fed his will to possess very large or famous things. By 2000 he was the biggest individual landowner in the United States, with the biggest herd of buffalo in the world, forty-five thousand head. He married the film and fitness icon Jane Fonda, and he won the America's Cup, sailing's greatest prize, twice.

Not surprisingly, Turner has been an irresistible subject for biographers. Four accounts have been published, not including his own, *Call Me Ted*—each retailing a seemingly inexhaustible stream of manic exertions: daring business strategies, sexual exploits, and a fierce competitiveness in all things. Chroniclers tend to credit, or blame, Turner's father, an abusive drunk who frequently whipped the young Ted with a coat hanger, even occasionally forcing the boy to do likewise to the father as a bizarre form of punishment.[2]

Such sordid details may seem irrelevant, but in fact they matter because they influenced how Americans get information. For as we have seen, the mogul makes the medium: the imprint of the personality inevitably informs it, often no less than the technology underlying it. Turner styled himself a heroic swashbuckler, an underdog fighting the brutal domination of the networks. And so network cable, when it finally took off, reflected the scrappy, overzealous character of its pioneer—wildly ambitious, bombastic, fearless, and always on the edge of total failure. Turner explicitly described himself rather as the industry's Alexander the Great: "I can do more today in communications than any conquerer ever could have done," he proclaimed, "I want to be the hero of my country."[3]

It was by this instinctive will to greatness, the desire for ownership of something big as a path to immortality in the fashion of the ancient epic heroes and emperors, that Turner transformed the cable industry. By dint of completely idiosyncratic programming choices, he turned around Channel 17, the failing little UHF station in Atlanta he had bought in 1970. But that wasn't enough: he didn't want to run a station; he wanted to run a major national network, like ABC, CBS, or NBC. For anyone who didn't already run one, that had been a laughable aspiration for most of the history of radio and television. But Turner was determined to find a way.

Even after the Nixon-era deregulation of the cable industry, the broadcast networks still had many lines of defense. A network, by definition, brought the same content to its affiliate stations around the country—and to do so they needed some means of moving massive amounts of information at high fidelity, a feat beyond the capacity of radio broadcasting except for the few high-powered, specially licensed clear channels. The answer to this necessity lay in access to AT&T's long distance network, and this would become the secret weapon of the broadcast networks. A show produced in New York would be carried around the country over AT&T's long lines or microwave towers: from the 1920s through the 1960s there was still no other way of constituting a network. And so, to a degree many have forgotten, the broadcast triopoly and the telephone monopoly were intimately linked.

Through the 1970s, running a network, then, meant paying AT&T's rates, which, while subject to government oversight, were still high enough to make any start-up a losing proposition. There was the cheap, low-tech alternative of taping shows and mailing the videos. That was how, for instance, Pat Robertson's Christian Broadcasting Network got its start. But one could scarcely be more than a bit player relying on the post office. For these reasons and others, the only national networks to enter the market from the 1930s through the 1970s were the government-funded Public Broadcasting System (launched in 1970) and ABC, which was fashioned by the FCC out of NBC's rib.[4]

Ted Turner did what had seemed impossible when, in 1976, he managed to create the first *cable* network—that is, the first station available on basic cable all over the country. He did it by substituting as his conduit the new technology of satellites for AT&T's long distance lines. With a satellite, you could take a single station signal, like that of Turner's WTCG in Atlanta, beam it up into space, and then beam it back down to a cable operator in, say, New Jersey or Michigan. What Turner created was technically not so much a network as a "superstation"—a station available around the country—yet it was, in effect, the prototype of our cable network.

To give credit where it is due, the use of satellites to carry television was not an idea that originated with Turner. The Home Box Office Network (HBO), under Gerald Levin and others, had done so since 1972 in order to offer so-called pay TV, which, for a premium, brought special content, such as Frazier-Ali boxing matches and feature films, to cable subscribers. But pay TV, however significant an innovation, was less an assault upon than a complement to the networks. Turner, in contrast, by making his WTCG available across the country on basic cable, was going head to head with the Big Three. For the many who depended on cable to get their network programming, his content was available on exactly the same terms. He started small—at first enlisting just four cable operating systems scattered across the country—but he made it big.

Turner's business model was simple. Overhead was limited, for the most part, to costs of operating his station—paying for access to content (say, old films), and leasing time on a communications satellite. Meanwhile, cable operators (e.g., companies like Cablevision) would pay Turner a fee in exchange for the right to make his channel available to subscribers. Turner would also sell advertising spots on his channel, based on the audience it could be expected to reach. Obviously the rates did not rival those of networks, but in a way, that was the point: more locally targeted advertising for sponsors who didn't have Coca-Cola's ad budget. As always, subscribers paid the cable operator for a slate of "basic" cable stations, offering as well the option to pay more for premium content (like HBO), typically without commercials. Then as now, there was some variation in what the cable operator paid to carry a channel (ESPN charging more than the Learning Channel, for example). That in essence is how he did it, and that is how the cable industry operates to this day.[5]

Sensible as Turner's business model was, however, in the late 1970s it was easier said than done. Regarding the challenge of drawing advertisers to his network, he recalled, "I knew it was going to be hard to convince the New York advertising community, but I had no idea how hard. My first team of salesmen ended up like the soldiers in the opening scene of *Saving Private Ryan.* They were mowed down to a man."[6]

While Turner described himself as a valiant liberator and cast the networks as oppressive scoundrels, his programming hardly offered a brave new world. Essentially, his was a network of reruns and specialty content like wrestling and cartoons. WTCG carried old sitcoms such as *I Love Lucy* and *Green Acres,* cartoons of yesteryear like *The Flintstones* and *Speed Racer* (the latter a Japanese import, dubbed into English), Hollywood movies from the 1930s and 1940s, and, of course Atlanta Braves baseball games (Turner bought the team in 1976, in part to gain content for his station). And so if Turner was an innovator in programming, he was an innovator of a rather paradoxical sort, finding an audience for classics of a bygone time, along with slightly "down-market" content like professional wrestling and music videos. Nonetheless he would find glorious terms even for retreads and junk, claiming to be pulling America back to television's golden age: "I want to get it back to the principles," he once said, "that made us good." Nostalgic, Manichaean, and bootstrappy: like programmer, like program.

By 1978, cable operators in fifty states were carrying Turner's station, giving him the national network he had dreamed of. With Turner's example having forged the way, the path was now clear for any number of new national cable networks, each dedicated to a specific type of content. Turner once said that if he had been born in another era, he'd have been an explorer, but failing that, he did open the undiscovered country of cable to an explosion of settlement in the late 1970s and early 1980s. For the first time since its invention in 1926, television was suddenly a wide-open medium, with all sorts of notions of what it was meant to be and do proliferating on the cable wire.

In less than a decade nearly a dozen cable networks were launched, including the Entertainment and Sports Programming Network (ESPN), Music Television (MTV), the Bravo channel (at the time of its launch in 1980 a commercial-free station dedicated to drama and the performing arts), Showtime (a competitor to HBO in showing recent feature films), Black Entertainment Television, the Discovery Channel (popular science), and the Weather Channel. Those actually are only the better known, by virtue of having survived; others, such as ARTS, CBS Cable, and the Satellite News Channel would fold or be acquired by other companies.

Turner himself joined the rush he had created, launching his second network, the Cable News Network, in 1980, on which his popular fame rests. The fact of his having launched it shows the marvelous plasticity of Turner's theories of programming. His original station carried almost no news, and he had once declared "I hate news. News is evil. It makes people feel bad." Yet with most of the nation still tuning in to the Big Three's evening news broadcasts, Turner spied an opportunity for profit (as well as mischief, perhaps) too great to ignore.[7]

Cable didn't turn out to be quite what its visionary backers had contemplated. Contrary to promoting a common interest and a shared vision of the good, cable television evolved to accommodate a universe of divergent special tastes. It catered to unserved or underserved interests and demographics of all kinds. The networks, from their beginnings, were aimed at the broad middle of American society, while cable pursued racial minorities (BET and Telemundo), perennial students (Discovery and History), news junkies (CNN), and people who didn't realize they were obsessed with the weather. This capacity to be all things to all people, however, has caused not a few to wonder whether cable has ultimately turned out to be a good or a bad thing.

The notion of the national audience as an amalgam of many diverse interests may seem obvious, particularly in the Internet age, yet the broadcast networks, since the 1920s, had been conceived according to the opposite principle. Committing themselves by design to the idea of a unified national community and culture, they espoused a more benign variant of what the Germans in the 1930s called a *volksgemeinschaft*. NBC was founded to serve as a *national* network not only in its reach but in its purpose of offering general content suited to everyone. The offerings of more sectarian or special-interest broadcasters, remember, were deemed "propaganda" by the Federal Communications Commission, who encouraged NBC and CBS in their mission of bringing the country together, the noble purpose of broadcasting for decades.

Since the 1980s, cable's appeal to the niche instead of the nation has thus, together with other factors, been blamed for the splintering, dividing, or clustering of America. There is a difference, of course, between recognizing and creating a social reality, but nevertheless the charge is not unwarranted. Obviously, there is a difference between a nation in which on any given night conservatives are watching Fox News, sports fans are tuned to ESPN, and teenagers are glued to MTV, as compared to the America of television's yesteryear, where the nuclear family would watch *I Love Lucy* and the *CBS Evening News* together, whether they would have preferred to or not. Indeed, the television writer Ron Becker observes, "cable networks and TV shows were designed not only to appeal to those in a targeted demographic group but also to send clear signals to unwanted eyes that certain media products weren't meant for them." The alienation was, in a way, the message, and the product.[8]

Critics like the law professor Cass Sunstein go so far as to describe the fragmenting powers of cable and other technologies, notably the Internet, as a threat to the notion of a free society. "In a democracy," writes Sunstein, "people do not live in echo chambers or information cocoons. They see and hear a wide range of topics and ideas."[9] There is a bit of a paradox to this complaint that must be sorted out. The concern is not that there are too many outlets of information—surely that serves the purpose of free expression that sustains democratic society. Rather, the concern is that in our society, we have been freed to retreat into bubbles of selective information and parochial concern (Sunstein's "information cocoons"), in flight from the common concerns we must address as Americans. And so if the networks suffered from being uninteresting in the effort to alienate no one, they did at least tend to feed a sense of a common culture, a common basis of the national self now lacking. There is little to talk about around the proverbial water cooler in a nation segmented by divides of gender, generation, political inclination, and so on.

There is clearly truth to this vision. The big picture encompassing the sweep of the past century, however, makes clear that cable television, and later the Internet, haven't so much splintered a heretofore united

America as returned us to the fragmentation of an earlier era that the rise of big concentrated media had suspended. In fact, the triopoly of NBC, CBS, and ABC itself marked an aberrant time in the history of America, or for that matter, of the world.

The age of "mass media" upended by cable television was actually a period of unprecedented cultural homogeneity. Never before or since the sixty-year interval from the 1930s to the early 1990s had so many members of the same nation watched or listened to the same information at the same time. In 1956, Elvis Presley's appearance on *The Ed Sullivan Show* attracted an unbelievable 83 percent of American TV households. A broadcast of the musical *Cinderella* in 1955 attracted 107 million viewers, nearly 60 percent of the entire U.S. population. Ken Auletta, *The New Yorker*'s media writer, could say as late as 1991 that "To most of us, television has always meant three institutions—CBS, NBC, and ABC. They have been our common church." The TV networks, around the world, were probably the most powerful and centralized information system in human history.[10]

Seen this way, the rise of cable and the Internet thereafter were less a revolution than a counterrevolution, a return to the more scattered pattern of American attention that once was. Before the rise of great media empires of the 1920s and 1930s, the United States, far from a united culture, was, almost by technological default, a house divided according to class, ethnicity, region, and other factors. Perhaps region most of all, for entertainment and culture were necessarily local. Early radio, before NBC, comprised hundreds of local stations, each generating its own content, however humble. Likewise, before the Hollywood studio gained ascendancy, local theaters decided for themselves what films to show. And of course before the telegraph, newspapers were necessarily local. In this sense, cable television's undoing of the mass, or national, media culture was merely the undoing of a transient twentieth-century invention. In the nineteenth century, as now, there was little common information for any two randomly selected American citizens to discuss, even if there had been a water cooler.

For the purposes of our narrative, the conclusion is clear: an open medium has much to recommend it, but not the power to unify the country. For a fully united national community, nothing succeeds like centralized mass media, a fact not lost on the Fascist and Communist totalitarians. With an open medium, one has diversity or fragmentation of content, and with it, differences among groups and individuals are accentuated rather than elided or repressed. None of this divisive power was plain to cable's early visionaries, who saw only good in a proliferation of diverse subjects and points of view rather than the diffusion of American identity that commentators like Sunstein lament.

Yet in one important sense cable was never truly diverse. From its birth, the medium was always as relentlessly commercial as the broadcast variant, and with a few exceptions (such as C-SPAN) it has never furnished many harbors for programming that doesn't need to make money, perhaps even fewer than broadcast television, which at least in the early days included so-called sustaining (or unsponsored) programming and before the ascent of cable had added the government-funded network, PBS.

Unfortunately, as it developed in the late 1970s, cable became wedded to the idea of niche marketing—of milking every audience segment with targeted advertising. It didn't have to be that way: cable television was never free as broadcast was, and could, in theory, have developed a model to support itself on fees alone (just as HBO did, and continues to do). Instead, it mostly developed in the cast of its founding mogul, driven by the incessant need to raise revenues and attract attention. And thus after Turner, cable gained both its enduring viability and its tendency toward pandering and crassness that have become its hallmark, giving the very word a whiff of the down market.

To understand how much difference the circumstances of creation can make, consider the contrasting example of the Internet. Cable was born commercial, while the Internet was born with no revenue model, or any need of one. Its funding came in research grants, making it, for a long time, the information media equivalent of a public park. And while today it can be used to make money, the network, being quite purely open, can still easily carry content that makes no financial sense, from personal blogs to sites like Wikipedia. Oddly enough, that's how many of the most lucrative Internet firms got their start.

Cable, on the other hand, with limited exceptions, has for its entire history been driven by a constant appetite for cash and for content that could deliver a sufficient audience to yield a return on capital. The scale needn't be vast; so long as there is someone with consumer needs and a credit card prepared to watch *Blue Crush* at 3 a.m., the model can work. But even with 1000+ channels, you could never just put on programming heedless of whether there was an audience. Cable operators carry channels because they believe their subscribers want them, and subscribers pay for the system. Even among a thousand choices, a channel with no viewers is a deadweight an operator can ill afford. That difference, along with certain structural ones, is still what distinguishes cable from the Internet today.

For all its shortcomings, there is no denying that cable shook up the way Americans get information and forever changed the face of television radically. As an object lesson in the way information networks can develop, it gives us occasion to consider what we truly want from our news and entertainment, as opposed to what sort of content we might be prepared to sustain, however passively, with our fleeting attention. For cable offered choices really only in the commercial range—enough, however, to suggest what a truly open medium could deliver to the nation, for better and for worse.

CHAPTER 17

Mass Production of the Spirit

On November 19, 1980, the hottest ticket in New York was the premiere of *Heaven's Gate*. Michael Cimino, acclaimed director of *The Deer Hunter*, had spent years and more than $35 million (over $100 million in today's dollars) on this masterpiece, his take on the American Western. Rising stars Jeff Bridges and Christopher Walken were featured, as well as a giant supporting cast and extensive period sets. Everything seemed set for a major triumph, yet in the rush to the premiere, almost no one, including its producers, had actually seen the film in final cut.

Cimino's effort followed in the footsteps of Francis Ford Coppola's *Apocalypse Now*. A hard-won triumph, that epic of Vietnam had creative romance about it: a masterpiece and a box office success that almost didn't get finished. *Heaven's Gate* was another director-driven vision, a symbol, even a high-water mark, of the second open era in American filmmaking that lasted from the late 1960s through the 1970s. This period brought Hollywood as close as it has ever come to fulfilling the ideal articulated by W. W. Hodkinson in the 1910s: directors enjoyed a new level of independence from studio preoccupations, and when those deemed blessed with inspiration asked for something, they got it. No studio embodied the spirit of New Hollywood more than United Artists, the only major independent to survive the purges of the 1920s, whose entire approach depended on finding and funding directors with a certain vision. In the 1970s, United Artists became the leading platform for new directors with big ideas, including not only Coppola and Cimino, but also Woody Allen, Martin Scorsese, and others.[1]

At the New York premiere, things got off to a bad start. The audience was oddly unresponsive during the first half of the film. Stephen Bach, a United Artists executive, later wrote that they were "either speechless with awe or comatose with boredom." During the intermission Cimino entered a subdued reception room and found that the glasses full of champagne had gone untouched. Bach relates the director's conversation with his publicist:

"Why aren't they drinking the champagne?"
"Because they hate the movie, Michael."[2]

The critics were brutal, particularly Vincent Canby of *The New York Times*. " 'Heaven's Gate,' " he wrote, "is something quite rare in movies these days—an unqualified disaster." It "fails so completely," wrote Canby, "that you might suspect Mr. Cimino sold his soul to obtain the success of *The Deer Hunter* and the Devil has just come around to collect."[3]

Indeed, *Heaven's Gate* would be ever after remembered as perhaps the greatest single bomb in film history, but not just in a financial sense. True, the film would fail to make much money, but that's not so unusual. This failure was deeper. Fairly or not, it would be understood as an indictment of United Artists' approach to filmmaking, the approach that had become emblematic of the art in the 1970s, based on prizing the independence of directors and glorifying artistic innovation. *Heaven's Gate* was the auteur film from hell and led directly to the financial collapse of United Artists and its subsequent sale to MGM, marking the beginning of the end for the second open age of film.

The fall of United Artists in the early 1980s and the second closing of the film industry represents a turn in the Cycle and the beginning of a new order that lasts to this day. This moment saw the triumph of a rival approach to production coinciding with the rise of the *media conglomerate*. Its greatest champion was a man named Steven Ross, a onetime funeral home director who pioneered a new way of organizing the entertainment industry. Unlike a freestanding firm like United Artists, Ross's firm, Warner Communications (today, Time Warner Inc.), held dozens of media concerns and other properties under a single roof. This vision would spread throughout the 1980s and 1990s to become the dominant industrial organization for movies, music, magazines, newspapers, book publishing—all forms of content once called "leisure."

There is no understanding communications, or the American and global culture industry, without understanding the conglomerate. Yet this industrial form, born originally to assist in creative accounting on the part of public corporations, remains surprisingly difficult to characterize, let alone justify. It is a hydra-headed creature whose operations and advantages have mystified lawyers and economists alike. As a 1981 paper in the *Bell Journal of Economics* put it: "Despite extensive research, the motives for conglomerate mergers are still largely unknown."[4]

Nonetheless, the conglomerate is the dominant organizational form for information industries of the late twentieth and early twenty-first centuries; here and abroad it is inseparable from the production of the lion's share of culture. Like the integrated Hollywood studio that preceded it, the conglomerate can be the worst enemy or the best friend of the cultural economy. With its hefty capitalization, it offers the information industries financial stability, and potentially a great freedom to explore risky projects. Yet despite that promise, the conglomerate can as easily become a hidebound, stifling master, obsessed with maximizing the revenue potential and flow of its intellectual property. At its worst, such an organization can carry the logic of mass cultural production to any extreme of banality as long as it seems financially feasible, approaching what Aldous Huxley predicted in 1927: a machine that applies "all the resources of science . . . in order that imbecility may flourish. . . ."[5]

DEFUSING BOMBS

The fact that a single big failure, *Heaven's Gate,* could take down an entire film studio made it starkly obvious that all the studios needed better ways of protecting themselves against disaster. As we shall see, defense against financial meltdowns was perhaps the driving reason for the rise of the media conglomerate. But the implications are broader than that: the shape and tenor of our current entertainment world are largely owing to the imperatives of risk management in a world where failure had become catastrophically expensive.

A few basic points about entertainment economics will make this clearer. The fundamental fact in the business is a high level of uncertainty as to the success of any given product, and a giant disparity between the rewards that come to those that succeed modestly, as against the real hits. Another way of saying this is that the entertainment industry is the classic, indeed definitive, example of what economists call a "hit-driven" industry.[6]

In such a context, a few hits will outperform the rest, sometimes by several orders of magnitude. The difference between number one and number twenty on any entertainment media chart is well captured by *Wired* magazine editor Chris Anderson in his book *The Long Tail* (itself a hit). While the book is famous for celebrating Internet business models, what Anderson also shows, using actual industry data, is that a relatively small number of hits account for the bulk of the revenue in those businesses. Hence the peculiar distribution of demand, as pictured below, which typically confounds and frustrates management consultants. In practical terms, to take book publishing as an example, this means that the seven Harry Potter books outperformed many thousands of other books combined. Likewise, in film, a giant blockbuster can outearn the combined receipts of hundreds of independent films, and so on.*[7]

The second peculiarity is that the hits are not so easy to predict. Sometimes a film comes out of nowhere and makes a killing, like the original *Rocky;* produced by a then unknown independent filmmaker (Sylvester Stallone) on a tiny budget, it became the top grossing film of 1976. On the other hand the big, obvious bets not infrequently do pay off, whether owing to classic mastery of craft, as with *Apocalypse Now,* or breakthrough technical wizardry and unsuspected thematic resonance, as with *Avatar.* But there is no long, expensive film that is so obvious a bet as not to be risky. The real fly in the ointment is those films like *Heaven's Gate,* or *Ishtar,* which, for whatever reasons, despite famous directors and actors and large budgets, prove complete financial failures. The megastinker can be as hard to forecast as the runaway hit. Even in the cases of *Titanic* and *Avatar,* the two top grossing films in history, many industry watchers predicted a financial bloodbath.

---

*In his book Anderson also pointed out that contrary to popular belief, there was as much revenue available in the "tail" as the "head" of customer demand; that is, a business could do well offering a great variety of less popular products instead of just a few highly popular ones.

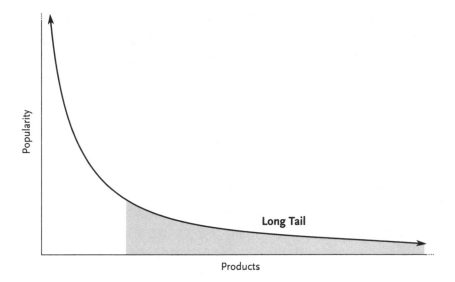

Demand in the entertainment industry

We might well ask why success in entertainment is so hard to predict. It's a tricky question, though one can start to answer it by looking to the nature of the demand. With any given entertainment product—as compared with, say, socks or beer—one is faced with selling something people don't ultimately *need;* they have to *want* it. They have to be inclined to invest time and money—ninety minutes with a film, twenty-five dollars for a book—without certainty of satisfaction or desired effect. To be sure, there are times when the desire for entertainment seems like a need—for instance, on a long flight. Yet even then the need is not unique; just about any decent film will do to pass the time. The upshot is that every book, film, or TV show launches amid the unsettling awareness that it could be a total and absolute flop. As the film industry economist Arthur S. De Vany writes, "every actor, writer, director and studio executive knows that their fame and success is fragile and they face the imminent threat of failure with each film they make."[8]

That uncertainty and variable demand at the heart of the entertainment industry has led to a wide range of countermeasures. As we shall see, the structure of the entertainment industries makes no sense apart from an understanding of the ways they manage risk. These range from the obvious—for instance, betting on well-known stars or directors (more typically the former) and the sequel (rerunning a past success in the hope that lightning will strike twice)—to somewhat esoteric systems of financial management and joint accounting aimed at diffusing success and failure over a broad balance sheet. All of these techniques have in common the way they end up altering the face of both American and global popular culture.

STRATEGY 1: CONGLOMERATIZATION

If unpredictability of success and failure was the chief problem in the entertainment industry, by the 1970s, a man named Steven Ross had the answer. As CEO of a company called Kinney National Service, Ross acquired the suffering Warner Bros.–Seven Brothers film and recording business for $400 million in 1969, and was on his way to becoming a player in entertainment. In two years, Kinney would be renamed Warner Communications, Inc., and over the next decade Ross would become the very first exemplar of a new archetype: the big media mogul, a breed apart from the studio heads and others who had preceded him in entertainment's corner office. He was the first pure businessman, a figure who bought and sold business properties, as opposed to a producer or exhibitor who'd hit it big. As such he thought about risk in entirely new ways. Ross is no longer a household name, but remains a pivotal figure, not merely on account of the corporate structure he pioneered, but also because his firm and his person would serve as the role model for other firms and their chiefs, among them Disney's Michael Eisner and Barry Diller at Paramount and Twentieth Century–Fox. Ross perfected the accounting practices that anchored the conglomerate, giving enough freedom to keep everyone happy, and the grand affect that would have made a Roman emperor seem mean.[9]

To be sure, as long as there's been a film business, the necessity to contain inherent risk has figured in the equation; even the old industry's modus operandi can be understood in relation to this imperative. The Edison cartel of 1908, for instance, fixed prices aggressively to make sure that, across the industry, costs would never exceed revenues. Likewise, the cartel's enforcement of a certain homogeneity of product—simple plots, short films, no stars, and a ban on most imports—had the effect of ensuring that one film was as good as (or as bad as) another; by making all their offerings "fungible," as an economist would put it, the cartel sought to iron out the discrepancy between the hits and flops, making the fate of any one film that much more predictable.

The vertical integration of the studios that arose in the 1920s with Paramount and the rest can also be understood as an attempt to minimize risk to their investments. By owning every single part of the production and exhibition process, from the actors down to the seats in the theaters, the studios were able not only to control costs, but also to guarantee the size of their audiences, to some extent. It didn't always work, of course, but it did achieve a certain stability for the industry.

In contrast, Ross's answer to the problem of entertainment failures was far more imaginative: he hedged the Warner Bros. film studio volatility with the steady revenues that came from unrelated businesses. Through the 1970s and 1980s, his acquisitions in the name of cash flow also included, at times, cleaning services, DC Comics, the Franklin Mint, *Mad* magazine, Garden State National Bank, the Atari video game company, and the New York Cosmos soccer team. Obviously, not every choice fit the rubric of "communications," but it was all in the name of "synergy."

There were, in fact, two forms of balance achieved by this weird portfolio. One, of course, was among various media. As we've said, record albums, television shows, and books are all subject to the vicissitudes of "hit-driven" industries. Collecting a group of media companies together is a means of sharing the risks and benefits across platforms, a bestselling novel helping to even out a movie dud to achieve something like a stable stream of revenue. But for real defense against *Heaven's Gate*–magnitude bombs, Ross's trick was to hedge the uncertainty of entertainment products as a whole with much more reliable sources of income. Under the Warner Communications umbrella sheltered not only films and music but parking lots, rental cars, and funeral parlors (his former métier).

What Ross was first to do with office cleaning and other services would be translated to the scale of heavy industry in the General Electric–Universal Studios merger of 2004.* Only now it was not the media mogul acquiring a pizza parlor but an industrial mogul, Jeff Immelt, buying a film studio. Universal would enjoy as much of a hedge as any

---

*The fact that Thomas Edison's General Electric now owns Carl Laemmle's Universal Studios is an irony appreciated only by film historians. It is the revenge of the Edison Trust, one century later.

entertainment firm could hope for. By 2008, GE had annual revenues of over $183 billion, while Universal had income of $5 billion, less than 3 percent of the total. With a holding company of that size, the prospect of losing millions on a single film, while not pleasant, is no existential threat. Here was the ultimate defense against even the biggest movie bomb: a corporate structure so titanic that the fate of a $200 million film can be a relatively minor concern.[10]

The conglomerate structure looked like a real boon to entertainment and culture. The capacity to absorb heavy losses could, in theory, provide breathing room for creative experimentation, a way to do the worthy, if riskier, projects. The profits from GE lightbulbs alone could keep dozens of great directors working indefinitely, or fund thousands of Sundance-style films. In fact, one could fantasize about the film studios supported by a tiny internal tax on lightbulbs, a sort of alternative to government funding of the arts. By defusing the bomb, the conglomerate held out the promise of ushering in a golden age of film and other entertainment subsidized by American corporate profit from other sectors.

In the 1970s, when the model was first deployed, with Ross and others, such as Gulf & Western, beginning to combine film studios with more staid businesses, conglomerates created exactly this stabilizing, creativity-enhancing effect. Like the independent studios, they tended to fund the more speculative and interesting films of auteur directors. As long as they broke even or didn't lose too much money, the conglomerate accepted the subpar return on capital. But this forbearance would not last indefinitely, and as one might fear, it was not long before the conglomerates would come to scrutinize their film divisions with the same green eyeshade they used for all their other products.

Before we consider how the values of conglomerates began to infect the business of cultural production, we might well ask why the conglomerates would have wanted any part of that business in the first place. It's clear what a deep-pocketed parent could do for a film studio, but as for a company like GE or Time Warner? Sure, they could well afford to fund plenty of films, and even accept heavy losses, but what was in it for them? Why would a bottom-line for-profit corporation seek exposure to a business as risky as 1970s director-centered film? It seems the sort of property most savvy businessmen would be seeking to dump. Over time, the conglomerates themselves and their frustrated shareholders would begin to ask such questions, eventually tightening the free rein of their studios. But in the 1970s, when Ross established his Warner, he and his cohorts enjoyed being corporate America's easy riders.

No honest account of the media conglomerate's rise could fail to concede the role of purely personal motivations, indeed vanities. For while throwing media properties together wasn't the likeliest path to profit, it provided Ross and his imitators with the chance to indulge some of the most primal pleasures known to the male of the species.

Being a corporate chief executive carries many rewards—above all, high salary and power—for suffering the loneliness at the top. But until relatively recently, those gratifications did not include being a national celebrity. Apart from those who actually owned their mighty companies—the J. P. Morgans and J. D. Rockefellers at the beginning of the twentieth century, the Bill Gateses at the end—the corner office was, for most of American history, a relatively anonymous place. When Ross ran funeral homes and parking lots he was very wealthy indeed but not famous, let alone glamorous. According to his biographer Connie Bruck, it was as a media executive that Ross found his passion, a life of such scale and drama as he craved. Running a conglomerate with media interests furnished Ross, and those who would imitate him, with the chance to befriend rock stars, date actresses, indulge in pet projects, and even influence public opinion.

These inducements are of course related to what economists call the will to "empire-build," but strictly speaking the phrase refers to an activity that is its own reward, the fulfillment of an innate desire as expressed by someone like Theodore Vail. While Ross certainly had such "pure" yearnings, he was also unquestionably drawn to what we might term imperial prerogatives, and these lures account for why he was attracted in particular to the film industry when he was already running a very solid business. It can be astonishing how much some executives prove willing to spend or sacrifice—particularly of other people's money—to enjoy visible proximity to and power over the world of the idolized. Few are the media executives who admit to the emotional need behind that vainglorious magnetism.

In available photos, Steven Ross is almost invariably arm in arm with the likes of Elizabeth Taylor, Barbra Streisand, and Dustin Hoffman. His obituary in *The New York Times* captures the giddy abandon with which he conducted life at the top:

> As Warner grew, lavishness became a company trademark. Generous gifts were doled out to employees at Thanksgiving and Christmas. If an executive wanted a face lift, the company paid. Stars were invited to corporate holiday homes in Acapulco, Mexico, and Aspen, Colorado. Invitations were thrown around to championship fights in Las Vegas, Nevada, and Warner's guests traveled on one of the company's half a dozen planes. On the flights, Mr. Ross would dole out candies, entertain stars and employees with card tricks, and play backgammon and dominoes. In some ways, he was a sugar-daddy, in others almost a child in the way he relished the pleasures available to him.[11]

His aggressive cultivation of celebrity friends sometimes cost Warner shareholders more than the odd trip to Acapulco. Donations to their favorite charities and expensive presents for their children were also in the sugar-daddy arsenal. As Connie Bruck relates, when courting Steven Spielberg, Ross spontaneously agreed to pay him over $20 million—ten times the reasonable value—for the rights to make a video game of his film *E.T.: The Extra-Terrestrial.* The result was the first major video game based on a movie—conglomerate synergy in action—but also a game generally acknowledged as the worst in video game history ("famously bad," according to *PC World*); after disappointing sales, untold millions of unsold *E.T.* game cartridges were buried in the desert near Alamogordo, New Mexico. Ross's disastrous deal damaged and may have wrecked Atari, whose role in the corporate portfolio was meant to be that of a cash cow, not glitzy loss leader.[12]

Of course it would be a gross oversimplification to say that the conglomerate represented a simple trade: the industrial mogul offering financial security in exchange for access to the Hollywood lifestyle. In addition to alleviating the volatility of cash flow in the movie business, and giving its master a new set of toys, the conglomerate served a more familiar purpose of empire building: material self-enrichment. For while most of the revenues of individual divisions might not justify truly outsize rewards, ganged together they represent a balance sheet that could justify the sort of compensation one would associate with an industrial or financial powerhouse today. Someone like Ross or Michael Eisner,

Disney's CEO, was well positioned to reward not only his friends and cronies but also himself. The fact of having made oneself a sort of celebrity creates some cover for such self-indulgence: Could a mogul fairly pay himself less than a movie star? In the 1990s, Eisner, for instance, famously awarded himself $737 million for five years of work.[13]

But there is an essential difference between a Ross and an Eisner. Ross was a service executive who bought into media, while Eisner was a born-and-bred media man. Though the latter would be forced out by a boardroom campaign orchestrated in part by Roy Disney, the founder's nephew, who would accuse Eisner of turning the family entertainment firm into a "rapacious, soul-less" company,[14] there is no denying that in his tenure, Eisner grew revenues by some 2,000 percent. Both men, for all their personal excesses, used the conglomerate structure in ways that were ultimately good for business, as a business, but detrimental to the variety and innovation in the production of content. With Ross we might say, generalizing broadly, that that effect was the result of being in the content business for all the wrong reasons, while with Eisner the problem was more nearly one of seeing content less and less as an end in itself. The Disney Company that Eisner inherited had pioneered the branding of content by way of various forms of merchandising, from theme parks to sweatshirts. But by the time Eisner would take the helm in 1984, the company, whose bread and butter was still the theatrical film release, was faltering and had barely survived takeover by corporate raiders. Eisner would turn the company around, making it for a while the largest media conglomerate in the world, but in ways that would seem to some, the founder's nephew among them, a betrayal of the company's founding devotion to content values above all. It was a common complaint against the media conglomerates rising in those years, and it stemmed from the other strategies typical of that corporate structure.

STRATEGY 2: INTELLECTUAL PROPERTY DEVELOPMENT

If Ross and Warner Communications pioneered the mixing of businesses to balance risk in the entertainment industries, through the 1980s and 1990s a complementary technique rose to prominence. As we've said, films and other entertainment products are risky investments, and the industry has historically structured itself to manage that risk. Among the oldest and perhaps the most intuitively apt strategies had been sticking with bankable stars.* A development engendered by the success of investing in star-driven films was the pursuit of film franchises. It had worked with the *Thin Man* films in the 1930s and has continued to work with the James Bond films since the 1960s, and more recently with the Bourne films. Slightly different but following the same basic logic is the sequel, which has given us multiple follow-ups of proven hits, among them *Jaws, Terminator,* and *Beverly Hills Cop.* In the 1980s yet another variation on this approach began to emerge, one by which films could come to be seen as, in effect, a delivery system for an underlying *property.*

By this approach, every film is anchored to an underlying intellectual property, typically a character, whether a primarily visual one drawn originally from a comic book, like Batman, or a literary one, like Harry Potter. The film is thus simultaneously a product in its own right as well as, in effect, a ninety-minute advertisement for the underlying property. The returns on the film are thereby understood to include not simply the box office receipts, but also both the appreciation in the property value and its associated licensing revenue—merchandise, from toys to movie tie-in editions and other derivative works. Since every film based on such a property can enjoy multiple types of return on investment, there is strong motivation to concentrate assets on these naturally diversifiable investments.

Consider the most expensive films of the 2000s:

2007	*Spider-Man III*	$258,000,000
2009	*Harry Potter and the Half-Blood Prince*	250,000,000
2009	*Avatar*	237,000,000
2006	*Superman Returns*	232,000,000
2008	*Quantum of Solace* (James Bond)	230,000,000
2008	*The Chronicles of Narnia: Prince Caspian*	225,000,000
2009	*Transformers: Revenge of the Fallen*	210,000,000
2005	*King Kong*	207,000,000
2006	*X-Men: The Last Stand*	204,000,000

---

*Strictly speaking, Hollywood's reliance on star-centered films was pioneered by Adolph Zukor in 1912 with *Queen Elizabeth* and continued with Mary Pickford; it was based on this strategy that Zukor called his production studios "Famous Players."

| 2007 | *His Dark Materials: The Golden Compass* | 205,000,000 |
| 2004 | *Spider-Man II* | 200,000,000 |

First, notice that with the exception of *Avatar*—the one flight of directorial fancy more like the high-risk gambles of days gone by—each of these films was a remake or a sequel. Even more telling, each was centered on an easily identifiable property with an existing reputation, appeal, and market value. And the power of merchandising is such that character would no longer appear to be entirely essential; one can be developed from something as inanimate as a toy, as with *Transformers*.

Let us now compare the most expensive films of the 1960s:

1963	*Cleopatra*	$36,000,000
1969	*Hello, Dolly!*	24,000,000
1965	*The Greatest Story Ever Told*	20,000,000
1969	*Paint Your Wagon*	20,000,000
1969	*Sweet Charity*	20,000,000
1962	*Mutiny on the Bounty*	19,000,000
1964	*The Fall of the Roman Empire*	19,000,000
1963	*55 Days at Peking*	17,000,000
1966	*Hawaii*	15,000,000
1960	*Spartacus*	12,000,000

From a twenty-first-century perspective, the problem with these films as a business proposition is clear: they don't build the value of an underlying property. A film like *Cleopatra* either makes money or it doesn't (it didn't—despite being the highest grossing film of 1963!). It doesn't leave the consumer with a desire for ancillary consumption once the experience is over. Stated in advertising terms, it wastes the audience's attention. In contrast, a film like *Transformers* or *Iron Man* doesn't just earn box office revenue, but it demonstrably drives the sale of the associated toys, comic books, and, of course, sequels.

You can't learn everything from looking at the most expensive films, but you can gain important insight into how the industry has changed, and how its energies and resources have come to be directed. The change has everything to do with the business's being part of conglomerate structures.

How does a film like *Transformers* suit the conglomerate in ways that even a money-making film like *Hello, Dolly!* does not? We can see the reason in broad terms, but a deeper understanding depends on considering both the economics of information and the law of copyright.

Unlike almost every other commodity, information becomes more valuable the more it is used. Consider the difference between a word and a pair of shoes. Use each a million times: the shoes are ruined, the word only grows in cachet. Every time you utter the word "Coke," "McDonald's," or "Lululemon," you are doing the brand owner a small service of marketing.*

One of the stranger consequences of the electronic age is that almost any word or image, reiterated a million, or a billion, times, can become a valuable asset. How likely does it seem that an odd-looking mouse with a squeaky voice and somewhat bland personality would become one of the world's most famous icons? Or that so many people would know who Paris Hilton is, and, even stranger, would seek to pay for her association with various products?

The key to capturing the economic potential of such phenomena is turning an image or a brand into a *signifier*—a symbol of something. A picture of Adolf Hitler has come to make most people immediately think "evil," although it seems just that no one profits from the association. By contrast, the image of Darth Vader leads just as directly to the same idea, but the character is a property owned by Lucasfilm. Snoopy the beagle has gradually become a platonic form, an image of fanciful fun, and that fact yields millions in licensing revenue every year for Snoopy's owner, United Media. As this suggests, the intellectual property laws of copyright and trademark are a way to own and profit by some signifiers. You cannot own Hitler or the idea of a giant squid or driven snow. But you can own Darth Vader, Batman, or the Pink Panther, thanks to the federal laws of intellectual property.

Strong, enforceable ownership of characters is actually a relatively new phenomenon in the development of copyright law. In the days when heroes were historical or drawn mainly from books, the courts often denied ownership of characters, even ones as distinctive as Sam Spade, protagonist of *The Maltese Falcon*.[15] But since the 1940s or so the

---

*Lawyers generally argue the opposite: that using a brand name too much destroys its value, because it makes the underlying trademark generic (as when one asks for a Kleenex or a Xerox, instead of a tissue or a photocopy), and therefore unprotectable as a matter of law. This helps explain why lawyers often don't get along with marketing people, as the latter always favor maximum exposure of the brand.

courts have generally been more accepting of ownership rights, provided the characters meet certain marks of minimal delineation or distinctiveness. You cannot own a stock character, like the barroom brawler. But give him claws of adamantium, a distinctive look, and a few other specific traits, and you have the comic character Wolverine, one of the most valuable properties in film.*

Legally speaking, it's hard to be precise about the standard for what kind of character can be vested with copyright, but suffice it to say the standard is not onerous. Unlike a patent, a character copyright doesn't require proof that you've invented something or effected a real innovation—minimal creativity will do. Nor is literary merit necessary—Grimace the Milkshake Monster resides safely in the protective realm of copyright. Characters are sturdy intellectual properties, whose ownership is protected by federal law. Their value can be measured and, by the device of film, increased.

So it was that in the first decade of the twenty-first century, many studios, almost in the manner of their contemporaries among real estate developers, would spend most of their time and money looking for properties ripe for redevelopment. They had tried selling stories, they had tried bankable stars. But by the 1990s, patience with plotting had reached a nadir and the salary demands of proven stars had reached a zenith. The coincidental revolution in digital technologies opened up the possibility for a much more notional type of film, in which dazzling effects and surreal imagery could serve just as well as, if not better than, absorbing narratives and memorable performances. By the twenty-first century, film would become much less predominantly a business of storytelling than it had been, and much more a species of advertisement, an exposure strategy for the underlying intellectual property. The exposure strategy also facilitated the globalization of entertainment media that had been under way for decades. While the export potential of the traditional sort of film, with its cultural particularity, was another unknowable quantity in the profit forecast, the new sort of film centered on literally cartoonish archetypes that traveled easily everywhere.

---

*As Judge Hand put the point, "If *Twelfth Night* were copyrighted, it is quite possible that a second comer might so closely imitate Sir Toby Belch or Malvolio as to infringe, but it would not be enough that for one of his characters he cast a riotous knight who kept wassail to the discomfort of the household, or a vain and foppish steward who became amorous of his mistress." *Nichols v. Universal Pictures Corp.*, 45 F.2d 119, 121 (2d. Cir. 1930).

And thanks to the accounting practices of conglomerates, the success of such a film was not reckoned based on its direct sales alone, but by its enhancement of the underlying property's worth. The film was, then, to some degree, a kind of giant business expense as well as a product in its own right. It was a clever concept, to be sure, but also a very different approach to culture, one that has proved unrecognizable to many people over thirty, let alone the question of what the founders of Hollywood would have made of it.

By splintering opportunities for return on investment, the intellectual property approach has served the conglomerates as the ultimate means of risk diffusion in their entertainment businesses. Edward Jay Epstein, an expert on the strange economics of the film industry, explains the system very clearly. He points out that since the 1990s or so, the studios have considered box office receipts as far from the most important measure of how a film "does." For it is outside investment partners that typically bear the risk at the box office. The revenue that matters to the studio, according to Epstein, is from everything else, including proceeds from the film itself in different media (DVD, cable video-on-demand, downloads, etc.) and in theatrical release around the world. But the studio's mother lode of profit depends on the character copyrights, coming from the merchandising, spinoffs, sequel rights, and other "derivative works," whose true value is never made public.[16]

That arrangement, Epstein suggests, makes the studio today more of a licensing operation than a filmmaking enterprise. It develops valuable properties and makes its money from licensing them in as wide a multiplicity of forms as possible. Such a view of filmmaking is a far cry from the essence of the medium since the rise of the studio. Whatever Michael Cimino may have had in mind when he created *Heaven's Gate*, burning a brand onto the popular consciousness was definitely not it.

STRATEGY 3: MINE FESTIVALS

In 1989, the brothers Harvey and Bob Weinstein were the owners of a small independent distribution company named Miramax, whose success was mainly in distributing concert films. That year, however, they made an important bet. Based on its reception at the Sundance and Cannes film festivals, they theorized that the low-budget film *Sex, Lies, and Videotape* would appeal to enough people to justify the costs of national distribution. It wasn't the traditional sort of Hollywood bet on a film to be made, but rather on one that already existed. But it was a highly speculative proposition all the same: with a production budget of $1.2 million, the film had a complicated plot centered on the subject of adultery and featuring a man who tapes women talking about their sexuality.

Sex, Lies, and Videotape was the test case of a third risk management strategy developed in the late 1980s and early 1990s, one that depended not on production but on discovering diamonds in the rough: films already in the can. The concept the Weinsteins pioneered looked primarily to film festivals—particularly Robert Redford's Sundance Film Festival in Utah, and for foreign productions, the Cannes and Toronto festivals—as a test market and hunting ground. Through leveraging, low-budget undervalued films would realize a very handsome rate of return on capital; but the festival approach was hardly as important an innovation in financial terms as the others we have discussed. Its true importance was rather in becoming perhaps the best means for foreign and artistically innovative film to reach a large audience.

The Miramax bet on *Sex, Lies, and Videotape* paid off, as the film made over $25 million in the United States alone. Of course, the Weinstein brothers were far from the first to turn low-budget efforts into gold. As we've already seen, it was a common operating move in the 1970s, when studios first funded productions by unknown independent filmmakers like Sylvester Stallone and Francis Ford Coppola. In fact, by the standard of *Rocky* or *The Godfather,* the success of *Sex, Lies, and Videotape* was quite modest. But since the failure of United Artists in the early 1980s and the rise of the media conglomerate, big investments in new directors had become increasingly rare.

The Weinsteins brought the model back, albeit with a twist. As we've said, they relied not so much on their own judgment, but on the collective judgment of critics, audiences, and industry insiders at the festivals. The institution of the festival—once the only exhibition chance for the potential art-house film, of which many were made but few were chosen—slowly evolved into something of a filter or wholesale market. Firms like Miramax, and its imitators like Sony Film Classics and Fine Line, began to see that by buying low and selling high, exposure to independent film could be profitable again.

There is a certain genius to the approach, which might be said to depend on the wisdom of crowds over the judgment of a single producer. Given the quarry, there may be no other realistic way of hunting for it: nearly ten thousand independent films are produced in the United States every year, and thousands more are made abroad. As with any other commodity, most are average or below average, but some will be brilliant; among that smaller cohort, an even smaller portion will have potential for popularity with the public. To find them oneself is to look for the needle in the haystack. The film festival, by happy accident of engineering, is a remarkably effective filter, with several layers of selection by festival staff, thinning the herd before the festival critics and filmgoers pronounce judgment. The process, obviously, isn't foolproof; audiences in Sundance, Cannes, and Toronto are not a perfect proxy for the real world. But the festivals do deliver enough information to justify relatively small bets on films that have already been made and, to some extent, tested.

Consider the classic example of a film that broke through the Sundance model, Kevin Smith's *Clerks*. The film portraying the life of a convenience store clerk and his friends was made on a budget of $27,000 and was shot, in black-and-white, in the very convenience store then employing Smith. Miramax was at first uncertain about the film's investment potential. At Sundance, however, it proved a huge hit, and so the Weinsteins decided to buy it, paying $227,000. In theatrical release, it went on to make over $3 million—not a large number by blockbuster standards, but a great return on capital. Later, Miramax would invest more in Smith's *Chasing Amy*, with even greater returns. The success of the follow-up of course meant further rental and residual revenues from *Clerks*. And meanwhile, in terms of cultural capital, Kevin Smith had been minted as an auteur, a bard of New Jersey culture.

The technique pioneered by Miramax in the 1990s was successful enough that within a few years, the media conglomerates began to adopt the model. Disney took the most direct route: it simply bought Miramax, while others built new boutique operations based on the same model: Sony Pictures Classics, Paramount Classics, and Fox Searchlight, among others. But as we have said, the primary benefit of this approach was artistic rather than financial. As with so many small-scale methods, once taken up by a conglomerate, the pressure was on for what in network engineering is called *scalability*. There are only so many films made for $20,000 that can be turned into a multi-million-dollar box office result. As a consequence, a certain amount of, as it were, ersatz Indie product has been marketed: pictures that resemble quirky successes without quite having a unique soul of their own.

THE MEDIA CONGLOMERATE IN THE TWENTY-FIRST CENTURY

By the year 2000, the form of the media conglomerate had reached its maturity and logical perfection. What had started as an impulse to group media concerns with other types of businesses had by virtue of the intellectual property revolution reconfigured the landscape of information industries exclusive of telecommunications. The homogeneous giant enterprises that dominated the first half of this book had, by the 1990s, given way to a gang of octopuses owning properties diversified mainly across media industries, typically holding a film studio, cable networks, broadcast networks, publishing operations, perhaps a few theme parks. The conglomerates added a management layer above the media firms, unifying their efforts by little more than a common name and the fact that, in some loose sense, they all trafficked in information.

Disney, once dedicated to its core brands (Mickey Mouse, Donald Duck, Snow White, and the rest), turned itself into a true media conglomerate, with completely unaffiliated holdings such as ABC, ESPN, and Miramax fulfilling an imperial dream of Michael Eisner while creating in Roy Disney's eyes a "rapacious, soulless" abomination, though in the long run a quite profitable one. General Electric, the industrial conglomerate founded by Thomas Edison, having sold off RCA in 1930 bought back NBC and its associated properties in 1986, launching CNBC in 1989. It would later take over Universal Studios, joining Disney and Time Warner as one of the three players with holdings in every major entertainment sector: films, characters, television stations, publishers, theme parks, and recording labels. Meanwhile, Gulf & Western, another 1960s conglomerate that had started in 1934 as the Michigan Bumper Company, bought Paramount, remaking itself in the image of Warner Communications as Paramount Communications, an effort that would founder, resulting in the company's sale to Viacom in 1994. In 1989, Sony bought Columbia Pictures and CBS Records, trying to create the first Japanese version of a Ross-style media/industrial conglomerate, with mixed results.

We have seen that size now and again attracts the notice of the federal government, and one might well wonder whether the Justice Department or FCC, noticing these new giants walking the earth, might have considered breaking them up. Once upon a time, the government had indeed been vigilant about discouraging, if not blocking, cross-industry ownership, but by the 1980s and 1990s, with the rise of antiregulatory sentiment, those days were over. Some restrictions did apply, as related to, say, acquiring concentrated holdings in a particular medium in one market. But as the conglomerates mainly sought holdings in unrelated markets—for instance, magazines and film—resulting in no price fixing or monopolies in any particular market, no Sherman Act alarms were sounded. The conglomerates therefore grew unmolested, with minimal oversight, through the 1990s, until they owned nearly everything. The only exception, as we shall see, was the world of the Internet and computing; but that seemed only a matter of time.

As a coda, let us consider one story of a better life through the corporate chemistry of the conglomerate, a twenty-first-century entertainment parable that avoided tragedy, indeed has a happy ending, thanks to creative risk management. Released in 2007 by Universal Studios, *Evan Almighty* concerns a man with a Noah complex: he is driven to build an Ark to save the world's animals from a coming flood. Like many a prudent production of its time, the film was a sequel; it was developed to get another bite of the apple that was *Bruce Almighty,* the highly profitable film starring Jim Carrey as a man given God's powers for a week. *Evan* starred Steve Carell, a sensible choice since he was by then a bankable star, having succeeded with *The 40-Year-Old Virgin* and, on television, with *The Office.* The script was written and rewritten by numerous writers, minimizing the risk of relying on the judgment of a single author. While the film, unfortunately, had no protectable intellectual property of which Universal could take possession, it was at least based, however loosely, on the Bible, a proven bestseller even if its copyright had lapsed. Based on these precautions, the studio ultimately invested $175 million in the production, making it, at the time, the most expensive comedy ever.

Unfortunately, the film had all the right ingredients but one: it wasn't any good. The influential critic Richard Roeper excoriated it as "a paper-thin alleged comedy with a laugh drought of biblical proportions, and a condescendingly simplistic spiritual message."[17] On Rotten Tomatoes, a popular website aggregating reviews, the film garnered an embarrassing 8 percent positive response. Despite an extensive marketing campaign, and a national opening on 5,200 screens, *Evan Almighty* made just $30 million or so its first weekend, pitiful by blockbuster standards. For all of these reasons, it would find distinction on *Rolling Stone*'s list of the worst films of 2007, and many other lists of notable bombs.

Yet here is the miracle: the bomb went off, but it did no damage. There was no collapse of Universal Studios, and few lasting consequences for anyone involved with the project. Life went on basically as before at Universal, and more important, at General Electric. The failure of *Evan Almighty*, a bona fide disaster for Universal, was but a rounding error in the performance of General Electric, with revenues of $168 billion that year.

The immediate failure, in short, went unpunished. Even more remarkable, over time, through DVD sales and foreign box office, *Evan Almighty* actually came close to the break-even point, this even though no one had anything good to say about it. Had it been the type of film that lent itself to merchandising and licensing income, it is possible that *Evan Almighty* could have made a healthy profit despite the fact of being, as a film, unequivocally lousy.

It is instructive, and rather disheartening, to compare the failures of *Evan Almighty* in 2007 and *Heaven's Gate* in 1980. As a consequence of *Heaven's Gate*, Michael Cimino was effectively exiled, never allowed to make another major film, and the director-centered system he and others of his stature had embodied was severely discredited. In contrast, despite the failure of *Evan Almighty*, the system that produces films like it carries on unperturbed, because in financial terms there was little real damage. *Evan Almighty*, in this sense, is proof of how secure the studio structure now is. Mediocrity safely begets mediocrity: behold the true miracle of the modern entertainment industry.

CHAPTER 18

The Return of AT&T

In 2002 President George W. Bush signed an executive order authorizing the National Security Agency to monitor telephone conversations and Internet transactions of American citizens without a court warrant.[1] The order was secret, as was its implementation, and even today the breadth of the domestic spying remains unknown. However, one thing was clear: the NSA could not have fulfilled the order alone. It needed help, most of all from the nation's telephone companies.

Four years later, in December 2005, the warrantless wiretap order was leaked to *The New York Times.* Senator Arlen Specter summoned Edward Whitacre, CEO of AT&T, to appear before the Senate Judiciary Committee.[2] In the Judiciary Committee's hearing room, with unusual intensity in his voice, Chairman Specter, a former prosecutor, questioned Whitacre precisely and slowly:

"Does AT&T provide customer information to any law enforcement agency?"

"We follow the law, Senator," answered Whitacre.

"That is not an answer, Mr. Whitacre. You know that."

"That's all I'm going to say, is we follow the law. It is an answer. I'm telling you we don't violate the law. We follow the law."

"No, that is a legal conclusion, Mr. Whitacre," said Specter, with evident rising anger. "You may be right or you may be wrong, but

I'm asking you for a factual matter. Does your company provide information to the federal government or any law enforcement agency, information about customers?"

"If it's legal and we're requested to do so, of course we do."

"Have you?"

"Senator, all I'm going to say is we follow the law."

"That's not an answer. That's not an answer. It's an evasion."

"It is an answer."

Whitacre's testimony in 2006 marked the first major public appearance of the resurrected AT&T. It was a moment that dramatized how much had changed since 1984: twenty-two years after the breakup, the Bell system was, in a word, back, and working closely once again with the U.S. government.

The inheritor of the great mantle of Theodore Vail, Ed Whitacre was a very different sort of man, and though he had ambitions similar to his predecessor's, his new AT&T was a different kind of company. Vail had been an idealist who believed earnestly in Bell's obligation to serve the country as a public utility and build the greatest phone system in the world. "We recognize a 'responsibility' and 'accountability' to the public on our part," wrote Vail in 1911.[3] By contrast, Whitacre was the product of a different corporate culture, whose credo was to maximize returns and minimize oversight. When a reporter asked Whitacre his vision for AT&T, he listed his top three priorities as follows: "I like to be the best. I like for our stock price to be the highest. I like for our employees to be the highest paid."

Whitacre's AT&T was a beneficiary of federal communications policy in the early twenty-first century, and a powerful reflection of the corporate ethos of that era. In the name of competition, it sought monopoly and power. Under the banner of libertarianism and small government, it manipulated regulatory regimes to eliminate competition. But it would be unfair to say that the benefits of the relationship flowed entirely one way. As Whitacre's testimony makes devastatingly clear, even while leaving it unspoken, AT&T found ways to be of use to the administration.

WHITACRE TAKES THINGS IN HAND

Edward Whitacre, Jr., the chief rebuilder of the AT&T system, is a man whose appearance makes a lasting impression. He is enormously tall, with a slow gait, an even slower way of speaking, and a textbook Texan drawl. As a former FCC staffer put it, "he was extremely intimidating, always polite, but you had the feeling that if you messed with him he would kill you." Whitacre, despite being head of a telecommunications company, cultivated a stubborn Luddite affect. He had no computer in his office and refused to use email. "I'm not computer illiterate," he once told reporters, "but I'm close."[4]

In 1999, *Business Week* put Ed Whitacre on a cover with the headline "The Last Monopolist."[5] The story sought to determine how Whitacre and his Bell company could survive the coming age of what was assumed would be ruthless competition. "Can Whitacre," asked *Business Week,* "a monopolist born and bred, survive without his monopoly?"

Whitacre had an answer: Why learn to cope with competition when you can eliminate it? Through the late 1990s and into the new millennium, despite or perhaps thanks to an official federal policy promoting "fierce competition," Whitacre would strangle nearly all of his competitors, largely reconstituting the Bell system that Theodore Vail had founded. By 2006, his resurrected empire would cover the whole country, excluding parts of the West and Northeast, those ruled by another giant born of reconsolidation, Verizon.

Whitacre, aptly termed "a monopolist born and bred," was certainly the man for the job. Having joined AT&T in 1963, during its pre-breakup heyday, he had stuck with the firm through the first stirrings of competition in the 1970s, and he would continue to rise through the ranks as Bell was dismembered in the 1980s. In the 1990s, still in Texas, Whitacre took charge as CEO and chairman of Southwestern Bell, then the smallest of the eight Baby Bells created by the breakup.

Whitacre, to his credit, was a man willing to take on a challenge. The task before him was immense. The regional Bell companies were, according to federal policy, supposed to lose business to their competitors. The Bell system was the corporate equivalent of a convicted felon, and the Baby Bells all operated under the direct supervision of both the FCC and Judge Harold Greene, who, now as "telecommunications czar," would oversee enforcement of the former monopoly's consent decree with the Justice Department. The punishment for the mother ship's mischief in the 1970s was a tight web of court-ordered restrictions and requirements cast over the Baby Bells' operations. FCC rules now obliged them to furnish all customers with a telephone jack to facilitate the attachment of "foreign" telephones and other devices. At their exchanges, the Bells were required to provide competing long distance carriers (MCI or Sprint) access to local callers. The decree—to which, for mysterious reasons, Bell had voluntarily submitted—also barred the Baby Bells outright from certain markets, including "online services."

Springing Ma Bell from this regulatory cage would amount to an escape from corporate Alcatraz. For the Baby Bells, survival was the order of the day; a return to monopoly control seemed a fruitless fantasy of those impractical enough to entertain it. If there was any hope of recovering even a bit of that former power, it lay in a painstaking long-term strategy.

But Whitacre and others with revanchist longings would bide their time. They knew that while Bell was officially a public menace, the old regime retained many loyalists and friends in Congress, federal agencies, and most of all, state and local governments. Southwestern Bell in particular had good working relationships with most of the Texas political class, since Whitacre and his company continued to do as they always had, supporting both parties with generous donations.

At a conceptual level, the Bells' lobbyists and policy strategists—mostly at Verizon (once Bell Atlantic, the East Coast Bell, which had long thought of itself as the most intellectual of the Bells)—began to rethink some of the fallen empire's long-held positions, particularly its attitude toward competition, which had always been regarded as anathema to the company whose credo was "One System, One Policy, Universal Service." Vail's writings of the 1910s are filled with denunciations of "the nuisance of duplication."[6] So close to the corporate core were these convictions that AT&T had upheld them as gospel from its beginnings in the 1880s right through the 1980s, by which time the idea of a regulated monopoly had long gone out of fashion. Like a man in a leisure suit, Bell strutted stubbornly into its last decade clad in its outmoded orthodoxy, refusing to abide even a whiff of "competition." It was on account of such stiff-necked absolutism that the Justice Department brought the lawsuit that led to the breakup.

But Bell's brain trust, a shadow cohort who, though scattered, never abandoned the cause of the empire, had an idea. With sympathetic academics, lobbyists, and some of Bell's best inside and outside lawyers in their ranks, these stalwarts came to understand that Washington's prevailing enthusiasm for competition and deregulation might actually be made to serve Bell's contrary interests. Paradoxical as it might seem, the ideology of competition itself could, they imagined, furnish the key for Bell's prison break. How was this possible for what had become the nation's most regulated businesses? For an answer, let us examine for a moment the history of competition's allure in America.

The perceived value of competition has varied considerably in American history. In the late nineteenth and early twentieth centuries, broadly speaking, many business leaders like Vail, as well as labor leaders and economists, thought that competition, particularly in operating utilities or other economic necessities, was wasteful and destructive. In such sectors, government regulation was deemed prudent to protect businesses serving a vital social function from the excesses of competition, assuring them of, if not monopoly, at least a reasonable degree of market share.

Competition hardly grew in favor during the 1930s, given the Depression and the New Deal. Private industry on the whole was suspect, as government began to trust more broadly to regulation as a means of achieving better results. It was only beginning in the 1960s and 1970s that this bias toward government control began to shift, inspired by a new generation of libertarian economists, mostly at the University of Chicago, among them Milton Friedman and George Stigler. Such analysts viewed the performance of government-regulated industries—essentially the New Deal paradigm—as disappointing, and they prescribed a dose of both deregulation and more competition as medicine for an economy by then ailing after its postwar expansion. Some went so far as to suggest that nearly any regulation was unnecessary, given a state of healthy competition.[7]

While such ideas were considered radical in the 1960s—sometimes dismissed as Goldwater economics—they began in the seventies to be applied experimentally across once regulated industries such as airlines, trucking, and energy. In communications, change began with Nixon, as we saw in the cable chapter, but continued under Jimmy Carter. Under Reagan, deregulation accelerated, and was regarded, along with tax cuts, as the key to economic growth. And so, at the time of the Bell system breakup, nothing could have been quite so anomalous as radically regulating a major industry.

In this ambience, the Bell partisans got thinking. If government regulation was an evil to be suffered only in the absence of competition, it followed that evident competition should render that evil unnecessary. If the most regulated company in the nation could somehow show that competition had indeed taken root in the telecommunications sector, there was a chance the Bells might yet shed most of their shackles.

The 1996 War of All Against All

The election of Bill Clinton did not turn back the deregulatory wave. Clinton would find himself forced to agree that "the era of big government is over," a sentiment that applied to the federal regulatory regime as well as the welfare state. In few places was this free-market vibe as strongly felt as at the FCC and among those paid to lobby the commission. Al Gore, the administration's point man on tech policy, believed as deeply in the power of competition as had the figures in Nixon's administration. For their part, in speech after speech, FCC officials touted a free market as the best means of achieving the social goals of communications policy. Gore's friend and FCC chairman Reed Hundt was a competition apostle as well. Speaking of a "national commitment to open markets, competition and deregulation" was boilerplate in the 1990s—as Reed explained, "competition in the communications markets will yield lower prices and more choices for consumers, rapid technological innovation and a stronger economy."[8]

Promoting competition wasn't a bad idea by any means. What didn't necessarily follow, however, was that the existence of competition in the most abstract sense could obviate all need for regulation—particularly regulation designed to prevent anticompetitive behavior! And how do we know when "competition" really is competition? The willingness of some to see an appearance of burgeoning competition as a reasonable substitute for regulation looked like an opportunity to the Bells, and so they changed their religion. Embracing the process of "competition" that was under way, the Bells prepared to make their comeback as a dominant player in a nominally open industry.

It was a perfect wedding of a new government ideology and a new corporate calculus when the Bells, AT&T, and the rest of the industry signed on to the Telecommunications Act of 1996.[9] The most sweeping legislative overhaul of the business since the Communications Act of

1934 was founded on the principle of "competition everywhere." The idea was to remove barriers to entry in all segments of the industry, a goal that the Bell companies (Bell Atlantic, Bell South, Pacific Telesis, Verizon, and the rest), the long distance firms (AT&T as well as MCI), and the cable companies all pledged to uphold. The Act was designed to encourage cable companies to enter the phone business, phone companies to offer TV service, long distance firms to build local networks, and so on. Officially, it was meant to create a Hobbesian struggle of all against all.

The law was hailed in 1996 as a sort of ultimate victory over the Bell monopoly and the dawn of a new age; in retrospect, the measure was hopelessly naïve. Its centerpiece was a complex regime whereby the Bell companies allowed their competitors to lease Bell's infrastructure to offer the same local telephone service as the regional Bells. Creating competition over the existing facilities might have worked in another industry, and it actually did seem to work in other countries. But somehow forgotten was the Bell company's hundred-year track record of annihilating or assimilating dependent competitors. The Bells were corporate America's reigning champs in the rope-a-dope game of keeping up appearances in the front office while quietly pummeling their rivals in the parking lot.

While not keen to share their toys under the Act's so-called unbundling rules, the Bells immediately understood that the deal was a win for them. What mattered most was one critical fact: the 1996 law superseded the consent decree that had ended the Bell antitrust lawsuit. With that decree abrogated, the Bells were now under the supervision of the FCC, as opposed to the hawk-eyed taskmaster Judge Greene. It was for them the catbird seat: there was no rival they couldn't handle, except for the federal courts and the Department of Justice.

There is a striking similarity between what followed the 1996 Act and what followed the Kingsbury Commitment of 1913. Both government interventions had sought to introduce permanent competition into the telephone market, and each was hailed at the time as an enormous victory over the Bell system. In fact, each would pave the way for a new age of Bell ascendancy. The difference was this: the old AT&T had pledged to operate as a public trust, and was as good as its word. The new AT&T had no such aspirations.

THE CAMPAIGN

Eliminating competition is rarely accomplished in a single grand stroke. The would-be monopolist does not round up rivals for a wholesale massacre; there are no corporate killing fields. Instead, the corporation seeking dominance behaves rather like a pest exterminator, setting poison bait traps, killing what he can see, and methodically decimating his foes by making their life a living hell. The monopolist's tools are lawyers and local statutes; his tactics are delays and court challenges, all deployed with an eye toward unraveling firms with lesser resources.

The 1996 Act enabled the Bells to blow the trumpet of competition while simultaneously eliminating all actual competitors. There were plenty: since the breakup, scores of new "competitive" phone and Internet companies had launched, hoping to take a piece of the Bells' billions in revenue for themselves. In part they were drawn by the more general tech boom and economic expansion of the 1990s, when it was easy to get funding. But the real rush began in 1996, after which telecommunications would account for an unprecedented share of GDP growth.

Each of the Bells did its bit to eliminate local challengers. Verizon, for its part, handled competitors in the Northeast, and managed to finally silence MCI, Bell's bête noire, forever. But the undisputed heavyweight champion was Ed Whitacre's Southwestern Bell, now renamed SBC. As *Network World* reported as early as 1997, "SBC, more than any other [Bell Company], uses a phalanx of lawyers and millions of dollars in lobbying efforts in a deliberate effort to thwart meaningful competition in its markets."[10] From the late 1990s into the following decade, SBC masterfully waged a war of attrition.

SBC's ground war was a guerrilla-style campaign devised to nullify concessions made in the 1996 Act. It would ultimately be copied by all the Bells, until in state capitals and in thousands of tiny local jurisdictions across the country, death would be inflicted by a thousand cuts to make the competition regret they'd ever thought of taking on a Bell. But it began in Texas, where, by 2003, SBC had nearly a hundred registered lobbyists working in Austin—as against the 181 members of the legislature.[11] Avowedly opposed as they had been to regulation, SBC and the others had long known that pressing for it on the local level was a handy tool against any challengers. When competitors had started cropping up in the early 1990s, SBC persuaded the Texas legislature to add some useful provisions to the Public Utility Regulatory Act of 1995 (PURA 95)—among the tweaks to the final bill, a hefty price on market entry. To offer service to a single customer, a would-be telephone company had to build physical lines reaching 60 percent of homes and businesses in a twenty-seven-square-mile radius. It was roughly like requiring that one build roads to every house in the area just to open a gas station.

There were other tactics beyond the legislative. To reach customers over SBC's lines, for instance, would-be competitors often needed to rent space in the local "central office," where the lines terminated. SBC was required to offer a lease, but the law didn't specify a rate. So in the late 1990s, when a 10x10 space in upstate New York was going for $10,000 a year, for instance, SBC would charge $500,000 for that much space in Texas. The aim of obtaining a more reasonable rent would oblige a competitor to file a lawsuit, and appeals—again, just to get started.

The industry trade magazines were full of similar stories of dirty pool in the late 1990s. By one account, SBC lawyers were dispatched to threaten an elementary school for choosing a rival phone company. In another case, SBC left the windows open in a space they operated where a competitor's switching equipment was housed; pigeons came in to roost, until eventually their excrement caused system failure. On other occasions, Bell simply flouted interconnect agreements until the competitors were forced to sue—just like the good old days! In such cases, the FCC would sometimes fine SBC and the other Bells, but to little effect. Over time the government would file a series of lawsuits under the Sherman Act, petitions superficially similar to MCI's case against AT&T in the 1970s, charging them with violating the spirit of FCC regulations by using its monopoly powers to destroy newly created competitors. The Baby Bell apples had not fallen far from the tree.

If the late 1990s and early 2000s departed from the 1970s in invigorating the promotion of competition, they also marked a different mood in the federal courts regarding assertiveness in enforcing the antitrust laws. Once competition, however nominal, was supposed to exist, there was little appetite on the bench for acts of interference in a "free market." *Verizon Communications v. Trinko* is the most instructive example. Broadly and on good evidence, Verizon was accused of interfering with competitors, and the matter went all the way to the Supreme Court. Justice Antonin Scalia, writing for the majority, held that violations of the Telecommunications Act did not create an antitrust problem; a conclusion opposite the one the courts reached in the 1970s, when AT&T was tormenting MCI. The decision reaffirmed that moving the Bells out of the line of antitrust fire was by far the most decisive result of the 1996 law.[12]

As Whitacre and the other Bells prosecuted the ground war by frustrating and sabotaging competitors, their lobbyists and lawyers launched an air campaign against the new Telecommunications Act itself. The Bells challenged nearly every aspect of the line sharing provisions in federal court. The Bells won some and lost some, but that wasn't the point, really. What mattered was tying up would-be competitors in years of complex and expensive federal litigation, thrusting their business model into a permanent state of uncertainty. At some level, the lawsuits were an end in themselves.

Engaging in complex litigation did not distract the Bells from their accustomed pursuit of influence. In 2000, Verizon appointed as its general counsel William Barr, the former U.S. attorney general and onetime CIA operative; his style was distinctive. On one occasion, angered by an anti-Bell vote cast by a FCC commissioner, Barr cooly allowed "I want his balls in a jar."

In 2000, George W. Bush became president, and within a few years most of the Bells' wishes came true. Unlike the Nixon and Reagan administrations, which had been serious about competition in communications, the Bush administration tended to agree that competition didn't necessarily require that there be any extant competitors. Within two years, a new FCC gutted the sharing rules,* and a market that since 1996 had been competitive in name only was now rushing headlong toward monopoly once more.[13] Indeed, with the sharing provisions gone, most of the firms that hadn't already entered bankruptcy were effectively doomed.

In the next few years, one after another of the Bells' would-be rivals

---

*Pursuant to a suggestion by the D.C. circuit court of appeals, the commission decided to eliminate the requirement that the Bells share the entire "platform" (the line plus switch and other necessary equipment), obliging them thereafter to share only the "line." That put on competitors, in all instances, the added burden of installing their own switching equipment within the Bell facilities.

withered and died, and all the while Bell representatives murmured about the challenges of surviving in a competitive industry. Indeed, the only companies that would manage to survive as challengers in telephony were the cable firms, who had wires of their own running into every home, and whom the Act of 1996 had freed to become an intermodal competitor, the only kind the Bells couldn't destroy. Nevertheless, within a decade after the Telecommunications Act of 1996, history had repeated itself, and the Bells once again ruled the telephone system unperturbed. The idea of inducing "fierce" competition over Bell's proprietary wires, like the fledgling companies that had taken the bait, was utterly dead.

Wiping out the competition was only half the dream, however, and during this time the Bells were reaching, none too discreetly, for something even bigger than collective control: the reconstitution of the great Bell system itself. In this, too, Whitacre was the ringleader. In 1990, he had controlled only Southwestern Bell, the smallest of the eight fragments of Bell's breakup. In the 1997, he bought the regional Bell companies that served California, Nevada, and the states of the Midwest: the Pacific Telesis Group. Finally, in 2006 he added Bell South. And so, after a decade of consolidation, his new Bell system covered most of the country.

Whitacre's greatest symbolic victory, however, had actually come two years earlier, in 2005, when he bought AT&T, beating out Verizon, his only rival in becoming the Bell of Bells. The main stated purpose of the

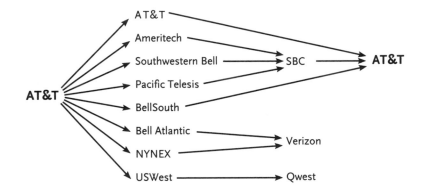

The Empire, long divided, must unite

1984 breakup had, after all, been to segregate the long distance company AT&T from the local carriers. With SBC's acquisition of the former flagship, a deal quickly approved by all levels of relevant parties in the Bush administration, the centerpiece of the breakup was no more. "The existence of separate local and long distance companies," wrote AT&T to the FCC in 2005, "no longer benefits consumers."[14] That was about the same time that Verizon bought out MCI, and like late Rome, the Bell system now existed as an eastern and a western empire—Verizon and AT&T (whose 24-karat name and logo Whitacre's company assumed)—but that was the only division; the vertical disintegration was abolished, and with it the second major era of openness and competition in telephony was over. That second era had lasted from 1984 to 2005, not even as long as the first age of competition, from 1894 to 1920.

Under the name that had signified unified telephony for more than a century, Whitacre's company became the largest communications firm in the world, just as its namesake had been. It had spent twenty-one years in the wilderness. But in name and in fact, AT&T was back.

Spying

"I flipped out," he said. "They're copying the whole Internet. There's no selection going on here. Maybe they select out later, but at the point of handoff to the government, they get everything."[15]

Mark Klein began his career as a young engineer at AT&T in 1982. Twenty-one years later, Klein was still at AT&T, working in the San Francisco offices, when he began to notice something odd. His fellow AT&T engineers were installing a raft of expensive hardware in little-used Room 641A, and access to that room was restricted.

Klein watched carefully and began to take notes. He noticed that the restricted room was connected to a larger one containing the high-speed fiber optic lines that went in and out of the building, the ones carrying Internet traffic to and from San Francisco, and the "peering links" to other major telecom providers. At some point—details are sketchy—Klein managed to get inside Room 641A. There he found an array of sophisticated networking equipment—crucially and in particular a semantic traffic analyzer, a special machine designed for intensive data mining and content analysis. After more than two years, Klein came to a most upsetting conclusion: AT&T had built a secret room to help the federal government spy on the Internet, and not just AT&T's customers but everyone's.

You might have wondered whether there had been any practical consequences to the return of AT&T, which, considering the drama of its dissolution, had been allowed rather quietly to regroup. It may seem that the average person, assuming he saw no spikes in his monthly phone bill, could afford to remain fairly indifferent to who runs the telephone system. But as Klein's story suggests, it can be a matter of very serious importance. It may be impossible to say for certain that the reconsolidation of AT&T fundamentally enabled the National Security Agency's surveillance program, but the need to involve so few companies in the conspiracy undoubtedly made things much easier. Suffice it to say, as the Cold War made clearest, the federal government has usually found an integrated telephone system more malleable to its needs than a fragmentary one.

In the early 2000s, when the spying began, SBC and the rest of the Bells had various merger deals pending before the Justice Department and the FCC. Again, while a direct causal link, much less a quid pro quo, is impossible to prove, this was obviously a very prudent time to be helping out the government.

Here is how Klein described the situation:

> In 2003 AT&T built "secret rooms" hidden deep in the bowels of its central offices in various cities, housing computer gear for a government spy operation which taps into the company's popular WorldNet service and the entire internet. These installations enable the government to look at every individual message on the internet and analyze exactly what people are doing.[16]

He described, in particular, the setup at his own place of work:

> In San Francisco the "secret room" is Room 641A at 611 Folsom Street. . . . High-speed fiber-optic circuits come in on the 8th floor and run down to the 7th floor where they connect to routers for AT&T's WorldNet service, part of the latter's vital "Common Backbone." In order to snoop on these circuits, a special cabinet was installed and

cabled to the "secret room" on the 6th floor to monitor the information going through the circuits.[17]

Mark Klein passed these declarations, along with pictures of one of the secret rooms, to the Electronic Frontier Foundation, a digital civil liberties group based in San Francisco. After holding a press conference, the EFF sued AT&T, alleging, on the basis of Klein's documents, violations of the Foreign Intelligence Surveillance Act (FISA), which at the time made it illegal for a private party to engage in electronic surveillance not authorized by statute.* The EFF declared, "We want to make it clear to AT&T that it is not in their legal or economic interests to violate the law whenever the president asks them to."[18]

When Klein made his startling allegations, however, it wasn't exactly clear that they were true. In its filings, AT&T was unresponsive, and the federal government wasn't about to admit it was spying on Americans. No one knew then about the secret order Bush had signed in 2002 authorizing domestic surveillance without a warrant, a contravention of FISA and of the administration's repeated claims that the NSA was spying only on foreigners. But by April 2006, the administration had entered the lawsuit, asking for its dismissal.[19] It was now clear that something was going on.

The resolution of this matter has little to commend it. In July 2008, during the presidential campaign, Congress passed a law granting AT&T and Verizon full and retroactive immunity for any violations of the laws against spying on Americans.[20] (That same measure, incidentally, also expanded the period that the FBI could spy on Americans without a warrant, now up to one week.) Tying the immunity provisions to a national security bill was crucial, for doing so terrified any congressmember, Republican or Democrat, who might have opposed the broad grant of immunity, lest he be accused of being weak on national security. Presidential candidate Senator Barack Obama, for example, while on record as opposing the immunity, nonetheless voted for the bill in order to defend his national security credentials; he described the bill as "improved but imperfect."

---

*The law, 50 U.S.C. §1809, makes it an offense if a person "discloses or uses information obtained under color of law by electronic surveillance, knowing or having reason to know that the information was obtained through electronic surveillance not authorized by statute."

With the bill's passage, the matter mostly faded from public view, like much else of questionable legality undertaken during these years in the name of national security. Just as surely as the iron fist of government had been the only power equal to dismembering the mighty Bell monopoly, the flameproof hand had again intervened, this time to pluck its sometime corporate adversary, but now, as in the good old days, its strategic partner, from the fires of legal jeopardy.

Thanks to the immunity grant, there is a good chance that the extent of government spying, past and present, will never be known. But the moral of the story is obvious. In an age more reliant than ever on telecommunications media, the more concentrated the power over information and communications, the easier it is for government to indulge its temptation to play Big Brother. With everyone in the country now connected, the fewer the parties that need to be persuaded to cooperate, the greater the risk. With the convergence of all communications by virtue of interconnected networks (aka intermodality), the reconstituted giants of telephony are closer to possessing a master switch than Vail himself could have dreamed. Those are the unremarked costs of the return of the empire.

THE CYCLE

By 2007, Ed Whitacre had fulfilled his mission, and his destiny. Most of the Bell system was back in place in the world's largest communications firm, with him at the helm. At age sixty-five, with nothing left to prove, he announced his retirement.[21] In accordance with the custom of the early twenty-first century, it would be the occasion for a very sizable payout, over $200 million, making it clear that even if money isn't the only motivation for building an information empire, it is certainly among the rewards.

In one way, more than any other phenomenon we have considered, the return of AT&T—that perennial phoenix—would seem to prove the irrevocability of the Cycle of information empires, their eternal return to consolidated order however great the disruptive forces of creative destruction. After all, despite the explicit wishes and mighty efforts of the FCC and Congress to maintain an open and competitive telephone market, within twenty years the national phone system was once again ruled by just a few companies, most of them parts of the old Bell system. Though such a view of inevitability perhaps makes for a tidier argument, no theory of history operates in isolation from the particularities of the times and the actors. Every consolidated entity may well have only until the next turn of the Cycle before being scattered, and everything scattered may await only its eventual imperial visionary.

PART V

The Internet Against Everyone

BY THE END OF THE FIRST DECADE of the twenty-first century, the second closing of the traditional information industries was complete. Most of the phone system was back in the hands of Bell, and its competitors were being steadily run down. A gang of conglomerates comfortably controlled film, cable, and broadcasting. And while this new order was by no means absolute, the industrial concentration had reached levels not seen since the 1950s.

The one great exception to this dominion of big business was the Internet, its users, and the industry that had grown on the network. Amid the consolidation, the 1990s also saw the so-called Internet revolution. Would it lead to the downfall of those consolidating superpowers? Some certainly thought so. "We are seeing the emergence of a new stage in the information economy," prophesied Yochai Benkler. "It is displacing the industrial information economy that typified information production from about the second half of the nineteenth century and throughout the twentieth century."

Unfortunately, the media and communications conglomerates didn't consult Benkler as their soothsayer. With aggregate audiences in the billions and combined revenues in the trillions, they had—in fact, have—a very different vision of the future: the Internet either remade in their likeness, or at the very least rendered harmless to their core business interests.

For even though its origins are distinct, the Internet by 2010 had become a fledgling universal network for *all* types of data: phone calls, video and television, data, a potential replacement for every single information industry of the twentieth century. Technologically this was a product of the Internet's design, conceived to be indifferent to the nature of the content carried, able to handle it all. But for the old media industries of the twentieth century, the shape-shifting nature of the Internet, its ability to be a phone, a TV, or something new, like Facebook, posed an existential threat. Hence the powerful desire to bring the network to heel, one way or another.

We now face squarely the question that the story told heretofore is meant to help us answer. Is the Internet really different? Every other invention of its kind has had its period of openness, only to become the basis of yet another information empire. Which is mightier: the radicalism of the Internet or the inevitability of the Cycle?

CHAPTER 19

A Surprising Wreck

On October 1, 1999, Steve Case and Gerald Levin sat together in a reviewing stand on Beijing's Tiananmen Square, chatting about the future as they watched the long procession of troops and tanks roll by. Tiananmen Square, where these celebrations of the Communist revolution's fiftieth anniversary were taking place, is surrounded by symbols of imperial power: the Forbidden City, home to generations of emperors, as well as the icons of the People's Republic, including the giant portrait of Mao Zedong. The occasion was a historic superimposition of imperial spirits, for these two men, like those whose ghosts hung all around them, were also emperors of a kind· Gerald Levin ran Time Warner, the world's largest media conglomerate, and Steve Case was president and CEO of America Online, then the single most successful Internet services firm in the world.[1]

The two men were in town for an event thrown by *Fortune* magazine commemorating the fiftieth anniversary of the Communist revolution, and watching the long parade, the two found they had much to talk about. In theory they were of rival clans, old media and new, but they found a connection. Each was an ambitious CEO with a taste for risk, but also a self-styled idealist. Both men of far-reaching vision and aspiration, they concurred in earnest that corporations, as Levin said, should be run in the public interest, to maximize long-term value, not short-term profit, the latter being the unfortunate obsession of too many American managers. Oh, and both felt they could see the future of the Internet.

Case and Levin had met briefly once before, in 1998, at a White House screening of the romantic comedy *You've Got Mail,* a Warner Bros. film that featured AOL product placement. Now as they spoke, a warm glow began to develop between a Montague and a Capulet who fantasized about all-embracing alliance between their seemingly irreconcilable houses. Theirs was to be a union that could move mountains—or least break down the old barriers and create a perfect new world.[2]

The two moguls plotting the future of the Internet had something else in common: neither was what you might call a natural computer geek, in the manner of Bill Gates or Steve Jobs. Entrepreneurs like Apple's Steve Wozniak got started by programming and soldering; Case was an assistant brand manager at Procter & Gamble in Kansas. He might have languished somewhere in upper middle management had he not resolved to grab the ring. Case took a job at a risky computer networking firm named the Control Video Corporation that had already failed twice. Three's a charm, however: by some miracle, that firm eventually managed to become America Online.[3]

Once a corporate lawyer, Levin during these years was working as a cable executive. He made his name in the business as an executive at HBO, the first premium cable network, persuading cable operators to install the satellite dishes that Ted Turner would later make use of. But unlike Turner, eternally entrepreneurial and advantaged at birth, Levin had worked his way up within the system, a slow and steady rise within Time Inc. as it became Time Warner. There he became a protégé of Steven Ross, who, before dying, anointed him as successor.[4]

Not long after that Beijing parade, Case telephoned Levin with a proposal. And within three months of that rendezvous on the reviewing stand, on January 10, 2000, they were holding a press conference to announce their own revolution: a $350 billion merger between the world's biggest media company and biggest Internet firm. AOL would be the engine that brought Time Warner's old media holdings—a treasury of what was becoming known as "content"—into the new world. It was an effusive spectacle. Levin said "We've become a company of high-fives and hugs." Ted Turner, the largest individual shareholder, likened it to "the first time I made love some forty-two years ago."[5]

It looked as if the future had indeed arrived. To many it seemed that the Internet would eventually belong to vertically integrated giants on the model of AOL Time Warner. Here's Steve Lohr writing in *The New York Times* in 2000: "The America Online–Time Warner merger [will] create a powerhouse for the next phase of Internet business: selling information and entertainment services to consumers who may tap into them using digital cell phones, handheld devices and television set-top boxes in addition to personal computers."[6] In time, it was envisioned, three or four consolidated firms—say, AOL Time Warner, Microsoft-Disney, and perhaps Comcast-NBC—would slowly divide up the juiciest Internet properties, such as Yahoo! and eBay, just as they'd already divided up the rest of the electronic media and entertainment world. In other words, Steven Ross's media conglomerate model now seemed poised to conquer the new frontier that was the online universe.

Let's stop for a moment to consider the world as it would look today had this expectation been fulfilled. The realm of information industries would be divided, essentially, in two: conglomerates on the one side and phone companies on the other, the rest serving as the business world's equivalent of condiments. If it had all worked out, the new century would have begun with the world of information far more consolidated than at any time in American history.

But it didn't work out.

Coming around the bend at full speed, AOL Time Warner ran right into a wall they didn't even see. Soon the very name became synonymous with "debacle." The share price plummeted, and within a short time, Case was forced out and Levin retired. Time Warner would carry on, though seriously enfeebled, while the AOL part of the operation would survive only as a zombie of its former self before being cut loose.

Ugly though the crash was, to this day, some still believe that AOL Time Warner was a good idea poorly executed. Steve Case makes this claim, as does Larry Kramer, a media analyst, writing in 2009, "One of the world's foremost content companies was merging with one of the largest distributors of online content. Content meets customers. Sounded perfect. . . . The idea that you could put world-class content in front of a huge audience wasn't a bad one."[7] Most take the opposite view—and they are far more numerous and vehement—but these critics have created a different sort of misunderstanding about the fiasco. For it wasn't, as generally reported, just a tale of epic personality clashes and bullheadedness, however much fabulous schadenfreude and great copy that story line made for. Far more than anyone realized at the time, the human errors plaguing the firm were less to blame than the very structure of the Internet in destroying whatever advantages the merger was meant to deliver. The principle of *net neutrality*, instilled by the Internet's founders, is ultimately what wrecked AOL Time Warner. And that now iconic wreck, if nothing else, would attest powerfully to the claim that the Internet was at last the great exception, the slayer of the Cycle we have been visiting and revisiting.

THE TWO SIDES

By the late 1990s, Time Warner had grown into an even larger giant than it had been under Steven Ross. It was now an empire encompassing almost every conceivable type of media. Yet as the turn of the century drew near, Ross's successor, Levin, became convinced he had a serious problem: the Internet, or interactive sector, represented a hole in the imperial panoply of Time Warner's media holdings; and were it not filled, it might one day be the giant's undoing.

Time Warner's first Internet forays proved disastrous and unintentionally hilarious: the original "Clown Co." Robert Wright, the journalist and bestselling author, once told me a story that captures their approach to the Internet in the 1990s. In 1995, he and the editor Michael Kinsley were looking for someone to back what was to become the first online magazine, Slate.com. Time Warner was an obvious possibility, and the two managed to get a meeting with Norm Pearlstein, *Time* magazine's editor in chief, and Walter Isaacson, recently appointed to establish Time Warner's presence online. Wright remembers:

> As we discussed the idea, it soon became clear there was something they didn't quite get. He [Pearlstein] kept asking about long distance charges. They seemed to think of the Internet as something like a fax machine, and that we were going to be sending our customers the physical magazine.

In 1995, abstractions like the World Wide Web, now familiar to the world as the Internet's content platform, had yet to penetrate Time Warner's upper-level consciousness.

What Time Warner was doing instead was investing in a cable alternative to the Internet—one not so different from Ralph Lee Smith's 1970 vision of what cable might become, with a pilot in Orlando, Florida. It was called the Full Service Network, an interactive form of television, allowing, for example, easy online shopping via the remote control and expanded on-demand options. But as an effort to reimagine both the computer and the Internet at once, the service never got out of Disneyland.

Meanwhile, Time Warner's first site on the World Wide Web, Pathfinder.com, was advertised as the one that would "conquer the digital world." It conquered only whatever doubts remained about how out of touch the conglomerate was. The idea, vaguely, was to create a "portal" whose drawing power would be access to Time Warner's content. It doesn't take a genius in hindsight to see the limitations of a portal lacking a search engine and featuring the content of only one firm. (Ironically, it billed itself as "Home of the Web's most complete news, information and entertainment site.") Needless to say, it was never to be a serious rival to Yahoo!, or later Google.[8]

If it had failed twice already, in a sense the merger with AOL represented a doubling down of Levin's will to take the Web. With this already full-grown success, Time Warner would at last have an "Internet strategy," whatever that meant. To be clearer, for Time Warner, the ultimate strategic goal of merger with AOL, the Internet's leading on-ramp, was to drive consumers toward TW's products. The Internet presented Levin with a great uncertainty: unlike the other channels of information, which TW owned, it was a vast new domain where eyes were free to wander toward any producer's content. By acquiring AOL's millions of users, TW could, in theory, expose them incessantly to TW offerings. With every click of the mouse, every AOL user would be, subtly or not so subtly, steered toward allied brands.

TW imagined taking its usual practice of steering watchers to its own content to a whole new level—to the medium destined to be the mother of them all, the Internet. It was logical in the abstract, and not without grand precedent, having been the basic model invented by Disney in the 1960s, when the Disney brothers began to realize that their films could drive customers to its merchandise and theme parks, which would in turn drive them back again to its films—a strategy Disney labeled "total merchandising." Gerald Levin referred to this as his theory of "touchpoints," and in business conglomerate jargon the idea went by names like "cross pollination."[9]

Levin had thought up a way to ensure Time Warner's long-term survival via a shrewd marriage. But unbeknownst to him, the bridegroom was already suffering from a progressive terminal illness. AOL had managed for the most part to keep the secret in the family, but Case and his colleagues were desperately aware that there was no logical place for AOL in the age of broadband, an age that was fast approaching. This point is technical; AOL's illness requires a bit of explanation.

They key fact about AOL is that it came of age *before* the mass Internet. In the early 1990s, you dialed up not to surf unfettered but to reach AOL itself, and to talk to other AOL users. This fact can be hard to grasp for those born and bred on the Internet: AOL was, in those early days, *the platform,* and, in the lingo, operated as a "walled garden" for its users. It dictated what content was available to users. For example, before The Motley Fool, the investor's guide, was a website, it was an AOL page.

In the 1990s, as the Internet began growing in popularity on college campuses, a few enterprising companies began offering home Internet service. Their popularity would eventually force AOL to change its business model, in a way that would give its millions of users direct access to the Internet, rather than merely to AOL's walled garden. It was a decision made reluctantly, for Case and his colleagues knew that Internet access would cost them control over their customers. Nevertheless, with the new Internet Service Providers (ISPs) nipping at their heels, the company changed from being primarily an online service company to primarily an Internet Service Provider.

At first, during the 1990s, the strategy worked brilliantly. Everyone wanted to see what the Internet was all about, and AOL was the easiest way to get there, because they put a CD in everyone's mailbox (free samples being one of the things Case learned about in his former career as marketing manager at P&G). In this sense AOL was a major part of the mass Internet revolution.

The problem for AOL as the 1990s became the 2000s was that a new way of reaching the Internet was on the horizon: broadband. The cable and phone companies had figured out how to get higher speeds out of the same old cable and phone lines, and they were offering customers a fast and direct Internet connection. Unfortunately for Case & Co., broadband service, by design, made AOL unnecessary. AOL's business model was premised on "dial-up"—on calling AOL to reach the Internet. On the phone company's new Digital Service Lines (DSL) and on cable broadband, the phone and cable companies offered customers the Internet directly, cutting out Internet Service Providers like AOL.

To understand what happened to AOL, imagine that the firm had been in the business of delivering pizzas by bicycle, until one day the pizza company bought its own fleet of cars. That is what the telephone and cable companies did to AOL. The diagrams below make clear the redundancy and inferiority of AOL's core business in the new technological order. If AOL was to have any chance of surviving, it would have to acquire its own fleet of cars, and fast.

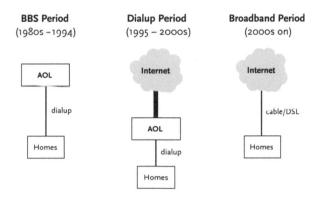

What AOL was

It is a strange thing to contemplate: in 2000, AOL was seen as the "new media" company, meant to reinvigorate the "old media" fossil Levin feared Time Warner would become. In fact, however, AOL was the dinosaur limping into the new age. Getting its own cable wires was a matter of life and death, and merging with Time Warner was a way to get those. More cynically, you might say that Steve Case, who understood AOL's problems, picked his moment to cash in on his company's literal and figurative stock at the moment when it could go no higher. Within a year of the merger, as the dot-com boom went bust, AOL would be worth considerably less than its lifetime high of $240 billion.[10]

In 2000, AOL and Time Warner were hardly alone in thinking that massive media integration would be the key to the future of the Internet. Microsoft, from the mid-1990s onward, was convinced of the imperative to own both distribution and content. From its vast cash reserves, the software giant invested heavily to take a 50 percent stake in a new cable channel, a joint venture with the peacock network, called MSNBC, conceived as a rival to CNN. It also began putting money into computer games via the Xbox, and it created an online service, MSN, to mimic the lustrous but doomed model of AOL, as well as developing new, signature content. Wright and Kinsey, for instance, would ultimately find their backer for *Slate* magazine not in a traditional media company but in Microsoft, who invested more than $20 million in the venture. Meanwhile Comcast, the cable provider founded in 1963, began an effort to merge with Disney to create a whole new Internet—media behemoth. The race was on.

Informed and well-capitalized opinion had it that history was repeating itself. Had the idea of these giants to stand on one another's shoulders panned out, the combined media-Internet world would have been dominated by the handful of conglomerates then already beginning to buy up the most important websites. The early 2000s might have turned into a war of accumulation among three vertically integrated great powers: Microsoft-GE (NBC's parent), AOL Time Warner, and Comcast-Disney. Eventually everything on the Internet would have been owned by one of them.

That would have made for a tidier information economy, centered mainly on two mega-industries: the media conglomerates and the telephone companies. But something went wrong. Microsoft stopped buying media. Disney rejected Comcast's merger offers. And AOL Time Warner became the textbook example of what not to do—as Ken Auletta calls it, "the Merger from Hell."[11]

What Happened?

The books about the AOL Time Warner saga are the work of business reporters and as such tend to focus on the personalities, clashes of corporate cultures, and terse boardroom encounters. Steve Case was wily enough to see the sand running out of AOL's hourglass, so he sold, or bought, when he could. Gerald Levin, head of Time Warner, was brash and impatient at the failure of his firm to conquer the Internet, and as a consequence was had. Some commentators blame Time Warner's units for being slow to adjust, as they were dragged kicking and screaming into a new media environment. (Perhaps some were still worried about incurring long distance charges!) While Case and Levin continued to get along well enough, they failed to see how the world of AOL might not meld with that of Time Warner—another leitmotif in every chronicle of a failed merger foretold.

It is true that organizational incompetence and the problems naturally attending megacombinations can't be discounted. Time Warner, for instance, continued to run a service that competed with AOL, named Road Runner. The reason was that, as a condition of the merger, the Federal Trade Commission, the antitrust enforcement agency, and the FCC required TW to provide consumers with a choice of ISPs; yet the consequence was that AOL's greatest competitor for cable internet customers was another division of its parent company.[12] There is nevertheless ample reason to believe that even if the two cultures had meshed like Cheech and Chong under an organizational genius to rival Jack Welch, much the same failure would have eventually resulted. There were deep structural problems that neither Case nor Levin nor many others fully comprehended at the time. It was, in a sense, a failure to understand the deep structure of the Internet.

The AOL–Time Warner fusion made sense only if the giant could find some way to induce among the existing customer base of each division a discriminatory preference for the other's products. In other words, consumers of Time Warner products had to be persuaded to become AOL users and AOL users made to pay for Time Warner content. In 2001, AOL had nearly 30 million subscribers, a huge headcount for any information commodity, described by *BusinessWeek* at the time as a "juicy prize."[13] The plan was that those subscribers would be directed and exposed to Time Warner's offerings: its content, its cable TV, and its Internet services. That experience would prove so agreeable that, in theory, it would result in a feedback loop, creating even more AOL subscribers. The only problem with this idea was the Internet. Neutral by design, it did not take naturally to the imposition of such a regime.

It might have worked back in the early 1990s, when AOL refused its subscribers access to the Internet beyond its walled garden. The original AOL could have simply restricted all comers to Time Warner content, rather as McDonald's serves only Coca-Cola products and not Pepsi's. But by 2000 AOL was less a destination in itself—*the* platform that it had been—than simply the most popular way to reach the Internet. While it could boast 30 million subscribers, it could exercise no meaningful control over them. Once online, a user could go wherever he wished, the Internet being set up to connect any two parties, whoever they might be. The rise of search engines and domain names would only exacerbate the problem. Type in Yahoo.com or Google.com, and the whole world of the Internet opens up before you; it scarcely matters by what ISP you have reached that point. At most, AOL could recommend Time Warner content to anyone logging on, but it was almost immediately clear that that dividend was not worth much (not much more than a pop-up ad, actually).

It may seem astonishing that anyone of Gerald Levin's considerable experience and business acumen could have failed to grasp this problem. Levin, moreover, did not consider himself an enemy of the Internet; he described it in 2010 as "a beautiful thing." Yet for many people, the Internet's structure was—indeed remains—deeply counterintuitive. This is because it defies every expectation one has developed from experience of other media industries, which are all predicated on control of the customer. Levin, an apostle par excellence of that control model, fell victim to Schumpeter's observation that "Knowledge and habit once acquired, become as firmly rooted in ourselves as a railway embankment in the earth." Unlike any medium Levin had known, the Internet abdicates control to the individual; that is its special allure, its power to be endlessly surprising, as well as its founding principle.

Other factors may well have clouded Levin's judgment. There are few intoxicants like the prospect of easy money, and that's what the billions in IPO dollars then raining down on nobodies in Silicon Valley must have looked like. Nor can be discounted the recurring theme of this book: the lure of information empire. In 2010, Levin described to me the condition of being a CEO as "a form of mental illness," with the desire for never-ending growth as a kind of addiction. As he said, "there's something about being able to say, 'I'm the CEO of the world's largest media company.' "

Was there any way for AOL Time Warner to work? The firm would have needed to change the nature of the Internet itself, transforming the network into one on which "foreign"—i.e., non-TW—content could be blocked or discriminated against. Alternatively, the company could have sought control over Internet's "openers"—namely, the search engines that were giving users what they wanted. AOL Time Warner needed to subdue Google, Yahoo! and their many cousins. In short, to be viable, the firm would have needed to overturn the net neutrality principles at the core of the Internet's design.

The only entity that has so far really succeeded in such a mission is the government of mainland China, as we saw in 2010, when it drove an exasperated Google out of its sovereign territory by demanding extensive control over what Google let users find. Indeed, the feat requires such power and resources as belong uniquely to the state: access to the very choke points of a nation's communications infrastructure, its Master Switch.[14] AOL Time Warner, however vast, did not have police power—it could not imprison Google's executives for failing to block Wikipedia or Disney content.

In any event, by 2000 the death spiral that Case had feared was already under way. AOL's huge customer base stopped growing and then started to decline as the cable and telephone companies began to deploy broadband services in earnest. Those who quit AOL for broadband had no reason to come back, and every defection pounded another nail into AOL's coffin. The company would make various desperate attempts at survival through self-reinvention. In 2004 it tried to differentiate its service with AOL Optimized 9.0, whose bells and whistles included personalized audible greetings for the subscriber. The next year, in a joint venture with a division of Warner, AOL founded the celebrity news site TMZ.com, attempting to identify its brand with the creation of content. The next year they stopped charging for email accounts to stanch the loss of customers to Hotmail, Yahoo! and other free providers.

But it was all for naught: on December 9, 2009, just one month shy of their tenth anniversary, the disastrous marriage of AOL and Time Warner ended in divorce.

In the aftermath of the calamity, both Levin and Case departed the company. Levin quit high-stakes business altogether, becoming director of the Moonbeam sanctuary, a spiritual wellness retreat for executives in Southern California. It seems to have had an effect, for to meet Levin nowadays is to meet a man with a steady, Buddha-like gaze and a slow voice. He told me that anyone running a media company should only do so "in service of a higher purpose."

Case stayed in Washington, D.C., where he oversees a private equity firm called Revolution LLC; its stated mission is in part "to drive trans-

formative change by shifting power to consumers and building significant, category-defining companies in the process." He remains one of
Time Warner's largest stockholders.

Angry recriminations would be voiced for some time on the Time Warner board; in particular by the never shy Ted Turner, who is reported to
have lost around $7 billion, most of his personal worth, to the merger.
Time Warner's people continue blame Levin for making such a bad
deal, and Steve Case for being the serpent in the old media garden.
For their part, AOL, as of this writing newly single and still adjusting
to the post-traumatic stress, has rebranded itself "Aol." It blames Time
Warner's inflexibility and hidebound reluctance to truly embrace the
online world.

This anger, though understandable and predictable, as with any
failed union, is ultimately misdirected. It rightly belongs with J.C.R.
Licklider and Vint Cerf. Without specifically intending to, the founders
of the Internet had foreordained by the radicalism of their conception
that Levin and Case's great image of the future would have—despite its
head of gold, its belly of brass, and legs of iron—feet of clay.

The design of the Internet blesses some companies and curses others. For if net neutrality destroyed the value of AOL Time Warner, it
would catapult to riches the likes of Google and Amazon, firms that, far
from discouraging or circumscribing consumer choice, would aim to
put everything one could want within easy reach. In this fulfillment of
the Net's dream of connecting any user with any other, comes the power
behind the great business success stories of the still young Internet age.
In such a world, the advantage of owning everything from soup to nuts
is far from evident; it may be no advantage at all.

In 2008, at Revolution headquarters, I asked Case whether he regret-
ted the merger. "Yes," he said, without hesitation. What would he have
done differently? Acknowledging that nonintegrated, or "pure play,"
firms like Google would in the end succeed where AOL failed, Case has
a different vision in hindsight: "I would have bought Google."

In some sense the work remaining to our story is to assess the merits
of that answer.

CHAPTER 20

Father and Son

Steve Jobs stood before an audience of thousands, many of whom had camped out overnight to share this moment. In his signature black turtleneck and blue 501s he was completely in his element: in perfect control of his script, the emotions of the crowd, and the image he projected to the world. Behind him was an enormous screen—another of his trademarks—flashing words, animations, and surprising pictures. In the computer world, and particularly among members of the cult of the Mac, the annual Jobs keynote at Apple's Macworld is a virtual sacrament. During this one, on January 9, 2007, Jobs was to announce his most important invention since the Apple Macintosh.[1]

"Today," said Jobs, "we're introducing three revolutionary new products. Three things: a widescreen iPod with touch controls; a revolutionary mobile phone; a breakthrough Internet communications device."

Loud cheers.

"An iPod, a phone . . . are you getting it? These are not three separate devices!"

Louder cheers.

"We are calling it iPhone!"

The audience rose to its feet. On the screen: "iPhone: Apple reinvents the phone."

The iPhone was beautiful; it was powerful; it was perfect. After demonstrating its many features, Jobs showed how the iPhone could access the Internet as no phone ever had before, through a full-featured real browser.

"Now, you can't—you can't really think about the Internet, of course, without thinking about Google. . . . And it's my pleasure now to introduce Dr. Eric Schmidt, Google's CEO!"

To more cheers, Schmidt came jogging in from stage left, wearing an incongruously long orange tie. The two men shook hands warmly at center stage, like two world leaders. A member of the Apple board, Schmidt thanked Jobs and began his comments with a perhaps ill-advised joke about just how close Apple and Google had become. "There are a lot of relationships between the boards, and I thought if we just sort of merged the companies we could call them AppleGoo," he said. "But I'm not a marketing guy."

Indeed, in 2007 Google and Apple were about as close as two firms could be. Schmidt was not the only member of both corporate boards. The two firms were given to frequent and effusive public acclamations of each other. Their respective foundings a generation apart, Google and Apple were, to some, like father and son—both starting life as radical, idealistic firms, dreamed up by young men determined to do things differently. Apple was the original revolutionary, the protocountercultural firm that pioneered personal computing, and, in the 1970s, became the first company to bring open computing, then merely an ideological commitment, to mass production and popular use. Google, meanwhile, having overcome skepticism about its business model at every turn, had by the new millennium become the incarnation of the Internet gospel of openness. It had even hired Vint Cerf, one of the network's greatest visionaries, giving him the title "Chief Internet Evangelist."[2]

Their corporate mottoes, "Think Different" and "Do No Evil," while often mocked by critics and cynics, were an entirely purposeful way of propounding deeply counterintuitive ideas about corporate culture. Both firms, launched out of suburban garages a few miles apart, took pride in succeeding against the grain. Google entered the search business in 2000, when searching was considered a "commodity," or low-profit operation, and launched a dot-com after the tech boom went bust. Apple's revolution had been even more fundamental: in the 1970s, still the era of central mainframe machines, it built a tiny personal computer and later gave it a "desktop" (the graphic user interface of windows and icons and toolbars that is now ubiquitous), as well as a mouse. The two firms also shared many real and imagined foes: Microsoft, mainstream corporations, and uptight people in general.

Back in San Francisco, Schmidt, done with his jokes, continued his presentation.

"What I like about this new device [the iPhone] and the new architecture of the Internet is that you can actually merge without merging. . . . Internet architectures allow you now to take the enormous brain trust that is represented by the Apple development team and combine that with the open protocols and data service that companies like Google [provide]."

Unnoticed by most, here was enunciated a crucial idea, a philosophy of business organization radical in its implications. Schmidt was suggesting that, on a layered network, in an age of open protocols, all the advantages of integration—the "synergies" and efficiencies of joint operation—could be realized without actual corporate mergers. Call it Google's theory of the firm. With the Internet, there was no need for mergers and exclusive partnerships. Each company could focus just on what it did best. Thanks to the Internet, the age of Vail, Rockefeller, and Carnegie, not to mention the media conglomerates created by Steven Ross and Michael Eisner—the entire age of giant corporate empires—was, according to this revelation, over.

But was it really? The warmth of Jobs's greeting concealed the fact that Apple's most important partner for the iPhone launch was not Google—not by a long shot—but rather one of Google's greatest foes. At the end of his speech, in an understated way, Jobs dropped a bomb. The iPhone would work exclusively on the network of one company: AT&T.*

"They are the best and most popular network in the country," said Jobs. "Fifty-eight million subscribers. They are number one. And they're going to be our exclusive partner in the U.S."

In entering this partnership, Apple was aligning itself with the nemesis of everything Google, the Internet, and once even Apple itself stood for.

. . .

---

*At the time of announcement, AT&T Wireless was operating under its old name, Cingular.

We don't know whether Ed Whitacre, Jr., AT&T's CEO, was listening to Eric Schmidt's speech at the iPhone launch. But we can be sure that he would have disagreed with Schmidt that the age of grand mergers was over. Just one week earlier, Whitacre had quietly gained final federal approval for the acquisitions that would bring most of the old Bell system back under AT&T's control. Unfazed by the arrival of the Internet, Whitacre and his telephone Goliath were practicing the old-school corporate strategies of leveraging size to achieve domination, just as AT&T had done for more than a hundred years. The spirit of Theodore Vail was alive and well in the resurrected dominion of the firm with which Apple was now allied.

Within two years of the iPhone launch, relations between Apple and Google would sour as the two pursued equally grand, though inimical, visions of the future. In 2009 hearings before the FCC, they now sat on opposite sides. Steve Jobs accused Google of wasting its time in the mobile phone market; a new Google employee named Tim Bray in 2010 described Apple's iPhone as "a sterile Disney-fied walled garden surrounded by sharp-toothed lawyers. . . . I hate it."[3]

As this makes clear, where once there had been only subtle differences there now lay a chasm. Apple, while it had always wavered on "openness," had committed to a program that fairly suited not just the AT&T mind-set, but also the ideals of Hollywood and the entertainment conglomerates as well. Despite the many missteps, including the AOL–Time Warner merger, the conglomerates were still at bottom looking for their entry point into the Internet game. By 2010, Apple would clearly seem the way—whether through its iTunes music store, its online videos, or the magic of the iPad. In fact, the combination of Apple, AT&T, and Hollywood now held out an extremely appealing prospect: Hollywood's content, AT&T's lines, and Apple's gorgeous machines—an information paradise of sorts, succeeding where AOL–Time Warner had failed.

For its part, Google would remain fundamentally more radical with utopian, even vaguely messianic, ideals. As Apple befriended the old media, Google's founders continued to style themselves the challengers to the existing order, to the most basic assumptions about the proper organization of information, the nature of property, the duties of the American corporation, and even the purpose of life. They envisioned taking the Internet revolution into every sector of the information realm—to video and film, television, book, newspaper, and magazine publishing, telephony—every way that humans send or receive information.

You might think that such splits are simply the way the capitalist cookie crumbles and one shouldn't dwell overmuch on the rupture between two firms. But these are not just any two firms. These are, in communications, the industrial and ideological leaders of our times. These are the companies that are determining how Americans and the rest of the world will share information. If Huxley could say in 1927 that "the future of America is the future of the world," we can equally say that the future of Apple and Google will form the future of America and the world.[4]

What should be apparent to any reader having reached this point is that here in the twenty-first century, these firms and their allies are fighting anew the age-old battle we've recounted time and time again. It is the perennial Manichaean contest informing every episode in this book: the struggle between the partisans of the open and of the closed, between the decentralized and the consolidated visions of a proper order. But this time around, as compared with any other, the sides are far more evenly matched.

APPLE'S RADICAL ORIGINS

Apple is a schizophrenic company: a self-professed revolutionary closely allied with both of the greatest forces in information, the entertainment conglomerates and the telecommunications industry. To work out this contradiction we need to return to Apple's origins and see how far it has come. Let's return to 1971, when a bearded young college student in thick eyeglasses named Steve Wozniak was hanging out at the home of Steve Jobs, then in high school. The two young men, electronics buffs, were fiddling with a crude device they'd been working on for more than a year. To them it must have seemed just another attempt in their continuing struggle to make a working model from a clever idea, just as Alexander Bell and Watson had done one hundred years earlier.[5]

That day in 1971, however, was different. Together, they attached Wozniak's latest design to Jobs's phone, and as Wozniak recalls, "it actually worked."[6] It would be their first taste of the eureka moment that would-be inventors have always lived for. The two used the device to place a long distance phone call to Orange County. Apple's founders had managed to hack AT&T's long distance network: their creation was a machine, a "blue box," that made long distance phone calls for free.

Such an antiestablishment spirit of enterprise would underlie all of Jobs and Wozniak's early collaborations and form the lore that still gives substance to the image long cultivated: the iconoclast partnership born in a Los Altos garage, which, but a few years later, in March of 1976, would create a personal computer called "the Apple," one hundred years to the month after Bell invented the telephone in his own lonely workshop.

In the 1970s this imagery would be reinforced by the pair's self-styling as bona fide counterculturals, with all the accoutrements—long hair, opposition to the war, an inclination to experiment with chemical substances as readily as with electronics. Wozniak, an inveterate prankster, ran an illegal "dial-a-joke" operation; Jobs would travel to India in search of a guru.

But, as is often the case, the granular truth of Apple's origins was a bit more complicated than the mythology. For even in the beginning, there was a significant divide between the two men. There was no real parity in technical prowess: it was Wozniak, not Jobs, who had built the blue box. And it was Wozniak who would conceive of and build the Apple and the Apple II, the most important Apple products ever, and arguably among the most important inventions of the later twentieth century.* For his part, Jobs was the businessman and the dealmaker of the operation, essential as such, but hardly the founding genius of Apple computers, the man whose ideas were turned into silicon to change the world; that was Wozniak. The history of the firm must be understood in this light. For while founders do set the culture of a firm, they cannot dictate it in perpetuity; as Wozniak withdrew from the operation, Apple became more and more concerned with, as it were, the aesthetics of radicalism than with its substance.

Steve Wozniak is not the household name that Steve Jobs is, but his importance to communications and culture in the postwar period merits a closer look. While Apple's wasn't the only personal computer invented

---

*Some may argue that the Macintosh was more significant than the Apple II; without discounting the importance of the former, the significance of personal computing seems categorically larger than the importance of adding the desktop interface to the personal computer.

in the 1970s, it was the most influential. For the Apple II took personal computing, an obscure pursuit of the hobbyist, and made it into a nationwide phenomenon, one that would ultimately transform not just computing, but communications, culture, entertainment, business—in short, the whole productive part of American life.

We've seen these moments before, when a hobbyist or limited-interest medium becomes a mainstream craze; it happened with the telephone in 1894, with the birth of radio broadcasting in 1920, and with cable television in the 1970s. But the computer revolution was arguably more radical than any of these advances on account of having posed such a clear ideological challenge to the information economy's status quo. As we've seen, for most of the twentieth century, innovators would lodge the control and power of new technologies within giant institutions. Innovation begat industry, and industry begat consolidation. Wozniak's computer had the opposite effect: he took the power of computing, formerly the instrument of large companies with mainframe resources, and put it in the hands of individuals. That feat, and every manifestation of communications freedom that has flowed from it, is doubtless his greatest contribution to society. It was almost unimaginable at the time: a device that made ordinary individuals sovereign over information by means of computational powers they could tailor to their individual needs. Even if that sovereignty was limited by the primitive capacities of the Apple II—48 KB of RAM, puny compared with even our present-day telephones but also with industrial computers of the time—the machine nevertheless planted the seed that would change everything.

With slots to accommodate all sorts of peripheral devices and an operating system that ran a variety of software, the Wozniak design was open in ways that might be said still to define the concept in the computing industries. Wozniak's ethic of openness extended even to disclosing design specifications. He once gave a talk and put the point this way: "Everything we knew, you knew."[7] In the secretive high-tech world, such transparency was unheard of, as it is today. Google, for example, despite its commitment to network openness, keeps most of its code and operations secret, and today's Apple, unlike the Apple of 1976, guards technical and managerial information the way Willy Wonka guarded candy recipes.

Put another way, Wozniak welcomed the amateur enthusiast, bringing the cult of the inspired tinkerer to the mass-produced computer. That ideology wasn't Wozniak's invention, but rather in the 1970s it was an orthodoxy among computing hobbyists like the Bay Area's Homebrew computer club, where Wozniak offered the first public demonstration of the Apple I in 1976. As Wozniak described the club, "Everyone in the Homebrew Computer Club envisioned computers as a benefit to humanity—a tool that would lead to social justice." These men were the exact counterparts of the radio pioneers of the 1910s—hobbyist-idealists who loved to play with technology and dreamed it could make the world a better place. And while a computer you can tinker with and modify may not sound so profound, Wozniak contemplated a spiritual relationship between man and his machine, the philosophy one finds in Matthew Crawford's *Shop Class as Soulcraft* or the older *Zen and the Art of Motorcycle Maintenance*. "It's pretty rare to make your engineering an art," said Wozniak, "but that's how it should be."[8]

The original Apple had a hood; and as with a car, the owner could open it up and get at the guts of the machine. Indeed, although it was a fully assembled device, not a kit like earlier PC products, one was encouraged to tinker with the innards, to soup it up, make it faster, add features, whatever. The Apple's operating system, using a form of BASIC as its programming language and operating environment, was, moreover, one that anyone could program. It made it possible to write and sell one's programs directly, creating what we now call the "software" industry.

In 2006, I briefly met with Steve Wozniak on the campus of Columbia University.

"There's a question I've always wanted to ask you," I said. "What happened with the Mac? You could open up the Apple II, and there were slots and so on, and anyone could write for it. The Mac was way more closed. What happened?"

"Oh," said Wozniak. "That was Steve. He wanted it that way. The Apple II was my machine, and the Mac was his."

Apple's origins were pure Steve Wozniak, but as everyone knows, it was the other founder, Steve Jobs, whose ideas made Apple what it is today. Jobs maintained the early image that he and Wozniak created, but beginning with the Macintosh in the 1980s, and accelerating through the age of the iPod, iPhone, and iPad, he led Apple computers on a fundamentally different track.

Jobs is a man who would seem as much at home in Victorian England as behind the counter of a sushi bar: he is an apostle of perfectibility and believes in a single best way of performing any task and presenting the results. As one might expect, his ideas embody an aesthetic philosophy as much as a sense of functionality, which is why Apple's products look so good while working so well. But those ideas have also long been at odds with the principles of the early computing industry, of the Apple II and of the Internet, sometimes to the detriment of Apple itself.

As Wozniak told me, the Macintosh, launched in 1984, marked a departure from many of his ideas as realized in the Apple II. To be sure, the Macintosh was radically innovative in its own right, being the first important mass-produced computer to feature a "mouse" and a "desktop"—ideas born in the mind of Douglas Engelbart in the 1950s, ideas that had persisted without fructifying in computer science labs ever since.* Nevertheless the Mac represented an unconditional surrender of Wozniak's openness, as was obvious from the first glance: gone was the concept of the hood. You could no longer easily open the computer and get at its innards. Generally, only Apple stuff, or stuff that Apple approved, could run on it (as software) or plug into it (as peripherals). Apple now refused to license its operating system, meaning that a company like Dell couldn't make a Mac-compatible computer. If you wanted a laser printer, software, or virtually any accessory, it was to Apple you had to turn. Apple thus became the final arbiter over what the Macintosh was and was not, rather in the way that AT&T at one time had sole discretion over what could and what could not connect to the telephone network.

Thus via the Mac, Apple was at once an innovative and a completely retrograde company. Jobs had elected the design principles that had governed the Hollywood studios, Theodore Vail's AT&T, indeed anyone who ever dreamed of a perfect system. He created an integrated

---

*Most notably at the Palo Alto Research Corporation, then owned by Xerox, whose labs, by 1975, had produced a computer closely resembling the Apple Macintosh. The Apple Lisa also, technically, came between the Macintosh but did not thrive.

product, installing himself as its prime mover. If the good of getting everything to work together smoothly—perfectly—meant a little less freedom of use, so be it. Likewise, if it required a certain restraint to create and market it, that was fine. Leander Kahney, author of *Inside Steve's Brain,* describes Jobs's modus operandi as one of "unrelenting control over his employees, his image, and even his customers" with the goal of achieving "unrelenting control over his products and how they're used."[9]

By the time the Macintosh became Apple's lead product, Wozniak had lost whatever power he had once held over Apple's institutional ideology and product design. One salient reason had nothing to do with business or philosophy. In 1981 he crashed his Beechcraft Bonanza on takeoff from Scotts Valley, just outside the San Francisco Bay Area. Brain damage resulted in pronounced though temporary cognitive impairment, including retrograde amnesia. He would take a leave of absence, but his return would not alter the outcome of a quiet power struggle that had been building since before the accident. Its resolution would permanently sideline "the other Steve," leaving the far more ambitious Jobs and his ideas ascendant.

Like all centralized systems, Jobs's has its merits: one can easily criticize its principles yet love its products. Computers, it turns out, can indeed benefit in some ways from a centralizing will to perfection, no less than French cuisine, a German automobile, or any number of other elevated aesthetic experiences that depend on strict control of process and the consumer. Respecting functionality, too, Jobs has reason to crow. Since the original Macintosh, his company's designs have more often than not worked better, as well as more agreeably, than anything offered by the competition.

But the drawbacks have been obvious, too, not just for the consumer but for Apple. For even if Jobs made beautiful machines, his decision to close the Macintosh contributed significantly to making Bill Gates the richest man on earth. No one would say it was the only reason, but Apple's long-standing adherence to closed design left the door wide open for the Microsoft Corporation and the many clones of the IBM PC to conquer computing with hardware and software whose chief virtue was combining the best features of the Mac and the Apple II. Even if Windows was never as advanced or well designed as Apple's operating system, it enjoyed one insuperable advantage: it worked on any computer, supported just about every type of software, and could interface with any printer, modem, or whatever other hardware one could design. After it was launched in the late eighties, early-nineties Windows ran off with the market Apple had pioneered, based mostly on ideas that had been Apple's to begin with.

The victory of PCs and Windows over Apple was viewed by many as the defining parable of the decade; its moral was "open beats closed." It suggested that Wozniak had been right from the beginning. But by then Steve Jobs had been gone for years, having been forced out of Apple in 1985 in a boardroom coup. Yet even in his absence Jobs would never agree about the superiority of openness, maintaining all the while that closed had simply not yet been perfected. A decade after his expulsion, back at the helm of the company he founded, Steve Jobs would try yet again to prove he had been the true prophet.

Just What Is Google?

In 1902, the New York Telephone Company opened the world's first school for "telephone girls." It was an exclusive institution of sorts. As the historian H. N. Casson described the qualifications for admission in 1910: "Every girl shall be in good health, quick-handed, clear-voiced, and with a certain poise and alertness of manner." There were almost seventeen thousand applicants every year for the school's two thousand places.[10]

Acquiring this credential was scarcely the hardest part of being a telephone girl. According to a 1912 *New York Times* story, 75 percent were fired after six months for "mental inefficiency." The job also required great manual dexterity to connect dozens of callers per minute to their desired parties. During the 1907 financial panic in New York, one exchange put through fifteen thousand phone calls in the space of an hour. "A few girls lost their heads. One fainted and was carried to the rest-room."[11]

People often wonder, "What exactly is Google?" Here is a simple answer: Like its harbinger the telephone girl, Google offers a fast, accurate, and polite way to reach your party. In other words, Google is the Internet's switch. In fact, it's the world's most popular Internet switch, and as such, it might even be described as the current custodian of the Master Switch.[12]

Every network needs a way to connect the parties who use it. In the early days of the telephone, before direct dial, you'd ask the telephone girl for your party by name ("Connect me with Ford Motors, please"). Later on, you'd directly dial the phone number, from either memory or the telephone directory, which seems rather a decline in service. Today, Google upholds the earlier standard, but on the Internet. Needing no address, you ask for your party by name (typing in "Ford Motor Company," for instance), and Google shows you the way to connect with them over the World Wide Web.

The comparison with Bell's telephone switchboard girls might sound a little anticlimactic to describe a firm with ambitions as grand as Google's, but this reaction betrays a lack of awareness of the lofty import the switch has in the information world. For it is the switch that transforms mere communications into networking—that ultimately decides who reaches what or whom. And at the superintending level, which most networks eventually do develop at some point, it is the Master Switch, as Fred Friendly reminds us, that will decide who is to be heard. However many good things the Internet has to offer—services, information resources, retail outlets—it hardly matters if you can't get to them.

There are, of course, some differences between Google and the switch monopolists of yesteryear, including the firm that is arguably its truest forerunner, Vail's AT&T. For one thing, Google is not a switch of necessity, such as the telephone company was, but rather a switch of choice. This is a somewhat technical point, but suffice it to say that Google's search engine is not the only Internet switch. There are other means by which to reach people or places on the Internet, as well as other points that might be described as "switches," like the physical routers that direct the flow of Internet traffic on the data packet level. There are plenty of ways around Google: you can use domain names to navigate the Internet, or use one of Google's competitors (Yahoo!, Bing, and the like), or for the truly hard-core, simply remember the IP addresses (e.g., 98.130.232.209), the way people once used to remember phone numbers. In fact, unlike AT&T, Google could be replaced at any time. And yet if by 2010 Google wasn't the only game in town, it was clearly the most popular Internet switch; by its market share of the search business (over 65 percent) it clearly qualifies as a monopoly.

In some ways, Google nevertheless enjoys a much broader control over switching than the old AT&T ever did. For it is not just the way that people reach one another to talk, but the way most people find all forms of information, across all media platforms, at a time when information is a far more prominent commodity in our national life and economy than it has ever been. Siva Vaidhyanathan's aptly titled book *Googlization of Everything* points out how much power this gives Google over global culture. As he writes, "for a growing (but not universal) portion of the Web and the world, we allow Google to determine what is important, relevant, and true."[13] As he suggests—and how many would disagree?—whatever shows up on the first page of a Google search is what matters in forming our sense of any reality; the rest doesn't.

To understand this unusual level of consumer preference and trust in a market with other real choices is to understand the source of Google's singular power. But is this a stroke of cold luck, or is Google something special? Quite enough has been written about Google's corporate culture, whether one looks to the cafeterias that serve free food, the beach volleyball, or the fact that its engineers like to attend the Burning Man festival in the Nevada desert.[14] Not that such things aren't useful inducements to productivity or the exception in corporate America, but they are more nearly adaptations of a general Silicon Valley corporate ethos than one particular to Google, a point the company readily admits.

Boiled down, the Google difference amounts to two qualities, rather than any metaphysical uniqueness. The first, as we've already remarked, is its highly specialized control of the Internet switch. We shall describe the nature of that specialization in more detail presently, but for now let us say it accords Google a dominance in search befitting an engine whose name has become a verb and synonymous with function. (No one Apples or AT&Ts a potential new boyfriend.) While the firm does have dozens of other projects, it is obvious (certainly from their individual direct contributions to cash flow, which are minuscule) that most, including the maps, the lavishly capacious Gmail accounts, even the hugely popular YouTube, are ultimately trial balloons, experiments of a kind, or a way of enhancing the primacy of the core business of search, whether by creating complementary information resources or simply engendering the goodwill that comes of offering cool stuff for free. The second Google difference is in its corporate structure. The firm, while having as many ventures as it has engineers, eschews vertical integration of these efforts to a degree virtually unprecedented for a communications giant. This structural distinction may be hard to grasp, so let's explain it more carefully.

·   ·   ·

A medieval architect looking at the skyline of New York City or Hong Kong would be astonished that the buildings manage to stand without flying buttresses, thick walls, or other visible supports. A nineteenth- or twentieth-century industrialist would feel much the same bewildered awe regarding a major Internet firm like Google. You might call Google a media company, but it doesn't own content. It is a communications company, but it doesn't own the wires or airwaves over which packets reach people. You might accept my characterization that Google is simply *the* switch, but a switch alone has never before constituted a freestanding company—what basis is that for value? Many credit Google with ambitions to take over the world, but an industrial theorist might well ask how such a radically disintegrated firm could long endure, let alone achieve global domination.

Compared with other giants, like Time Warner circa 2000, or Paramount Pictures circa 1927, or AT&T's original and resurrected incarnations, Google is underintegrated and undefended. The business rests on a set of ideas—or more precisely, a set of open protocols designed by government researchers. But that is the point: it is the structure of the Internet, much less than anything particular to the firm itself, that keeps Google standing. It traffics in content originated by billions of people, none of them on salary, who build the websites or make the home videos. It reaches its customers on wires and over airwaves owned by other firms. This may seem an improbably shaky foundation to build a firm on, but perhaps that is the genius of it.

If that seems a bit abstract, it is well to remember that Google is an unusually academic company in origins and sensibility. Larry Page, one of the two founders, described his personal ambitions this way: "I decided I was either going to be a professor or start a company." Just as Columbia University effectively financed FM radio in the 1930s, Stanford got Google started. With its original Web address http://google .stanford.edu/, the operation relied on university hardware and software and the efforts of graduate students. "At one point," as John Battelle writes in *The Search,* the early Google "consumed nearly half of Stanford's entire network bandwidth."[15]

Google's corporate design remains both its greatest strength and its most serious vulnerability. It is what makes the firm so remarkably well adapted to the Internet environment, as a native species, so to speak. Unlike AOL, Google never tried to resist or subdue the Internet's essential structure. It is a creature perfectly suited to the network as its framers intended it. In this sense, it is the antithesis of AOL.

Google's chief advantage, as we have suggested, can be summarized in a single word: *specialization.* Companies like AT&T or the big entertainment conglomerates succeed by being big and integrated—doing everything, and owning everything. A company like Google, in contrast, succeeds by doing one (well-chosen) thing, but doing it better than anyone else. It's the trait that makes Google the fox to so many others' hedgehog. The firm harvests the best of the Internet, organizing the worldwide chaos in a useful way, and asks its users to navigate this order via their own connections; by relying on the sweat of others for content and carriage, Google can focus on its central mission: search. From its founding, the firm was dedicated to performing that function with clear superiority; it famously pioneered an algorithm called PageRank, which arranged search hits by importance rather than sheer numerical incidence, thereby making search more intelligent. The company resolved to stand or fall on the strength of that competitive edge. As Google's CEO, Eric Schmidt, explained to me once, firms like the old AT&T or Western Union "had to build the entire supply chain. We are specialized. We understand that infrastructure is not the same thing as content. And we do infrastructure better than anyone else."

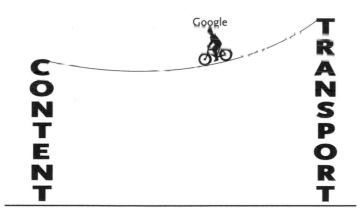

Google, between content and transport

Unlike AOL Time Warner, Google doesn't need to try to steer users anywhere in particular. They need only focus their resources on helping you get wherever you want to go, whether you know where that is or not. Needless to say, it is a great plus not to be involved in trying to persuade anyone to consume, say, Warner Bros. content. Such was the inherently corrupting project of AOL when Steve Case joined his company to Time Warner. Case had assumed that any Internet company would need control of both wires and content to succeed in the 2000s. He was wrong.

That's the advantage. On the other hand, Google's lack of vertical integration leaves it vulnerable, rather like a medieval city without a wall.* He who controls the wires or airwaves can potentially destroy Google, for it is only via these means that Google reaches its customers. To use the search engine and other utilities, you need Internet access, not a service Google now provides (with trivial exceptions). To have such access, you need to pay an Internet Service Provider—typically your telephone or cable company. Meanwhile, Google itself must also pay for Internet service, a fact that, conceptually at least, puts the firm and its customers on an equal footing: both are subscription users of the Internet. And so whoever controls those connection services can potentially block Google—or any other site or content, as well as the individual user, for that matter.

Nor is this matter of infrastructure the firm's only weakness. A concerted boycott among content owners—website operators or other sources—could achieve the same choking effect. Under long-established protocols, any website can tell Google that it doesn't want to be indexed.† In theory, Wikipedia, *The New York Times,* CNN, and dozens of other websites could begin telling Google, "Thanks, but no thanks," or conceivably strike an exclusive deal with one of Google's rivals.

---

*As we've seen, vertical integration serves as often as a means of corporate defense as efficiency. By combining related functions, the integrated entity can prevent rivals from depriving it of some essential component, as for instance when the Hollywood studios acquired movie theaters to prevent theater owners from shutting out studio products. Interesting, but beyond the scope of this book, is whether this defense function suggests an alternative explanation to the prevailing theory of the firm as shaped by the relative efficiency of internal and external contracting, which the economist Ronald Coase articulated in 1937.

†Technically, this is achieved by placing a "robots.txt" file on the root directory of the Web server in question. Google, for its part, could ignore the robots.txt files; in the United States that would foreground an unsettled copyright question, namely, whether expressly involuntary indexing is copyright infringement.

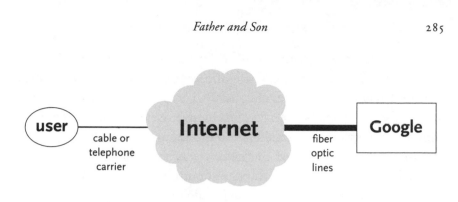

How Google reaches customers

Both of these vulnerabilities are a direct consequence of Google's corporate design, of the fact it owns no connections and no content. As we shall see, the firm's most determined enemies have begun to understand and exploit these frailties.

In Chicago in 2005, AT&T's Ed Whitacre took a break during a typical day of empire building to grant *BusinessWeek*'s Roger Crockett an interview.[16] In the midst of his campaign to reunify the Bell company, the CEO was refreshingly clear about his strategy. "It's about scale and scope," he told Crockett a few times, "scale and scope."

Crockett asked, "How concerned are you about Internet upstarts like Google, MSN, Vonage, and others?"

Whitacre immediately homed in on their weakness. "How do you think they're going to get to customers? Through a broadband pipe.

"Cable companies have them. We have them," he continued. "Now what they would like to do is use my pipes free, but I ain't going to let them do that."

From this it was clear that AT&T had identified precisely the soft underbelly of Google and the rest of the Internet industry: "How do you think they're going to get to customers?" Whitacre understood that he, allied with the cable industry and the other parts of Bell, was strategically positioned to choke the Internet industry into submission.

Such comments make vivid just why the ideal of net neutrality and the government's enforcement of it by statute or regulatory rules have become such urgent concerns for Google and the rest of the Internet industry, as well as increasingly for a great many individual users. If one allows that the Internet is our key means of conveyance, the "common medium" of our national life and economy, net neutrality is the twenty-first century's version of common carriage. Just as with the operator of the sole ferry between an island and the mainland, proprietorship of any part of the Internet's vital infrastructure rightly obliges one to carry the whole Internet, without discrimination or favoritism, in accordance with one of the oldest assumptions of our legal tradition. To be entrusted with a utility of such unique public importance comes with responsibilities such as AT&T assumed in 1910. In the case of the Internet, common carriage under the name of net neutrality amounts to an FCC rule that bans any degree of blocking individual sites, transmission of data (whether according to size, sender, time of day, or any other factor). Put most simply, net neutrality is what prevents the telephone and cable industry from killing Google, Amazon, Wikipedia, blogs, or anything else that might incur their displeasure.

In 2006, when Whitacre made his remarks, it seemed a plausible inference that AT&T and its allies might undertake—if not imminently, at least gradually—to subjugate the Internet and thereby the firms that depend on it, aiming to accomplish with long-honed lethal efficacy what AOL Time Warner had bungled. The initial step would be subtle: AT&T would begin offering, for a fee, a "fast lane" by which to reach consumers, inspiring the cable firms and Verizon to do the same. The precedent was Vail's policies of the 1910s, a system of preferential treatment with an eye toward creating vassals of the dependent industries. Of course AT&T would offer its ever-ready excuse: management of the network in the name of better service. The effects at first would be small. But it doesn't take a genius to realize that if AT&T and the cable companies exercised broad discretion to speed up the business of some firms and slow down that of others, they would gain the power of life and death over the Internet.

Google's advantage in being obliged to promote no one's product is double-edged: the ingenious idea of depending entirely on others for content leaves one entirely at the mercy of others for access to it. Google itself owns almost nothing: no movies, no websites, videos, or texts of any significant interest. In most instances, content owners have been only too happy to allow Google to lead its customers to them. On the other hand, the company's commitment to liberating content, making it accessible to as many as possible, has also left it the object of copyright-holder grievance and exposed it to potential lawsuits as well as threats of organized boycott.

Sometime in 1996, when Google began operations,* it made a copy of the entire World Wide Web in order to prepare a search index. In retrospect, no one really knows whether that copying was legal—whether a massive copyright violation occurred at the birth of the firm, confirming Balzac's observation that behind every great fortune is a great crime. As a matter of law, copying generally requires permission, something that Google never asked for, and indeed never has requested, for to do so comprehensively would be impracticable. Today, most copyright scholars would agree that Google has implied permission to copy the Web—no one brought suit against them for having done so, and so a new norm has been suggested. It is also likely an instance of "fair use," though, given the uniqueness of the act, there is little case law quite supporting such an assertion. Certainly at the time, the legality of what was done wasn't entirely clear; and truth to tell, if a copyright lawyer had been among Google's founders, it's doubtful the thing would have gotten off the ground.

Since its audacious birth, Google has never been completely at peace with the owners of content upon which it depends. The dispostion of those owners has varied according to what was in it for them in each instance. Some, of course, love, or at least respect, Google as the primary means by which their content gets found. For the small-scale business and those struggling to be heard without a major platform, Google's engine is a godsend, for it tends to equalize giant and one-man retailers, new bloggers and those who write for highly capitalized publications. Thanks to Google's proprietary algorithm, an entry on the nonprofit Wikipedia consistently outranks any official site related to a search term. A search for McDonald's also turns up McSpotlight, a page dedicated to exposing the misdeeds of the restaurant chain.

In contrast, owners of "valuable" content have a far more ambivalent relationship with the great Internet switch. In the United States, Google receives a daily stream of notices demanding that it remove links to copyright-infringing materials (YouTube accounts for the lion's share).

---

*At the time, it went by the name "BackRub."

Many, especially in New York's old-media conglomerates and publishing industries, hold Google in deep suspicion, a feeling that persists no matter how many earnest professions of benign intent are offered by Google's employees. Those professions, in fact, tend to make matters worse, as they leave the old content generators feeling Google doesn't appreciate how a dollar should be made in the information game.

When such anxiety boils over, it is expressed through lawsuits. When in 2004 Google proposed a system for searching books modeled on its search engine for the Web, it was promptly sued by a consortium of publishers and authors. YouTube, similarly, was subject to a deck-clearing lawsuit in the 2006, funded by Viacom, the entertainment conglomerate. By the first decade of its existence, Google's legal department had accumulated a large collection of copyright experts, and they needed every one of them.

Google has so far managed to settle many of the most serious claims, thanks in part to its lawyers, but a different sort of danger looms in the form of threatened content boycotts. Rupert Murdoch, owner of the News Corp. conglomerate and a master of exploiting the structural weaknesses of other firms, started complaining in 2009 about sites like Google that "steal" newspaper content.[17] Here is a portion of a television interview he gave on the subject (reproduced as a matter of fair use):

Murdoch: [The problem is] the people who just simply pick up everything, and run with it, who steal our stories . . . Google, Microsoft, Ask.com. . . .

Interviewer: Their argument is that they are directing traffic your way. . . . Aren't they helping you?

Murdoch: What's the point of having someone come occasionally, who likes a headline they see in Google? . . . We'd rather have fewer people coming, and paying.

Interviewer: The other argument from Google is that you could choose not to be on their search engine, you could simply refuse . . . so that when someone does a search, your websites don't come up—why haven't you done that?

Murdoch: Well, I think we will, but that's when we start charging.

While Murdoch doesn't go so far as to announce or promise a boycott, his implication is perfectly clear—as is the risk to Google owing to extreme specialization. To persist in doing what it does, Google, though a powerful monopoly, needs information industries disposed to play nice, cooperate, and share—to let the world's greatest organizer of information index their content and make it accessible over their wires. Unfortunately, playing nice has never been common practice in the information industries, as this book should already have made clear. Something about the intangible nature of information products seems to make everyone only more cutthroat than the average widget manufacturer.

THE BATTLE FOR TERRITORY IN THE 2010S

In Hindu mythology, deities and demons assume different incarnations to fight the same battles repeatedly. At the beginning of the 2010s, as a chasm opened between Google and its allies like Amazon, eBay, and nonprofits like Wikipedia on the one side and Apple, AT&T, and the entertainment conglomerates on the other, it was obvious that what loomed was just the latest iteration of the perennial ideological struggle into which every information industry is eventually swept. It is the old conflict between the concepts of the open system and the closed, between the forces of centralized order and those of dispersed variety. The antagonists assume new forms, the generals change, but essentially the same battles are fought over and over again. It is the very essence of the Cycle, which even a technology as radical and powerful as the Internet seems able at most to moderate but not to abolish.

For the information industries that now account for an ever increasing share of American and world GDP, the coming decade will be given over to a mighty effort to seize territory, to bolt the competition from its habitat. But this is not a case of one pack of wolves chasing another out of a prime valley. While it may sound fanciful, the contest in question is more like one of polar bears battling lions for domination of the world. Each animal, insuperably dominant in its natural element—the polar bear on ice and snow, the lion on the open plains—will undertake a land grab where it has no natural business being. The only practicable strategy will be a campaign of climate change, the polar bears seeking to cover as much of the world with snow as they can, while the lion tries to coax a savannah from the edges of a tundra. Sounds absurd, but for these mighty predators, it's simply the law of nature.

For the past few years, Google, together with Amazon, eBay, Facebook, and nonprofits like Wikipedia, has generally been trying to convert as much of the world as they can into something that looks like the Internet: a clear, free path between any two points, with no hierarchy or preferential treatment according to market capitalization, number of paid lobbyists, or any other prerogative of size and concentration. Meanwhile, AT&T, the entertainment conglomerates, and the rest are trying to succeed where AOL Time Warner failed, and bring the Internet to heel. They envision a rational regime of access and flow of information, acknowledging that the network is not some renewable natural resource but a man-made structure, one that exists only owing to decades of infrastructure building at great cost to great companies, entities that believe they ultimately are entitled to a say. For the telephone and cable companies it is a matter of respecting the ownership of the Internet's sine qua non: the wires, bandwidth, and cable. Naturally allied to such respect for ownership are copyright holders, whose just due they fear is being lost in the giddy idealistic effort to make everything available to everyone without limit, and as often without compensation. There is, the partisans of this side argue, a cost to building a bridge, a cost to writing a novel. An information economy, so called, cannot ultimately be sustained without acknowledging such hard facts. Information may "want" to be free, but we cannot expect it to be moved or created if we drive down to nothing the incentives for performing either function. If this side has its way, the twenty-first-century world of information will look, as much as possible, like that of the twentieth century, except that the screens that consumers are glued to will be easier to carry.

This, in essence, is our present war for information, one being waged on multiple fronts in ways subtle and not so subtle. Let us consider now the face of battle.

Apple's Challenge to the Computer

In 2006, Professor Jonathan Zittrain of Harvard made the startling prediction that over the next decade, the information industry would undertake a determined effort to replace the personal computer with a new generation of "information appliances."[18] He was, it turned out, exactly right. But the one thing he couldn't forecast exactly was the general who would lead the charge. How indeed could anyone have guessed that Apple Inc., the creator of the personal computer, would be spearheading the effort to replace it? Unlikely though it was, beginning in 2010, Apple, allied with the entertainment conglomerates, became the key firm in a broad challenge to the whole concept of the personal computer.

When, in 1997, following another boardroom coup, Steve Jobs took back control of Apple, it was clear he had not changed or abandoned his basic ideas; to the contrary, he had intensified them, taking his whole ideology to, as it were, the next level. In doing so he repudiated, now decisively and forever, Steve Wozniak's vision of the firm. The transformation would be symbolized by the moment in 2007 when Jobs renamed Apple Computers "Apple Inc."—and at roughly the same time, as a personal flourish, refused to write a foreword for his old friend's autobiography, *iWoz.*[19]

By the dawn of the decade, the cornerstone of Jobs's strategy seeking perfect control over product and consumer had been laid; it took form as a triad of beautiful, perfect machines that have since won the allegiance of millions of users. Usurping the throne of the personal computer, in their order of succession, came the iPod, the iPhone, and the iPad. These would be, if all went according to plan, the information appliances of the 2010s.

On the inside, the iPod, iPhone, and iPad are actually computers. But they are computers that have been reduced to a strictly limited set of functions that they are designed to perform extremely well. It's easy to see this with the iPod, which, rather obviously, is designed solely and optimally for playing music and watching videos. The limitation is much harder to see on the iPhone and the iPad, both of which can do things like make phone calls, send email, surf the Web, and allow one to read books, in addition to the seemingly unlimited variety of functions that can be acquired through the "app store." But even if invisible to many consumers, the inescapable reality is that these machines are closed in a way the personal computer never was. True, Apple does allow outsiders to develop applications on its platform—the defeat of the Macintosh by Windows taught Jobs that a platform completely closed to outside developers is suicide. Nevertheless, all innovation and functionality are ultimately subject to Apple's veto, making these devices antithetical to the Apple II and all the hardware development it inspired.

Apple's new generation of devices are user-friendly, but also what you might call "Hollywood-friendly." They are engineered with an eye to complicated deals the firm has made with the existing entertainment conglomerates, deals securing access to content that Apple's rivals have had trouble matching. In exchange for this access, Apple generally, if not quite perfectly, guards the intellectual property of its partners. The devices, in a similar way, are also "telecom-friendly": designed to operate with one carrier only, they reinforce the favored telephone company's power—for a price.

The veto that Apple maintains over functionality and specific applications is not notional, but one wielded in service of its partnerships' interests. The first major exercise was in blocking Skype, the voice-over-IP firm whose software lets users call each other over the Internet for free, eating into AT&T's long distance margins. Later, during the summer of 2009, Apple bodychecked an application written for the iPhone by Google. The product, named "Google Voice," was designed to make a single phone number, when dialed, ring on all one's phones at once. The rejection of this service six months after its submission for consideration was not another effort at protecting the telephone partner (all GV users would place and receive calls over their carriers' networks and not, as some had feared, over the Internet). Rather, this rejection of a widely anticipated function appears to have been motivated by perceived competition with existing Apple applications such as the dialer, voice mailbox, and others. This, in a way, makes the move seem pettier. And the pattern would continue. As Tom Conlon of *Popular Science* would write when the iPad was unveiled, "How long before it [Apple] blocks movies, TV shows, songs, books and even web sites? Scoff now, but don't be so naïve as to believe that this isn't possible."

Lest these examples be taken amiss, let me speak plainly: These are amazing machines. They make available an incredible variety of content—video, music, technology—with an intuitive interface that is a pleasure to use. But they are also machines whose soul is profoundly different from that of any other personal computer, let alone Wozniak's Apple II. For all their glamour, these appliances are a betrayal of the inspiration behind that pathbreaking device, which was fundamentally meant to empower its users, not control them. That proposition may appeal to geeks more than to the average person, but anyone can appreciate the sentiment behind putting enormous power at the discretion of any individual. The owner of an iPod or iPad is in a fundamentally different position: his machine may have far more computational power than a PC of a decade ago, but it is designed for consumption, not creation. Or, as Conlon declared vehemently, "Once we replace the personal computer with a closed-platform device such as the iPad, we replace freedom, choice and the free market with oppression, censorship and monopoly."

Google's Countermove

Throughout the summer of 2007, rumors flew that some kind of Google phone was in the works. At the Googleplex, the firm's storied campus, a suspicious statue, a human-robot, or android, with red eyes showed up in a nondescript building across from the main campus. Finally, on November 5, 2007, Google effectively announced the Gphone—by letting it be known that there was no such thing.

In contrast with the unveiling of the the iPhone, there was no stadium event, no screaming crowd, and most important, no product. Instead, there was just a blog post entitled "Where's My Gphone?"[20] An employee named Andy Rubin wrote the following: "Despite all of the very interesting speculation over the last few months, we're not announcing a Gphone. However, we think what we are announcing—the Open Handset Alliance and Android—is more significant and ambitious than a single phone."*

Here it was: Google's first real foray into the world of the telephone, as distinct from the computer and the Internet. The significance cannot be overstated. Until 2007, the Internet industries had, in the main, been playing defense—attempting to preserve the status quo of net neutrality and limit the power of their rivals among other information enterprises. Now, coming out of this defensive crouch, Google took the fight to its adversaries, attempting to plant the flag of openness deep in the heart of telephone territory, Bell's holy land since the 1880s.

---

*Notable members of the alliance at its launch included China Mobile, Intel, NTT DOCOMO, Sprint/Nextel, T-Mobile, HTC, LG, Samsung, and Motorola.

Project Android has puzzled many industry observers, for it has no obvious revenue model. Google distributes Android for free, as it does most of its other products. Mind you, what Google is giving away is not a telephone or even a telephone service—users must still buy those—but rather an operating system for telephones, based on the Linux kernel, the Ur–free and open software beloved of tech geeks. By giving away a version adapted for telephony, Google was distributing a free set of tools for programmers of any affiliation to write applications.

Given what we understand about Google, it should be obvious that this move was, like so many other initiatives, a means to an end rather than an end in itself. Project Android is a hearts-and-minds effort, a use of soft power to "convert" the mobile world into territory that is friendly to Google rather than to its enemies. It is, to return to our wild kingdom analogy, an effort to extend the world of ice and snow, where the polar bear cannot be defeated. And of course it is a long shot. As I wrote at the time, in *Slate* magazine, "Google is making its deepest foray yet into a foreign territory where its allies are few. It faces the challenge of not just entering the wireless world but also converting its inhabitants. Provided that Google has the nerve and resources to try to remake wireless in its image, it'll either prove its greatest triumph or its Waterloo." A high-stakes long shot, but one that Google's adversaries have given it little choice but to take.

Not surprisingly, Android has made the already strained relationship with former pal Apple downright hostile. Perhaps stung by the memory of Windows and what had followed, Jobs was quick to trash Android in *The New York Times.* "Android hurts them [Google] more than it helps them," said Jobs. "It's just going to divide them and people who want to be their partners."

This quote illustrates a crucial difference of mind-set. Since 2000, Jobs's innovations have depended on making the right deals. The success of his iTunes store has had less to do with the technology than with his being the first to get the music industry to consent to online downloads. Having been CEO of Pixar Animation Studios during his years in the Apple wilderness, Jobs is one of the few players who can move with ease between Hollywood and Silicon Valley. And he was able to extend his dealmaking reach beyond that corridor to work with the world's largest telephone company, making the iPhone the ultimate expression of his partnership mentality.

Schmidt and Google, meanwhile, have taken a different view. Their partnerships are few, and rarely, if ever, exclusive. For at bottom the firm believes, almost as an article of faith, that open protocols obviate the need for big combinations. As Schmidt puts it, the "interconnection makes you appear as one company while operating as two." In other words, why incur the burdens of marriage when you can have friends with benefits? Implicit in this view is the basic conception of the Internet and Wozniak's idea of the computer as worlds that minimize the need for permission.* The very same idea animates the Android.

Android may be the most significant of Google's territorial maneuvers, but it is not the only one. In the winter of 2010 the firm announced plans to build its own fiber optic connections, another bold incursion into the lands of telephone and cable and its first real flirtation with vertical integration. The full scope of the motivations isn't clear—Google insists it means only to create a "showcase" designed to spur the telephone and cable companies to expand broadband penetration, in which America lags the developed world despite having invented the technology. More startling were reports in the *New York Times* in the summer of 2010 that Google was on the verge of a deal with Verizon to align their policy positions and launch special "managed services." Google's close relationship with Verizon—its first friendship with a Bell—is hard to interpret. Eric Schmidt and Google believe they are converting Verizon to the side of openness and sundering the Bell Empire for good. But we've heard that before, and it is not completely clear who is converting whom. Verizon/Google makes for a powerful vertical combination. Verizon, formerly known as Bell Atlantic, is a seasoned monopolist, having held parts of its domain since the capitulation of Western Union in the 1870s. And so it may be Google that is learning from the master.

For now, we may still view Google and its Internet industry allies as locked in a complex, slow-moving struggle with AT&T and cable, the entertainment conglomerates and Apple. But while there are two sides in the broadest terms, the underlying reality is not so simple, as the firms

---

*Permission is a fundamental feature of what we call "property," and in this sense you can understand that at some level the entire struggle is between a world with more or fewer property rights.

have a web of allegiances as complex as those of nineteenth-century Europe. Verizon, a former Bell, for instance, has been in Google's camp for some time, having become an Android convert in 2009. It likes to declare itself an apostle of an open wireless future,[21] which presents the odd prospect of the reborn Bell system split into an open and a closed half. Meanwhile, some Internet firms, including Yahoo!, have long allied themselves with the centralizers, if only as a hedge against Google. Nevertheless, no one denies that the future is to be decided by one of two visions.

If the centralizers—AT&T, Hollywood, and Apple—prevail, the future will be informed by a marriage of twenty-first-century technology and twentieth-century integrated corporate structure. The best content from Hollywood and New York and the telephone and networking power of AT&T will converge on Apple's appliances, which respond instantly to ever more various human desires. It is a combination of undeniable power and attraction. And not least among its virtues, the worst of the Internet—the spam, the faulty apps, the junky amateur content—is eliminated. Instead, the centralizers pledge to deliver what Lord Reith promised from the BBC: "the Best of Everything."

For its part, the openness movement, of which Google has been the leader, despite whatever sort of pact may loom with Verizon, is based on a contrary notion of virtue, one that can be traced back to the idealism of 1920s radio and of course the foundation of the Internet itself. At some level, the apostles of openness aspire to nothing less than social transformation. They idealize a postscarcity society, one in which the assumption of limited resources dictating traditional economic theory are overturned, a world in which most goods and services are free or practically free, thereby liberating the individual to pursue self-expression and self-actualization as an activity of primary importance.[22] It may sound fantastical, but our lives are already full of manifestations of this idea. Digitization, for example, by eliminating most of the expenses associated with activities like making a film or distributing a recording, has enabled virtually anyone to prove his worth as a filmmaker or a singer. But the feasibility of such a quasiutopian information economy depends on an open communications infrastructure that facilitates individual expression, not mass conformity.

There is, as with all competing visions of the good, a downside to each. More specifically, as ever in the history of information networks, something is lost in seeking the benefits of an open system, just as there is in adopting a closed one. Each side of course imagines its preference as offering us more than it denies us. Apple and the conglomerates think it perfectly sensible to identify popular desires and then to fulfill them. As Jobs put it, "We figure out what we want. And I think we're pretty good at having the right discipline to think through whether a lot of other people are going to want it, too. That's what we get paid to do." Such cultural surrogacy does deliver an extremely polished product, both as content and as delivery system, indeed one very widely desired. But inevitably it is not to every taste. The champions of openness propose an untidier world of less polish, less perfection, but with more choice. It is, in that side's view, choice, the freedom to figure out what one wants, that people prize most. In Eric Schmidt's words: "The vote is clear that the end user prefers choice, freedom, and openness."

And so we have the essential alternatives: a world of information that looks much like the twentieth century's, only better—more beautiful and more convenient. Or a revolution in the very means by which information is produced and consumed.

The conflict is familiar in its contours; we have now seen it several times before, as the Cycle has worked its way through the film, radio, and telephone industries. The difference now, however, is this: In the 1920s and 1930s, there was a sense that the progress toward centralized, integrated models was somehow inevitable, simply the norm of industrial evolution. In the time of Henry Ford, Theodore Vail, and the rest, it had seemed quite natural, in a Darwinian way, that the big fish ate the little ones until there were only big ones trying to eat one another. All the power would thus come to reside in one or two highly centralized giants, until some sort of sufficiently disruptive innovation came along and proved itself a giant killer. Small fry would then enter the new decentralized environment, and the natural progression would start all over again.

The twenty-first century begins with no such predilection for central order. In our times, Jane Jacobs is the starting point for urban design, Hayek's critique of central planning is broadly accepted, and even governments with a notable affinity for socialist values tout the benefits of competition, rejecting those of monopoly. Nor does the new century partake of the previous one's sense of what is inevitable. Technology has reached a point where the inventive spirit has a capacity for translating inspiration into commerce virtually overnight, creating major players with astonishing speed, where once it took years of patient chess moves to become one, assuming one wasn't devoured. The democratization of technological power has made the shape of the future hard to know, even for the best informed. The individual holds more power than at any time in the past century, and literally in the palm of his hand. Whether or not he can hold on to it is another matter.

CHAPTER 21

The Separations Principle

An empire long united, must divide; an empire long divided,
must unite. Thus it has ever been, and thus it will always be.

—LUO GUANZHONG
The Romance of the Three Kingdoms

Luo Guanzhong's fourteenth-century novel captures the perennial
alternation between concentrated and dispersed power that has
shaped most of human history. Aside from a few patches of relative
enlightenment, our own among them, the course of political power has
continued on this winding way all over the world for most of recorded
time. Nowadays, we sometimes like to think we have progressed past
the cyclical rise and fall of centralized power, but in truth, even in the
absence of an actual Caesar or Khan, the human ambition to build
and overthrow empires lives on, however adapted to new forms and
contexts. It has been the aim of this book to show that our information industries—the defining business ventures of our time—have from
their inception been subject to the same cycle of rise and fall, imperial consolidation and dispersion, and that the time has come when we
must pay attention.

Living in a contemporary democracy can lull us into regarding concentrated power as a historical problem we have more or less solved.
The American Constitution was designed above all in the awareness of
the danger of centralized power, and its response to that danger was, as

Justice Kennedy once wrote, to "split the atom of sovereignty." By this he was referring specifically to our federal system, by which we distinguish powers reserved to the states from federal powers, but it could equally refer to any number of essential separations that the Constitution enshrines: separations of executive, legislative, and judicial powers; separations of the government's enumerated powers from those reserved to the individual, the latter protections found in the Bill of Rights, from which also derives the separation of church and state. Not that the framers invented the notion. The various divisions of power found in most of the world's constitutional governments today are based on an idea as old as ancient Greece and continued under the Roman Republic. Behind the very notion of separation is a theory of countervailing power. Separations are an effort to prevent any single element of society from gaining dominance over the whole, and by such dominance becoming tyrannical.

The American political system is designed to prevent abuses of public power. But where it has proved less vigilant is in those areas where the political meets the economic realm, where private economic power comes to bear on public life. We seem loath as a society to acknowledge the historical coincidence of the two, even though historians such as Arthur Schlesinger, Jr., have persuasively described our history as an ongoing contest between public and private power. We like to believe that our safeguards against concentrated political power will ultimately protect us from the consequences of accumulated economic power. But this hasn't always been so.

This relative indifference to the danger of private power is of complex origin. It owes in part to the Lockean sanctification of private property as enunciated by Jefferson. It owes as well to the nature of our constitutionalism: the American system reserves to the individual or, as the case may be, to the individual states any powers not explicitly granted to the federal government. The federal government's right to interfere with free enterprise is derived mainly from the Commerce Clause, and the extent of that right has never been uncontroversial. As a consequence, while popular demand for regulation has waxed and waned, American economic life has been built mostly on freewheeling capitalism.

This tradition has bequeathed to us an economic history far more spasmodic and cyclical than American political history (aside from the obvious exception of the Civil War). While the U.S. Constitution has proved relatively sturdy and adaptable, it is America's economic life that has been subject to a dynamic of imperial rise and fall akin to that in *The Romance of the Three Kingdoms*. The rise of an explicit political empire was successfully forestalled by the Constitution; and, as if in response, American history became, in no small part, a chronicle of commercial empires, including the industrial dominions of men like Carnegie and Rockefeller as well as those described in this book. The reason this displacement of energies could even have occurred is simple: while our political theology seeks to tame the state of nature, our economic orthodoxy submits to it. And so most influential economic thought, from Smith to Keynes to Schumpeter, accepts as intrinsic to a free-market system the ravages of boom and bust, as well as the various consequences of imperial growth and overreach, recommending that government policy should seek, at most, to moderate the resulting tremors.

It would be quite radical today even to contemplate imposing on the economy the kind of safeguards that the Constitution places on the political system, though such ideas have occasionally been proposed in our history, for instance by Justice Louis Brandeis and President Andrew Jackson. The latter, who fought and destroyed the Second Bank of the United States, warned in 1837 that without control over private power, "you will in the end find that the most important powers of Government have been given or bartered away, and the control over your dearest interests has passed into the hands of these corporations." But in our times, it goes without saying that economic vitality—innovation, growth, and opportunity—depends on the freedom of the economic system to rise and fall, crash and burn.

The difference in the American approaches to political versus economic power is a subject too vast to be done justice here. Suffice it to say that one must recognize it in order to understand the course of American industrial history. But there is a further difference to bear in mind, whatever we might think of the special treatment of industrial power in general; and that concerns the special case of concentrated power over the creation, transmission, and exhibition of information.

It is an oft-repeated assertion, but one that nevertheless always bears repeating: information industries, enterprises that traffic in forms of individual expression, can never be properly understood as "normal" industries, ones dealing in virtually any other sort of commodity.* Hence, the problem of cognitive entrenchment—a problem for any part of society—is much more serious when we speak of an industry fundamental to democracy. For humans, speech—in the broad constitutional sense extending beyond simple oral or even verbal communication—has effects and purposes that transcend mere transactional utility. To offer it and to consume it can take on a spiritual dimension that ensures that a television or mobile phone can never be remotely considered, as it were, a toaster that doesn't toast but happens to present pictures and sound. Whether we have in mind a song, a film, a political speech, or a private conversation, we are considering forms that have the potential to alter sensibilities, change lives. Every one of us has read or watched something that has made an indelible impression, impossible to quantify in relation to production and distribution costs. For such a reason did Joseph Goebbels describe radio as "the spiritual weapon of the totalitarian state." Indeed, for such reasons is there almost always, behind every political revolution or genocide, a partnership with some kind of mass medium. That kind of claim can't be made of orange juice, heating oil, running shoes, or dozens of other industries, no matter their size.

Today, the information industries are collectively embedded in our existence in a way unprecedented in industrial history, involving every dimension of our national and personal lives—economic, yes, but also expressive and cultural, social and political. They are not just effectively integral to every transaction; they also decide who among us gets heard or seen and when, whether it be the aspiring inventor, artist, or candidate. And that creates a challenge for an American system used to a clean split between the treatment of political and economic power, a strict control of the former and only moderate regulation of the latter. Among the great questions of our time is whether our approach to the power of information should be informed by a sense of that power's political consequences, subject to our ingrained habit of balancing and checking any great power. Or should we simply follow our approach

---

*This point might be described as axiomatic in communications scholarship, and indeed the justification for the communications departments found at many universities. It is, for example, the whole premise of Harold Innis's *Empire and Communications* (1950), which held, rather boldly, that the nature of various civilizations from the Egyptians onward was much the product of their communications systems.

to economic power in general, in which we tolerate, and even reward, aggrandizement?

While perhaps not immediately obvious, such questions are in fact at the heart of the ongoing struggle between the armies of open systems and closed, represented in the last chapter in the battle between Google and Apple but manifest elsewhere as well and destined to outlast that rivalry. The two defining firms of our time have come to represent, respectively, the utopia of openness (the dream of the Internet's founders, of which the early days of telephony, radio, and film offered a foretaste) and the perfection of the closed system (Vail's dream). The same question of how to treat information industries is also raised by a less well reported alignment now shaping up: by the FCC's own reckoning, the cable companies will soon enjoy an uncontested monopoly over broadband Internet in much of the United States beyond the East Coast, and they are also seeking control of more Hollywood studios and television networks.

To come at these problems afresh in the twenty-first century is to be struck by an obvious reality: information has become exceptional as an industrial category even in relation to that industry's own history. To consider the extent of that reality is to recognize immediately that the purely economic laissez-faire approach, the old television-as-toaster thinking that prevailed in the late twentiety century, is no longer feasible. To leave the economy of information, and power over this commodity, subject solely to the traditional ad hoc ways of dealing with concentrations of industrial power—in other words, to antitrust law—is dangerous. Without venturing into the long, rancorous debate over what, if any, kind of antitrust policy is proper in our system, I would argue that by their nature, those particular laws alone are inadequate for the regulation of information industries.* One reason is fairly simple: historically, the application of those statutes has been triggered by manipulation of consumer prices and certain other very particular abuses of market power; but those aren't the most troubling problems in this context. More subtly, there is the problem of taking an after-the-fact approach to

---

*More broadly, it seems clear to me that a pure antitrust approach is inadequate for any of the main "public callings," i.e., the businesses of money, transport, communications, and energy.

a commodity so vital to our basic liberties: a framework that has worked well enough for oil and aluminum is ultimately unsuited to an industry whose substrate is speech.

The rejection of a narrow economic approach might seem to propose a high degree of regulatory involvement, along the lines taken by the New Deal agencies. And indeed, following the logic that the information and news media peform a vital public function, most nations in the twentieth century, even liberal democracies, have simply made broadcasting, telephony, and the news media either actual or de facto parts of government. The United States came close to this model during the years of AT&T's regulated monopoly and the golden age of the television networks, but has since reverted to the idea that industry is industry.

Yet this approach is also wrong. What I propose is not the sort of nationalization that found favor in Western Europe and briefly in the United States during the 1930s. Far from it. For history shows that in seeking to prevent the exercise of abusive power in the information industries, government is among those actors whose power must be restrained. Government may function as a check on abusive power, but government itself is a power that must be checked. What I propose is not a *regulatory* approach but rather a *constitutional* approach to the information economy. By that I mean a regime whose goal is to constrain and divide *all* power that derives from the control of information.

Specifically, what we need is something I would call a Separations Principle for the information economy. A Separations Principle would mean the creation of a salutary distance between each of the major functions or layers in the information economy. It would mean that those who develop information, those who own the network infrastructure on which it travels, and those who control the tools or venues of access must be kept apart from one another. At the same time, the Separations Principle stipulates one other necessity: that the government also keep its distance and not intervene in the market to favor any technology, network monopoly, or integration of the major functions of an information industry. Such interference—often to preserve an industry that figures mightily in the national economy (in a sense, too big to fail)—is ultimately destructive of both a free society and the healthy growth of an information economy or any other kind.

Like the separation of church and state, the Separations Principle means to preempt politics; it is a refusal to take sides between institutions that are historically, even naturally, bound to come into conflict, a refusal born of society's interest in preserving both. Thus the First Amendment's church and state separation has been used by secularists and the religiously minded alike in the defense of their respective causes. Such refusal to favor is the essence of how a liberal society preserves itself as such while availing itself of the dynamism of diverse, sometimes disruptive, perspectives and ideas.

And like the separation of powers, the Separations Principle accepts in advance that some of the benefits of concentration and unified action will be sacrificed, even in ways that may seem painful or costly. An autocracy may make the trains run on time, and in the information world, a perfectly unified Bell system might be able to guarantee a good connection 99.999 percent of the time. But those satisfactions come at too high a price.

As we have seen, power can be concentrated both by monopolistic control of a technology (such as telephony or film) and by the integration of industrial functions (as when a single entity controls every stage of creating and delivering the product). Such concentration through horizontal monopoly and/or vertical integration typically finds its license, its basis for societal acquiescence, in a specific kind of consumer gratification that size and centralization make possible: reliable, universal telephone service (the Bell system), radio shows backed by advertising (the networks), big-budget movies (the Hollywood studios and the media conglomerates), a dazzling device that seems to put the world in the palm of your hand (Apple and its collaborators). To see what is sacrificed to such efficiency, polish, and convenience, however, takes work, and to some it may forever remain invisible. It requires appreciating the merits of systems in which, so to speak, the trains do not always run on time. It requires an appreciation of the forms of speech and technical innovation that are excluded in the name of perfection and empire.

More than anything else, the preceding chapters chronicle the corrupting effects of vertically integrated power. A strong stake in more than one layer of the industry leaves a firm in a position of inherent conflict of interest. You cannot serve two masters, and the objectives of creating information are often at odds with those of disseminating it. That is the very first reason for the Separations Principle. Broadcast witnessed a dramatic winnowing of content with the introduction of the advertising-based model by the Radio Trust. Film, for its part, was subject to two regimes of severe private censorship. The first, under the Edison Trust, was a matter of simple monopoly: a patent on technology, restricting its application. But the second was entirely a result of Zukor's and his fellow studio heads' efforts to protect their empire by acquiring the industry's means of exhibition. As technological monopoly, film was a boring, underdeveloped medium. But as a unified, fully integrated industry, film was vulnerable to the efforts of a few private individuals to enforce the Production Code, a regime of censorship without equal in American culture and entirely insulated from First Amendment challenge.

By fighting vertical integrations, a Separations Principle would remove the temptations and vulnerabilities to which such entities are heir. It represents the difference between free speech as an abstract ideal and the habit of fostering a practical environment in which the ideal can be realized. It is a recognition that the disposition of firms and industries is, if anything, more critical than the actions of the state in controlling who gets heard. The public square is a fine conceit, but in an information society it matters little that one is free to speak one's mind in public; the public square, if it exists, is an information network nowadays.

The second broad justification for a Separations Principle may be derived mainly from the story of the AT&T monopoly that is woven throughout this book. Communications by wire, an incredibly dynamic market at the turn of the century, became a stagnant, oppressive industry under decades of AT&T rule. The sector began to resemble a small-scale version of the planned economies of the Soviet Union. We like to imagine that in the United States, we've never had such a manifestly socialist industrial regime. But of course we have, only with ultimate power in the hands of regulated monopolists in partnership with government planners. The Separations Principle protects entrepreneurial freedom by preventing stagnation and repression of business innovation, especially repression abetted by the state. It also promotes vitality and innovation in different parts of the information economy by preventing one layer from smothering the others.

There was, as I've allowed, much to admire about the internal efficiency of AT&T, particularly in its early incarnations, and the achievements of the Bell Laboratories cannot be doubted. But this does not negate the pernicious effects of Bell's having gained control over too many layers of the industry and having blocked every way forward inconsistent with its consolidated vision of progress. Everything is a matter of degree. Had the monopoly limited itself, say, to local telephony, the trade-off between quality and innovation might have been far more tolerable. But it became a menace when it sought to control every single aspect of "the System"—all handsets, long distance, data communications—ultimately making itself the gatekeeper for all innovation. As a consequence, inventions from magnetic recording and electronic television to packet networking and fiber optics, developments feasible long before the moment with which they are associated, were squelched. The consequences of such action for economic growth and further innovation are incalculable: imagine trying to determine the effect on GDP growth if the broad rollout of email had been delayed for ten years to suit one company.

This brings me to the inadequacy of traditional efficiency calculations in regulating information industries. As we have observed, it has been the habit of the Justice Department to identify failures of competition by their effect on prices. In practice, however, not all dangerous arrangements inflate prices. The Edison Trust, one will remember, kept prices low by preventing more sophisticated product development. AT&T reaped handsome rewards by undercutting its competition with lower prices. The real problem with AT&T was in fact evident only after the government took decisive action to break up the telephone monopoly: as wave after wave of new services crashed on the market, beginning with voice mail and ending with the Internet, it became clear how drastically the Bell system had retarded progress. And when the government did take its long-deferred action, the suit was triggered not by objective calculation of malfeasance but by Bell's increasing arrogance. With no objective or automatic standard of response to anticompetitive behavior, the application of the Sherman Act, a relatively rare and extreme step, is largely discretionary, unlike most responses to the violation of federal law. A Separations regime would take much of the guesswork and impressionism, and indeed the influence trafficking, out of the oversight of the information industries.

Finally, the third justification of the Separation Principle derives from an awareness of the historical role of government in the information industries. That awareness leads to an inescapable conclusion that what is sauce for industry should be sauce for the state as well.

Again and again in the histories I have recounted, the state has shown itself an inferior arbiter of what is good for the information industries. The federal government's role in radio and television from the 1920s through the 1960s, for instance, was nothing short of a disgrace. In the service of chain broadcasting, it wrecked a vibrant, decentralized AM marketplace. At the behest of the ascendant radio industry, it blocked the arrival and prospects of FM radio, and then it put the brakes on television, reserving it for the NBC-CBS duopoly. Finally, from the 1950s through the 1960s, it did everything in its power to prevent cable television from challenging the primacy of the networks.

It isn't merely that government has been slow to act against the bad; it has quite often misconstrued the good. Time and again it has stood beside concentrated power against the underdog at the expense of economic dynamism. Government's tendency to protect large market players amounts to an illegitimate complicity, whether by reason of the firm's involvement in important government concerns (such as AT&T's work with the NSA) or a general sense of obligation to protect big industries irrespective of their having become uncompetitive.

Most of the federal government's intrusions in the twentieth century were efforts at preventing disruption by new technologies in order to usher in a future more orderly, less chaotic. That might sound like a sensible objective, but the effort can easily be perverted into serving special interests. The simplest expression of the Separations Principle as it relates to the state is that government's only proper role is as a check on private power, never as an aid to it. To grant any dominant industrial actor the protection of the state, for whatever reason, is to arrest the Schumpeterian dynamic by which innovation leads to growth, an outcome that is ultimately never in the public interest.

The Separation Principles I've called for require a certain breadth and ambition in its application. I've described it as more a constitutional than a regulatory framework, the former sort generally understood as being implemented by multiple institutions, including those restrained by it.* The norms found in the U.S. Constitution work not because

---

*It is critical to understand that I do not mean a constitutional principle in the formal sense, that is, an amendment to the U.S. Constitution. Rather, what I mean by "constitutional principle" is a norm taken as axiomatic or generally accepted to such an extent that to the degree it regulates, the regulation is a matter of self-regulation.

the Supreme Court, the system's final arbiter, is inherently, let alone supremely, powerful; in fact the Court can do nothing but opine. Rather, the system works because the president, the armed forces, and Congress swear fealty to it and the way the Court interprets it, generally observing constitutional principles. It is on such consensus that the Separations Principle depends, a sort of informal compact between the people and their government, an acceptance on the part of the three estates of government as well as the governed—that is, the information industry, and most of all, the people.

Let us talk about each party in turn, starting with government and the Federal Communications Commission, which has day-to-day authority over the information industries, the duty to specify the basic rules under which they operate. The commission's birth was ignoble, and in recent memory its abolition has been called for by those across the political spectrum (including Peter Huber of the Manhattan Institute and Larry Lessig of Harvard University). And yet whatever its beginnings, from the 1970s through the 1990s, it effected some extremely successful policy, some of it arguably a prototype for just the sort of dispensation I am recommending.

It was through the FCC's power that the Nixon administration implemented the first and still the most fundamentally important extant separation: that between carriage and services. It took the form of the FCC's separating AT&T's phone system from all the new services that had begun to operate on that network, from computer networking, through the Internet. The commission's second great separation parted the phone networks from the equipment that attached to them, thereby creating a market not only for telephones but also for answering machines, faxes, and modems. The work continued in this vein in the 1990s, when, under President Clinton, Chairman Reed Hundt protected thousands of new Internet Service Providers from being bled to death by the cash demands of the Bell companies.

It is worth noting that the divergent political dispositions of the Nixon and Clinton administrations were of no matter in the course of this progress. In this realm, both subscribed the same essential principles: that a free market would foster economic growth, and that government's only proper role in the market was to ensure opportunity, not to favor entrenched interests. The subsequent history speaks for itself. True, phenomena like the infotel revolution of the late twentieth century are complex macroeconomic events, and this one resulted from an incalculable combination of many factors, from the "peace dividend" created by the end of the Cold War to certain technological advances. But it would be impossible not to count among the foremost what was, in effect, the FCC's extensive pilot program for a Separations regime— a use of federal regulatory power not to limit freedom (as it is popularly believed to do) but to promote it. It is, in other words, a case of regulation achieving the good we commonly ascribe to deregulation.

There is a persistent misconception in the annals of American information industries that the radical transformation of the sector beginning with the rise of the computing Internet in the 1970s and continuing to pulse ever since was essentially owing to a return of laissez-faire, a purer free-market capitalism that had fallen out of favor after the Great Depression and was slow to regain its natural place. If the stories in this book tell us anything, however, it is that the free market can also lead to situations of reduced freedom. Markets are born free, yet no sooner are they born than some would-be emperor is forging chains. Paradoxically, it sometimes happens that the only way to preserve freedom is through judicious controls on the exercise of private power. If we believe in liberty, it must be freedom from both private and public coercion.

What was understood in the 1970s, and what needs to be understood again, is the role of such restrictions in preserving both the free market of goods and services and the free market of ideas. While the idea of regulation as a safeguard of freedom in any sense has come to seem incomprehensible—at least in the politics of sound bites—it is an idea perfectly at home in any serious understanding of the nature of law and of government. What is the First Amendment, or the Fourth, if not law that restricts power for the protection and promotion of freedom? The controls on private power to protect individual freedom are no different. Whether the state restricts a corporation from dumping toxic waste in a river or toxic assets in the financial markets, would one more reasonably regard this as an abridgment of freedom of the malefactors, or a protection of the freedom of individuals and businesses that would be adversely affected?

The implementation of a working Separations Policy, then, falls in the first instance to the Federal Communications Commission, where it finds its expression mainly in two classes of regulations. The first class comprises antidiscriminatory or common carriage rules, the ancient laws meant to govern how firms that operate or own essential infrastructures treat those who use those infrastructures. As we've seen, since antiquity, certain functions have been recognized by the state as being essential to the economy and commerce and therefore necessarily subject to nondiscriminatory policies. For such firms, also described as "public callings," freedom and opportunity for profit come with responsibility as well. In the American information industries, such duties were first imputed to the telegraph and telephone companies by the Taft administration in 1910; once it is recognized that a network has passed from a novelty to a necessity, the ancient justifications for common carriage reappear, even if under different names.

Such as "net neutrality," a concept I have espoused in other contexts, which is essentially the application of the idea of common carriage to a twenty-first-century industry. By this specific nomenclature I mean to add only a somewhat more specific understanding of how information networks function, as compared to, say, the operation of the only ferry to the mainland. Discrimination can take various forms when a network traffics in information packets. For while the boatman may fail in his obligations by refusing you passage or charging you more than the next passenger, the keeper of an information network may also speed up or slow down your transmission, or give right of way to one over another stream of traffic, among other manipulations. The Internet's nature affords many options, but whatever may be the justification, a vibrant information economy cannot countenance discrimination at a level so basic as transmission on a public network. If the carrier is determined to capture greater profits, the carrier ought to be obliged to do so by expanding his capacity, not by charging similar parties different prices, bestowing on the favored a competitive advantage.

The second essential component of a Separations Policy concerns industrial structure, which, I have argued, is the ultimate determinant of the scope of expressive freedom in our time. Here, the priorities must be both the prevention and dissolution of large-scale vertical mergers in the communications industry, a stricture perfectly within the commission's legal authority to impose. Under such a rule, a merger of Comcast, the emerging broadband monopolist for much of the nation, with NBC or Disney—a combination obviously resulting in the sort of conflicts of interest a Separations Principle is meant to prevent—would simply be out of the question; it would thus not be subject to the customary gaming of the commission's approval process whereby applicants offer marginal concessions in exchange for extravagant license. It is a rule the FCC can and should effect without delay. The histories we have examined make clear enough their power to do so, and also the unsavory consequences of allowing the creators of content to be conjoined with its disseminators.

Despite some good work done by the FCC that I have acknowledged, entrusting to that one agency total responsibility for the nation's information policy would be a serious mistake. The FCC is inevitably close to the action—sometimes too close to be perfectly impartial, and always in danger of capture in ways obvious and subtle. History shows what problems can result. Despite some finer moments, the agency has on occasion let itself become the enemy of the good, effectively a tool of repression. And so what is needed is not only an FCC institutionally committed to a Separations Principle but also a structural arrangement to guard against such deviations, including congressional oversight as well as attention and corrections from other branches of government.

Here is where antitrust law becomes so important to communications policy. It is inevitable that the FCC will occasionally fail in its mission, and for this reason the government's competition authorities, the Justice Department's Antitrust Division and the Federal Trade Commission, are necessary as a backup. Notwithstanding my earlier criticism of the antitrust system's narrowness of focus, the DOJ and FTC have a vital role to play generally, and particularly in one pernicious situation: when a private power has become so closely affiliated with government that only the government can take action against it. We should at least be able to depend on antitrust as a last safeguard against the FCC's lapses.

As things stand, the American antitrust regime, unlike its European counterpart, is virtually dormant respecting the entertainment and communications industries. That's not necessarily a bad thing, for these are fast-moving industries and Sherman is a slow-moving law. And no one would deny that the awesome power of the law should be used sparingly, working more as a deterrent than a ready remedy. Nevertheless, its application must be a far more credible eventuality in those relatively rare instances when an industry has manifestly defeated normal efforts to place reasonable constraints on it, and specifically whenever it has somehow managed to circumvent the FCC. To fulfill such a mandate, the antitrust law must be responsive to its own discrete criteria, not deferring to the FCC's oversight of the industry, which of course is not inerrant. An antitrust law preempted by FCC discretion is no safeguard at all. And in a constitutional democracy we simply cannot do without such a line of defense. For once it is entrenched in our national life, particularly once an industry has virtually merged with the state, such a power can be dislodged in no other way.

Reasonably effective though I believe the FCC/Antitrust model can be, given the force of the Cycle it would hardly be prudent to rely on government institutions exclusively to ensure a durable compliance with a Separations Principle. But how else do we achieve such a goal?

It may sound improbable, if not hopelessly naïve, but one place to apply pressure is among members of the industry itself. If legal scholarship over the past few decades has proved anything, it is that we have little choice. The better part of compliance with rules of all sorts actually depends on the power of self-regulation, not the threat of force, though of course that threat can help. Both church and state (or at least individual politicians) may occasionally feel motivated to push the boundaries of their coexistence, but overall both institutions tend to accept the wisdom of the divide between them, which is why it works. The consent of the governed is not strictly necessary, but it helps.

Likewise the information industries, whatever their actions may suggest to the contrary, are much closer to an acceptance of the Separations ideal, at least in theory, than one might imagine. It should be recognized that there are uncodified norms governing the behavior of infotel firms in the twenty-first century, ones that did not exist decades ago, such as the norms that stigmatize site blocking, content discrimination, and censorship, broadly defined. Consider that when phone or cable companies have been accused of blocking an Internet site, their tendency has been to deny it, or to blame a low-ranking official, rather than to baldly defend a right to block or censor, as for instance the Edison Trust once did. While not always resulting in a practical difference, the change nonetheless suggests that such behavior has become *malum in se*. And the consensus to this effect is a powerful force.

Consider all the mischief that the information firms could undertake but choose to eschew. Cable operators, though not obliged by law to do so, generally carry channels that a cruder calculus would motivate them to block. Likewise, Apple, the maker of the iPhone, has been, in effect, shamed into allowing apps, such as Skype or Line2, that compete with its own services. Meanwhile Verizon, a born-and-bred Baby Bell, gains public applause by publicly declaring itself an "open" company. And Google, one of the great corporate hegemons of our time, does likewise under its banner "Don't Be Evil." Whatever its missteps and shortcomings, that firm has, so far, done more than any other to promote what we have been describing as a constitutional policy of separations for the information industry. And while the extent of Google's commitment has been exceptional, the basic impulse is not.

In fact, rare is the firm willing to assert an intention and a right to dominate layers of the information industry beyond its core business, an ambition that someone like Theodore Vail, Adolph Zukor, or David Sarnoff would have proclaimed with unabashed glee. Now of course, some of these high-minded professions we hear might be cynical constructs of public relations departments, a scrim behind which a company can hatch some diabolical master plan. But I find little evidence of that level of conspiracy. (Generally, the Cycle moves in a manner more akin to the classical invisible hand than to Strangelovian machinations.) However insincerely embraced, corporate norms have in many ways proved to be a far more powerful deterrent to misconduct than regulations, which in corporate psychology exist only to be circumvented, preferably though not necessarily by legal means. Certainly this is so for the financial services sector. Perhaps in this way, too, information industries show their exceptionalism.

Anyone who would discount the power of such norms might care to know what a world without them would look like. For that, one need search no farther than China, where blocking, discrimination, and censorship of the Internet are perfectly routine and in no way stigmatized. What is so striking (although, I would argue, not surprising) is that the vast preponderance of Chinese censorship is private, undertaken voluntarily, rather than enforced by actions of the state. In that society, it is as if the American commandment "Thou shalt not block" has had a minus sign placed before it; there exists a diametrical inversion of our norm, an orientation influenced, to be sure, as ours is, by codes both legal and extralegal, but to the opposite effect.

Many would consider it simply a foolish denial of capitalism's red tooth and claw to look for virtue among corporate titans. But as our narrative has also shown, the urge to dominate is never one of simple greed or the warped megalomania of a James Bond villain. It is in fact heartening to discover in the history of the information industries a recurring strain of idealism, even if it occasionally comes to unwholesome fruition. The motivations of information moguls can almost never be exhaustively described in terms of simple greed and vanity. Were it otherwise, we as a culture might be irretrievably lost. For ultimately, no matter how many regulatory fetters we may succeed in placing on them, the men and women who run the information empires of today and tomorrow will inevitably have enormous power over the extent of our free expression. Their values will always be the first line of defense, but so, too, will their vices be the most immediate source of public outcry. Whatever external system of controls might be created, there is no substitute for self-control. Put another way, we have hardly managed to improve on the Roman conception of virtue in governance.

If, as I've suggested, corporate norms can provide a critical basis for self-regulation, the question naturally arises: Where, exactly, do such norms come from? The answer is, quite simply: From the general sentiment, the popular sense of right and wrong. And so this is where the ordinary citizen becomes involved in the Separations cause. I'm not suggesting that every American need become an avid follower of FCC proceedings. But the population's general "information morality," so to speak, is decisively important. In any industry, corporate behavior that strikes most people as wrong can bring a great cost to the perpetrator. But information commerce, as we have repeatedly observed, is more entangled with daily life than any other sort of commercial enterprise. Even the misdeeds of an industry as vitally important as health insurance do not have the potential to provoke such instantaneous reaction as the blocking or impeding of network traffic. The ever-growing wired majority is a particularly vociferous one, quick to adopt and exploit every new application for self-expression.

Can we really depend to any degree on a popular groundswell to accomplish anything significant? In fact, every existing principle of separation, every effective limit on power in the American system, manages to be upheld precisely because of a broad consensus actively favoring it. Laws may continue on the books indefinitely, even after falling into desuetude. But the laws that continue to bite are those for which there continues to be a strong consensus. Thus, for instance, years before the Supreme Court struck down antisodomy statutes in the fifty states, prosecution of violators was already rare in nearly all jurisdictions because of a lack of strong consensus. Democracy expresses itself not only in the erection of walls but in the enforcement of prohibitions.

In this way, a successful Separations regime ultimately depends less on the enactment of useful laws—although we need these, too—as on the cultivation of a popular ethic concerning our society's relation to information, an ethic consistent with the importance of information in our individual and collective lives. A strong general conviction that it is wrong to block sites on the Internet, wrong for studios to censor films that deal with controversial problems, can do more to secure our freedom than an army of regulators. For the Cycle may have enormous force. Those who lobby government on behalf of the information industries may be legion. But in a society such as ours, they should be fairly matched by a generally elevated awareness of the imminent perils of a closed system.

THIS TIME IS DIFFERENT

Here at the very end it behooves us to return to the two questions posed at the beginning of this book. First, why should you care? Second, is the Internet different?—or, put another way, have we seen the last of the Cycle? The two questions are in fact intertwined, and to answer the second is to leave little doubt about the first.

Notwithstanding what may seem the slow, progressive realization of information dystopias à la Aldous Huxley, the outright repressions of speech, of innovation, and of entire industries might seem a relic of the twentieth century and its totalitarianisms, nothing to fear in our day and age. It cannot be denied that the Internet has ushered in a time of unprecedented diversity and ease of communication and commerce, a broadly available way of reaching millions of people. And each of those millions of networked parties can in turn claim the role of what was once called, with appropriate distinction, a "broadcaster." Beyond the Universal Network, cable television carries hundreds of channels, our mobile phones exceed the communicators of *Star Trek* in functionality, and even mature industries such as print journalism and book publishing have sought renewal in opening up to an unprecedented variety of voices, sensibilities, and forms. While the decline of many once proud industries is cause for real sorrow, we do, I think, live in what is in some ways an informational golden age. Television, the Internet, film, and mobile devices each force one another to become better. The breathtaking diversity of content in our age has actually engendered in us an anxiety perversely contrary to the one that plagued our ancestors: it is not that there's too little produced to meet demand, but that there's way too much to sustain all our would-be writers, reporters, and thinkers in a world of content so cheap and abundant.

Yet if we generally like the way things are now, we must also ask whether our current situation is really so different from the open ages of radio, film, or the telephone. Might it not have also seemed in those times that the orgy of limitless entrepreneurism would never end? The point is that we are near the high end of a pendulum arc that, so far, has always begun to swing in the opposite direction—toward greater integration and centralization—with a force that can seem inexorable. So let us evaluate the basis for suggesting that "this time is different."

The cornerstone of this view is that with the coming of the Internet, we have been, at least as makers and consumers of information, "saved"; now, as with the Resurrection, things can never be the same again. The Internet inaugurated a principle so fundamental and powerful that it cannot be abolished; ever after, all will agree that open beats closed. It is an attractive notion; but in fact it is an article of faith in a domain of experience where fact, not faith, should guide us. It is true that the Internet naturally harnesses the power of decentralization and defies central control, but in the face of a determined power, that design alone is no adequate defense of what we hold most dear about the network.

The simple fact is that the Internet is simply not the infinitely elastic phantasm that it is popularly imagined to be, but rather an actual physical entity that can be warped or broken. For while the network is designed to connect every user with every other on an equal footing, it has always depended on a finite number of physical connections, whether wired or spectral, and switches, operated by a finite number of firms upon whose good behavior the whole thing depends.

There is a dark underbelly to the diversity of content and services that the Internet has brought us, one that leaves it more vulnerable to centralization, not less. The Internet with its uniquely open design has led to a moment when all other information networks have converged upon it as the one "superhighway," to use the 1990s term. While there were once distinct channels of telephony, television, radio, and film, all information forms are now destined to make their way increasingly along the master network that can support virtually any kind of data traffic. This tendency, once called "convergence," was universally thought a good thing, but its dangers have now revealed themselves as well. With every sort of political, social, cultural, and economic transaction having to one degree or another now gone digital, this proposes an awesome dependence on a single network, and a no less vital need to preserve its openness from imperial designs.

Where might the next domineering empire come from? It is impossible to predict, though history offers some good guesses. It could arise from a takeover of content by the great carriers of our time, a future whose harbinger might be the takeover of NBC-Universal by Comcast, an even vaster effort to realize what AOL Time Warner failed to be. It might arrive through some further melding of Hollywood with AT&T in the devices marketed by Apple and friends. Or it could begin on the day that mighty Google, still the greatest corporate champion of openness, decides that its survival has come to depend on integration and the elimination of whatever competition it has. Whatever the source, the prospect of a new imperial age, even if only partially visible now, seems to me as likely as it ever has been at this point in the Cycle. This time *is* different: with everything on one network, the potential power to control is so much greater.

It is also possible that we could undergo such a consolidation blissfully unaware. Dazzled by ever newer toys, faster connections, sharper graphics, and more ingenious applications, we might be sufficiently distracted from the consequences of centralized control. After all, many still recall living perfectly productive and contented lives in the age of Hollywood's Production Code or the years when a long distance phone call was an expenditure to give one pause. With systems and industrial orders changing faster and faster, however, and with virtually everyone nowadays—not only hobbyists as in days past—enjoying an astonishing variety of venues for self-expression and entrepreneurship, it is difficult to imagine a new order not coming as a very rude awakening.

There is no escaping the reality that we have evolved into a society in which electronic information represents the substrate of much of daily life. It is a natural outcome of our having advanced past the mechanical age. And just as our addiction to the benefits of the internal combustion engine led us to such demand for fossil fuels as we could no longer support, so, too, has our dependence on our mobile smart phones, touchpads, laptops, and other devices delivered us to a moment when our demand for bandwidth—the new black gold—is insatiable. Let us, then, not fail to protect ourselves from the will of those who might seek domination of those resources we cannot do without. If we do not take this moment to secure our sovereignty over the choices that our information age has allowed us to enjoy, we cannot reasonably blame its loss on those who are free to enrich themselves by taking it from us in a manner history has foretold.

Acknowledgments

Many people helped me with this book. George Andreou is the best editor I have ever worked with and a fine prose stylist. Tina Bennett, my literary agent, understands writers better than they understand themselves. My dean David Schizer's support made this book possible, and I thank the entire faculty of Columbia Law School for their support and tolerance. I thank the editors at *Slate* magazine, particularly Jacob Weisberg, Dahlia Lithwick and Josh Levin, for giving me room to try out many of the ideas that went into this book.

Research assistants at Columbia Law School and the New America Foundation provided indispensable help with this book. Their ranks initially included Hailey DeKraker, the lead research assistant, Alex Middleton, who dug out the Hush-A-Phone hearings, and Luis Villa. Later help came from Anna-Marie Anderson, Kendra Marvel, and Judd Schlossberg, who provided research rescue at a critical hour. Faith Smith at New America and her team of researchers found things I wouldn't have thought existed, and I also thank the UCLA library, site of the Hodkinson papers. The multitalented, long-suffering Stuart Sierra did the diagrams. I also thank the reference librarians at Columbia Law School, who provided timely access to everything, the library staff at Stanford Law School, and Lily Evans at Knopf.

Kathryn Tucker made important early suggestions and helped crystal-lize the idea of a Cycle at the center of the book. Scott Hemphill gave me especially useful comments, twice, and constant feedback on economic questions; other helpful assistance, ideas, and feedback came from Larry Lessig, Chris Libertelli, Charles Sabel, Derek Slater, Michael Heller, Andrew McLaughlin, Jennifer 8. Lee, Siva Vaidhyanathan, Hal Edgar, Diana Sanchez, Robert Wright, Richard Posner, Judith Judge, David Wu, and Louis Wolcher. I also feel deeply indebted to a series of authors, some of whom I have never met, who have written particularly helpful histories of communications and the media, including Paul Starr, Katie Hafner, Matthew Lyon, Milton Mueller, Connie Brooks, Lawrence Lessing, Thomas White, Ken Auletta, Herbert N. Casson, and many others. I presented early versions of this book at the New America Foundation, Columbia Law School, the University of Washington, the Stanford Communications Department, the Institute of International and European Affairs in Dublin, and the West Virginia School of Law.

Finally, I wish to thank my family, particularly my mother, who broke our budget to buy an Apple II+ in 1982 and thereby started this book, and my in-laws, who helped me finish it. And finally thanks to Kate Judge, who helped straighten out the logic and waited patiently every time I embarked on those long train rides meant to make this book happen.

Notes

Introduction

1. The description of the banquet is in "Voice Voyages by the National Geographic Society: A Tribute to the Geographical Achievements of the Telephone," *National Geographic* XXIX (March 1916): 296–326. Another account can be found in Albert Bigelow Paine's biography, *Theodore N. Vail: A Biography* (New York: Harper & Brothers, 1921).

2. Alan Stone, *How America Got Online* (New York: M. E. Sharpe, 1997), 27; *Annual Report of the American Telephone and Telegraph for 1910* (New York, 1911), 34; Albert Bigelow Paine, *In One Man's Life: Being Chapters from the Personal and Business Career of Theodore N. Vail* (New York: Harper & Bros., 1921), 213–14; and Allan L. Benson, "The Wonderful New World Ahead of Us," *The Cosmopolitan* (February 1911): 294, 302.

3. Nikola Tesla, "The Transmission of Electrical Energy Without Wires," *Electrical World and Engineer*, 1904. D. W. Griffith is quoted in Richard Dyer MacCann, *The First Film Makers* (Metuchen, NJ: Scarecrow Press, 1989), 5. Sloan Foundation, *On the Cable* (1971). Tom Stoppard's character Jackson makes this remark and then exclaims, "Electricity is going to change everything! Everything!" Tom Stoppard, *The Invention of Love* (New York: Grove Press, 1998), 53.

4. Authors Frank Manuel and Fritzie Manuel discuss the age of the Utopia Victoriana in *Utopian Thought in the Western World* (Cambridge MA: Belknap Press, 1979), 759. The other great influence on Vail's time was Frederick Taylor—his theories of scientific management and the concept of the "one right way." The classic is Frederick W. Taylor, *The Principles of Scientific Management* (New York: Harper & Bros., 1911); see, generally, Bernard Doray, *From Taylorism to Fordism: A Rational Madness* (London: Free Association Books, 1988).

5. See *Annual Report of the American Telephone and Telegraph for 1910*, 36; and "Public Utilities and Public Policy," *Atlantic Monthly* (1913): 309. These articles are reprinted in Theodore N. Vail, *Views on Public Questions: A Collection of Papers and Addresses of Theodore Newton Vail* (privately printed, 1917), 111.

6. *Annual Report of the American Telephone and Telegraph for 1910*, 36.

7. Ibid.; Henry Ford with Samuel Crowther, *My Life and Work* (Garden City, NY: Garden City Publishing Company, 1922), 2.

8. Aldous Huxley's diary of his long journey through several countries and his experience

reading Henry Ford's autobiography along the way is in Aldous Huxley, *Jesting Pilate: The Diary of a Journey* (London: Chatto and Windus, 1930).

9. These observations on human equality and the social order may be found in Henry Ford's autobiography, *My Life and Work* (New York: Garden City Publishing Co., 1922), 10, 3.

10. Huxley's early impressions of American culture and the revolutionary changes being wrought by advances in communications technology may be read in Aldous Huxley, "The Outlook for American Culture: Some Reflections in a Machine Age," *Harper's Magazine,* August 1927.

11. Joseph Goebbels, "Der Rundfunk als achte Großmacht," *Signale der neuen Zeit. 25 ausgewählte Reden von Dr. Joseph Goebbels* (Munich: Zentralverlag der NSDAP, 1938), 197–207. On the Lucy ratings, see Gary Edgerton, *The Columbia History of American Television* (New York: Columbia University Press, 2007), 134.

CHAPTER 1: THE DISRUPTIVE FOUNDER

1. There are, unsurprisingly, many useful histories of the Bell Company and AT&T. For the pre-1984 history, see Herbert Newton Casson, *The History of the Telephone* (Chicago: A. C. McClurg, 1910), 24–25; N. R. Danielian, *AT&T: The Story of Industrial Conquest* (New York: Vanguard, 1939); Arthur Page, *The Bell Telephone System* (New York: Harper & Brothers, 1941); Horace Coon, *American Tel & Tel: The Story of a Great Monopoly* (New York: Books for Libraries Press, 1939); Sonny Kleinfeld, *The Biggest Company on Earth: A Profile of AT&T* (New York: Holt, Rinehart, and Winston, 1981); John Brooks, *Telephone: The First One Hundred Years* (New York: Harper & Row, 1976).

2. William W. Fisher III, "The Growth of Intellectual Property: A History of the Ownership of Ideas in the United States," in *Intellectual Property Rights: Critical Concepts in Law,* vol. I, 83, David Vaver ed., (New York: Routledge, 2006).

3. The controversy over the invention of the telephone has engendered a small industry, including four volumes written in the twenty-first century. It begins with "How Gray Was Cheated," *New York Times,* May 22, 1886; see also A. Edward Evenson, *The Telephone Patent Conspiracy of 1876: The Elisha Gray–Alexander Bell Controversy* (Jefferson, NC: McFarland, 2000); Burton H. Baker, *The Gray Matter: The Forgotten Story of the Telephone* (St. Joseph, MI: Telepress, 2000); Seth Shulman, *The Telephone Gambit* (New York: W. W. Norton, 2008); Tony Rothman, *Everything's Relative* (Hoboken, NJ: Wiley, 2003). For the full Drawbaugh case, see *Dolbear v. American Bell Tel. Co.,* 126 U.S. 1 (1888).

4. Malcolm Gladwell, "In the Air," *New Yorker,* May 12, 2008. Gladwell explores these themes further in *Outliers: The Story of Success* (New York; Little, Brown, 2008).

5. Clayton M. Christensen, *The Inventor's Dilemma: The Revolutionary Book That Will Change the Way You Do Business.*

6. Casson, *History of the Telephone,* 24–25.

7. See Evenson, *Telephone Patent Conspiracy of 1876,* 65.

8. Joseph A. Schumpeter, *The Theory of Economic Development: An Inquiry into Profits, Capital, Credit, Interest, and the Business Cycle* (New Brunswick, NJ: Transaction Publishers, 1983), 84, 86.

9. One account of this first successful telephonic communication is found in Charlotte Gray, *Reluctant Genius: Alexander Graham Bell and the Passion for Invention* (New York: Arcade, 2006), 123–24.

10. A history of the Associated Press relationship with Western Union and the influence of wire-communicated news may be found in Menahem Blondheim, *News Over the Wires: The Telegraph and the Flow of Public Information in America, 1844–1897* (Cambridge, MA: Harvard University Press, 1994).

11. The text of this advertisement is in Brooks, *Telephone*, 60.

12. According to some accounts, quite possibly apocryphal, Orton simply chuckled when offered the Bell patents, asking pleasantly, "What use could this company make of an electrical toy?" This exchange appears in Casson, *History of the Telephone*, 58–59.

13. During this period, it appears that "Western Union's strangle hold [on the industry] began to tighten into a death grip," as described by historian John Brooks in *Telephone*, 70.

14. Jonathan Zittrain, *The Future of the Internet—And How to Stop It* (New Haven: Yale University Press, 2008), 106.

15. Bell's depression and subsequent hospitalization are recorded in Casson, *History of the Telephone*, 74.

16. Schumpeter's ideas are central to this book. Two of his works are particularly important for this work: *The Theory of Economic Development*; and *Capitalism, Socialism, and Democracy* (New York: Routledge, 2006) (1942). For more about his work see Robert Loring Allen, *Opening Doors: The Life and Work of Joseph Schumpeter, Volume One— Europe* (New Brunswick, NJ: Transaction Publishers, 1991); on his up-and-down life, see Richard Swedberg's *Schumpeter: A Biography* (Princeton, NJ: Princeton University Press, 1991); Thomas K. McCraw, *Prophet of Innovation: Joseph Schumpeter and Creative Destruction* (Cambridge, MA: Belknap Press, 2007).

17. Albert Bigelow Paine, *In One Man's Life: Being Chapters from the Personal & Business Career of Theodore N. Vail* (New York: Harper & Bros., 1921) 114.

18. Schumpeter, *Theory of Economic Development*, 93; Paine, *In One Man's Life*, 27.

19. This letter is reprinted in Casson, *History of the Telephone*, 67.

20. Brooks, Coon, and Casson all give full accounts of the agreement between Western Union and Bell.

21. Coon, *American Tel & Tel*, 41.

CHAPTER 2: RADIO DREAMS

1. An account of this boxing match at Jersey City, the first mass sportscast event, can be found in "Voice Broadcasting the Stirring Progress of the 'Battle of the Century,'" *The Wireless Age*, August 1921, 11–21. *The New York Times* also covered the event, in the article "Wireless Telephone Spreads Fight News Over 120,000 Miles," *New York Times*, July 3, 1921, 6.

2. For a description and photographs of Jack Dempsey, the reigning heavyweight champion in 1921, see Randy Roberts, *Jack Dempsey, the Manassa Mauler* (Baton Rouge: Louisiana State University Press, 1979).

3. The Radio Corporation of America and David Sarnoff are key characters in this book—although famously historians have been hoodwinked by Sarnoff's misrepresentations, so read with care. On the man and the firm, see Gleason Archer, *Big Business and Radio* (New York: American Historical Company, 1939); Tom Lewis, *Empire of the Air: The Men Who Made Radio* (New York: HarperCollins, 1991); and Kenneth Bilby, *The General: David Sarnoff and the Rise of the Communications Industry* (New York: Harper & Row, 1986). Descriptions of Sarnoff's life and work in his own words—a notably unreliable source—can be found in *Looking Ahead: The Papers of David Sarnoff* (New York: McGraw-Hill Book Co., 1968).

4. *Webster's Revised Unabridged Dictionary* (G. & C. Merriam Co., 1913), available online at http://machaut.uchicago.edu/websters.

5. A. Frederick Collins, *The Book of Wireless* (New York: D. Appleton and Company, 1916).

6. A description of Lee De Forest's radio station in the Bronx can be found in Brian Regal, *Radio: The Life Story of a Technology* (Westport, CT: Greenwood Press, 2005), 59–60.

7. QST Magazine continues to be the publication of the American Radio Relay League. The magazine's estimate is in "De Forest Wireless Telephone," *QST Magazine,* April 1917, 72.

8. See "Voice Broadcasting the Stirring Progress of the 'Battle of the Century,' " *Wireless Age,* August 1921, 11.

9. Todd Lappin discusses radio's dramatic rise in popularity in the early 1920s, which he compares to the personal computer's rise in the 1980s, in "Déjà Vu All Over Again," *Wired,* May 1995.

10. Lee De Forest's encouragements are in *How to Set Up an Amateur Radio Receiving Station* (New York: De Forest Radio Telephone and Telegraph Company, 1920), 2–7.

11. Waldemar Kaempffert, "Signalling and Talking by Radio," *Modern Wonder Workers: A Popular History of American Invention* (New York: Blue Ribbon Books, 1924), 351, 378.

12. Todd Lappin quotes this *Radio Broadcast* column in "Déjà Vu All Over Again," *Wired,* May 1995.

13. Alfred Goldsmith spoke about the potential cultural benefits of the radio in an interview with Edgar Felix, "Dr. Alfred N. Goldsmith on the Future of Radio Telephony," *Radio Broadcast,* May 1922, 42, 45.

14. Mark Caspar, "Radio Broadcasting," *Radio Dealer,* June 1922, 42–45. William Boddy references this article in *New Media and Popular Imagination: Launching Radio, Television, and Digital Media in the United States* (New York: Oxford University Press, 2004).

15. Lee De Forest recommends radio as a hobby in *How to Set Up an Amateur Radio Receiving Station.*

16. An example of the radio station "lists" is Armstrong Perry, "What Anyone Can Hear: Complete List of Broadcasting Stations in U.S." *Radio News,* March 1922, 814. An

excellent source describing the role of the radio in the history of jazz music is Clifford Doerksen, *American Babel: Rogue Radio Broadcasters of the Jazz Age* (Philadelphia: University of Pennsylvania Press, 2005).

17. David Sarnoff's remark is quoted in Kenneth Bilby, *The General: David Sarnoff and the Rise of the Communications Industry* (New York: Harper & Row, 1986), 65.

18. John Reith's diaries were edited by Charles Stewart and published in *The Reith Diaries* (London: Collins, 1975). See also Asa Briggs, *The History of Broadcasting in the United Kingdom,* vol. I (Oxford: Oxford University Press, 1995), 126; Derek Parker, *Radio: The Great Years* (Devon, UK: David & Charles, 1977). For an inside look at the history of the BBC, see Arthur Richard Burrows, *The Story of Broadcasting* (London: Cassell and Co., 1924).

19. Gale Pedrick's description of Savoy Hill can be found in Briggs, *History of Broadcasting in the United Kingdom,* 193.

20. John Reith expressed his view of the radio in his own book, *Broadcast over Britain* (London: Hodder and Stoughton, 1924).

21. Walter Lippmann, *The Phantom Public* (New York: Macmillan Company, 1927).

22. Reith wrote this in an internal BBC memorandum in November 1925. It has become a rather well-known quote, and is discussed in Burton Paulu, *Television and Radio in the United Kingdom* (Minneapolis: University of Minnesota Press, 1981), 8.

23. This statement by Reith comes from Paulu, *Television and Radio in the United Kingdom,* 300.

24. Asa Briggs relates this anecdote about George Bernard Shaw in his *The BBC: The First Fifty Years* (Oxford: Oxford University Press, 1985).

25. Ibid., 54.

26. Reith's assessments of Churchill can be found in various entries in his diaries, which were later edited by Charles Stewart and published in *The Reith Diaries.* Reith's opinion of the government was also discussed in a *Time* magazine article four years after his death: "Britain: Lord Wrath," *Time,* October 6, 1975.

27. This sentiment by Reith's contemporary at BBC can be found in Briggs, *The BBC: The First Fifty Years,* 156.

28. Asa Briggs describes the language advisory committee in *History of Broadcasting in the United Kingdom,* 221–22.

29. Reith confessed this in Stewart, ed., *The Reith Diaries,* 68.

30. Reith explains his conception of BBC as a public utility in *Broadcast over Britain.*

31. Briggs, *The BBC: The First Fifty Years,* 49.

CHAPTER 3: MR. VAIL IS A BIG MAN

1. For an interesting discussion of Edmund Burch and the development of these early Independent telephone lines, see Michael L. Olsen, "But It Won't Milk the Cows: Farmers in Colfax County Debate the Merits of the Telephone," *New Mexico Historical Review* 61:1 (January 1986).

2. Scientific magazines and rural newspapers frequently discussed the mechanics of establishing local telephone lines. The two that inspired Edmund Burch, according to Michael L. Olsen's account cited above, were "A Cheap Telephone System for Farmers," *Scientific American,* 1900; and a piece in the *Rural New Yorker* from June 11, 1903, at 437, quoted in Ronald R. Kline, *Consumers in the Country: Technology and Social Change in Rural America* (Baltimore: Johns Hopkins University Press, 2000), 28.

3. One early description of the Independent telephone movement can be found in *The Independent Telephone Movement: Its Inception and Progress* (s.n., 1906).

4. The Independents described themselves in populist terms, as evidenced by this quote found in W. A. Taylor, "The Art of Cable Splicing," *Sound Waves,* January 1907, 61, 64. Their dedication to achieving American industrial independence is referenced in Henry A. Conrad, "An Ohio Company's Splendid Record," *Sound Waves,* March 1907, 107, 108.

5. The circumstances surrounding Vail's first departure from Bell can be found in Horace Coon, *American Tel & Tel: The Story of a Great Monopoly* (Freeport, NY: Books for Libraries Press, 1939), 66–67.

6. This description of the impact of the telephone on farm life was first published in *Raton Range,* March 10, 1904; quoted in Olsen, "But It Won't Milk the Cows," 1.

7. Ronald L. Klein describes the uses to which the early phones were put in *Consumers in the Country,* 43.

8. This campaign is detailed in Paul Latzke, *A Fight with an Octopus* (Chicago: Telephony, 1906), and sporadically in *Sound Waves,* infra; the reliability of these sources, of course, can be questioned.

9. This account can be found in Norton E. Long, "Public Relations Policies of the Bell System," *Public Opinion Quarterly,* vol. 1, no. 4, 5, 12.

10. This quote, along with fascinating Independent perspectives on the early era of telephony, can be found in *Sound Waves* vol. XIII, no. 1, December 1906.

11. *Sound Waves* vol. XIII, no. 4, March 1907.

12. Latzke, *Fight with an Octopus,* 12.

13. For the history of Vail's interaction with J. P. Morgan, as well as his return to Bell as president of the newly formed AT&T, see *In One Man's Life: Being Chapters from the Personal & Business Career of Theodore N. Vail* (Harper & Brothers, New York, 1921); Coon, *American Tel & Tel;* from Morgan's perspective, see Jean Strouse, *Morgan: American Financier* (New York: Random House, 1999), 563.

14. "Universal service" in Vail's mind did not mean serving every home in the country with a telephone, but rather was a slogan calling for the elimination of competition from dual service and the grand unification of telephony under AT&T's authority. An excellent description of the underlying impetus behind the adoption of this strategy may be found in Milton Mueller, *Universal Service: Competition, Interconnection, and Monopoly in the Making of the American Telephone System* (Cambridge, MA: MIT Press, 1997), 96.

15. This observation of the effect of price cutting on competition may be found in David Ames Wells, *Recent Economic Changes* (New York: D. Appleton and Co., 1889).

16. One excellent means of gauging Vail's industrial philosophy is by consulting records

of Vail's public communications during his lifetime. Most can be found in Theodore N. Vail, *Views on Public Questions: A Collection of Papers and Addresses of Theodore Newton Vail* (privately printed, 1917).

17. AT&T's takeover of Western Union is well documented in George P. Oslin, *The Story of Telecommunications* (Macon, GA; Mercer University Press, 1992), 262.

18. Vail's strategy to establish control over the Independents through carrotlike incentives is described in Mueller, *Universal Service*, 107.

19. William Doan, "Manager's Duty to the Public and to Himself," *Sound Waves* vol. XIII, no. 2, January 1907, 69.

20. The Mesa Telephone Company sellout to Bell is described in Olsen, "But It Won't Milk the Cows," 13.

21. The theory that J. P. Morgan secretly undermined the Traction Kings and the Telephone, Telegraph, and Cable Company of America is in Noobar Retheos Danielian, *A.T.&T.: The Story of Industrial Conquest* (New York: Vanguard Press, 1939), 47. Danielian himself was drawing on a three-volume FCC document entitled *Telephone Investigation: Special Investigation Docket, Report on Control of Telephone Communication, Control of Independent Companies* (1936–37).

22. The settlement is discussed in Mueller, *Universal Service*, 130.

23. The government reaction to the agreement is covered in Brooks, *Telephone*, 136.

24. Bork's opinion on the irrelevance of corporate intent can be read in Robert H. Bork, *The Antitrust Paradox* (New York: Basic Books, 1978), 38–39.

25. This quote is in Theodore Vail, "Some Observations on Modern Tendencies," *Educational Review* vol. 51, February 1916, 109, 129.

26. Mueller, *Universal Service*, 146.

CHAPTER 4: THE TIME IS NOT RIPE FOR FEATURE FILMS

1. For a more detailed account of this initial meeting (and an excellent history of the rise of the American film industry), see James Forshcr, *The Community of Cinema: How Cinema and Spectacle Transformed the American Downtown* (Westport, CT: Praeger Publishers, 2003), 30–32.

2. Two interesting histories of French cinema and its early industry dominance are Richard Abel, *The Ciné Goes to Town: French Cinema, 1896–1914* (Berkeley and Los Angeles: University of California Press), and W. Stephen Bush, "The Film in France," *Moving Picture World*, July 12, 1913.

3. This vivid description of the early theatergoing experience from *Moving Picture World* magazine may be found in Eileen Bowser, *The Transformation of Cinema* (New York: Maxwell Macmillan International, 1990), 3–4.

4. There are conflicting accounts of how much Zukor truly paid for the rights to *Queen Elizabeth;* some scholars place the figure at $18,000, while others, going by what Zukor claimed in his later years, estimate closer to $40,000—either amount a huge sum at

the time. See Forsher, *Community of Cinema,* 33, for one account. The quote regarding Zukor's intentions is in Anthony Slide, *Early American Cinema,* 2nd ed. (New York: Rowman & Littlefield, 1994), 60.

5. The exchange between Kennedy and Zukor is recounted in Forsher, *Community of Cinema,* 30–32.

6. Ibid., 32.

7. As related in Evan L. Schwartz, *The Last Lone Inventor: A Tale of Genius, Deceit, and the Birth of Television* (New York: HarperCollins, 2002), 132.

8. The history of Carl Laemmle's entry into the film industry, and this scene in particular, may be found in John Drinkwater, *The Life and Adventures of Carl Laemmle* (London: W. Heinemann, 1931, reprinted 1978), 63; see generally Neal Gabler, *An Empire of Their Own: How the Jews Invented Hollywood* (New York: Crown Publishers, 1988).

9. Drinkwater, *Carl Laemmle,* 64.

10. Ibid., 65.

11. Laemmle declared his film company "independent" in *The Sunday Telegram,* April 18, 1909, as described in Drinkwater, *Carl Laemmle,* 67.

12. Drinkwater, *Carl Laemmle,* 69–70.

13. This observation is drawn from Upton Sinclair, *Upton Sinclair Presents William Fox* (Los Angeles: self-published, 1933), 39.

14. One excellent history of the Warner brothers in Hollywood can be read in Cass Warner Sperling, Cork Millner, and Jack Warner, *Hollywood Be Thy Name: The Warner Brothers Story* 2nd ed. (Lexington: University Press of Kentucky, 1998).

15. The international alliance to break the film trust is described in Rosalie Schwartz, *Flying Down to Rio: Hollywood, Tourists, and Yankee Clippers* (College Station: Texas A&M University Press, 2004), 163.

16. For a description of Hodkinson's role, and his belief that producing and distributing higher quality and more expensive films made economic sense, see Morris L. Ernst, *Too Big* (New York: Little, Brown, 1940, reprinted 2000), 142.

17. Drinkwater, in *Carl Laemmle,* 73, describes the founding of IMP, later Universal Studios.

18. One account of the industry's first European-style films, and another excellent resource on the history of American cinema, is Joel Waldo Finler, *The Hollywood Story* (London: Wallflower Press, 2003), 115.

19. Drinkwater, *Carl Laemmle,* 102–3.

20. This classic account of the industry's early lawless years and the growth of the Hollywood empire may be found in Lewis Jacobs, *The Rise of American Film: A Critical History,* 2nd ed. (New York: Harcourt, Brace, 1947), 85.

21. Bardèche, Brasillach, *The History of Motion Pictures* (New York: W. W. Norton/Museum of Modern Art, 1938), 61–62.

22. Balio insists that the "hop-skip-and-jump to the Mexican border" tale should be put to rest. In actuality, claims Balio, "Trust producers led the way [to Los Angeles]. . . . By 1910 most MPPC producers had sent companies to the area where they were shortly joined by such newcomers as Bison, Nestor, Lux, Éclair, Fox, and IMP." Tino Balio, *The American Film Industry* (Madison: University of Wisconsin Press, 1985), 108–9.

23. Balio, *American Film Industry,* 143. Stephen Prince also interestingly catalogs the history of film censorship of violence in *Classical Film Violence: Designing and Regulating Brutality in Hollywood Cinema, 1930–1968* (New Brunswick, NJ: Rutgers University Press, 2003).

24. Bardèche, 60–61.

25. Paul Starr, *The Creation of the Media: Political Origins of Modern Communications* (New York: Basic Books, 2004), 309.

26. Gabler, *Empire of Their Own,* 23, 60.

27. Balio, *American Film Industry,* 150.

28. Starr, *Creation of the Media,* 310.

29. Grimmelmann's observation was made in the context of the recent Google Books lawsuit and proposed settlement; James Grimmelmann, "The Google Book Search Settlement: Ends, Means, and the Future of Books," *American Constitution Society Issue Brief 5,* April 15, 2009.

30. The full text of the opinion dissolving the trust may be read at *U.S. v. Motion Picture Patents Co.,* 225 F. 800 (E.D. Pa., 1915).

31. This explosion of film diversity is well explored by Steven J. Ross, *Working-Class Hollywood: Silent Film and the Shaping of Class in America* (Princeton, NJ: Princeton University Press, 1998), 61.

32. *Reno v. American Civil Liberties Union,* 521 U.S. 844 (1997).

CHAPTER 5: CENTRALIZE ALL RADIO ACTIVITIES

1. Statement by the secretary of commerce at the opening of the radio conference of February 27, 1922, reprinted in, among other places, Fred Friendly, "Retrieving a Lost Rocket: How Television Went Haywire and What We Can Do About It—Part II," *Life,* March 24, 1967, 70–83.

2. Hoover described his governance ideas in *American Individualism* (Garden City, NY: Doubleday, Page & Co., 1923); see also Vincent Gaddis, *Herbert Hoover, Unemployment, and the Public Sphere: A Conceptual History 1913–1933* (Lanham, MD: University Press of America, 2005).

3. "Report of the Department of Commerce Conference on Radio Telephony," III.E, *Radio Service Bulletin,* May 1, 1922, 23–30. A complete description of the first conference can be found in Hugh Richard Slotten, *Radio and Television Regulation: Broadcast Technology in the United States* (Baltimore: Johns Hopkins University Press, 2000), 15–17. McQuiston's quote is in *Radio News,* August 1922, 232, 332–34.

4. This advertisement is reprinted in many places, including James Twitchell, *Adcult USA: The Triumph of Advertising in American Culture* (New York: Columbia University Press, 1996), 83–84. Whether it is actually the first radio advertisement is open to question, yet its significance lies in the fact that it was the first AT&T advertisement, and therefore the seed for all that followed.

5. For these statistics on AT&T's National Broadcasting System in 1924, see Douglas

P. Craig, *Fireside Politics: Radio and Political Culture in the United States, 1920–1940* (Baltimore: Johns Hopkins University Press, 2000), 28.

6. The full advertisement, along with a description of AT&T's plan for commercially sponsored broadcasts, can be found in Gleason L. Archer, *Big Business and Radio* (New York: Stratford Press, 1939), 54. For another source on AT&T and radio broadcasting, see Leonard S. Reich, "Research, Patents, and the Struggle to Control Radio: A Study of Big Business and the Uses of Industrial Research," *Business History Review* 51: 2 (Summer 1977), 208–35.

7. The Gillette advertisements discussed evolving fashions in facial hair and the benefits of a safety razor. For more information about early radio advertising, see Marc Weinberger et al., *Effective Radio Advertising* (New York: Lexington Books, 1994), 3–4.

8. The early sponsored programs are well described in Erik Barnouw, *A Tower in Babel: A History of Broadcasting in the United States to 1933*, vol. 1 (Oxford: Oxford University Press, 1966), and also John Dunning, *On the Air: The Encyclopedia of Old-Time Radio* (Oxford: Oxford University Press, 1998), 1, 159, 235.

9. For more information about the development and operation of radio broadcasting in different parts of the world, including the use of radio under the Third Reich, see Walter B. Emery, *National and International Systems of Broadcasting: Their History, Operation and Control* (East Lansing: Michigan State University Press, 1969).

10. A. H. Griswold gave this speech in 1923, at a meeting for AT&T executives across the country. The quote is reprinted widely; see, e.g., Michele Hilmes, *Hollywood and Broadcasting: From Radio to Cable* (Champaign: University of Illinois Press, 1999), 19.

11. The author has a copy of the original manual of the AT&T (actually Western Electric) radio on file. On the presenting of President Coolidge with a radio unit in 1924, see Barnouw, *A Tower in Babel*, 161.

12. For a full explanation of the creation of the RCA, see Archer, *Big Business and Radio*, 4–7, and Christopher H. Sterling and John M. Kittross, *Stay Tuned: A History of American Broadcasting*, 3rd ed. (London: Lawrence Erlbaum Associates, 2002), 57–58.

13. Sarnoff played a leading role in the rise of radio broadcasting; see Kenneth Bilby, *The General: David Sarnoff and the Rise of the Communications Industry* (New York: Harper & Row, 1986). Sarnoff's writings were published in *Looking Ahead: The Papers of David Sarnoff* (New York: McGraw-Hill, 1968).

14. Sarnoff had, in fact, trained in his youth to become a rabbi. See Daniel Stashower, *The Boy Genius and the Mogul: The Untold Story of Television* (New York: Random House, 2002), 32.

15. This quote, which evidences the obvious tension between AT&T and Sarnoff, can be found in Bilby, *The General*, 77.

16. Essentially, AT&T agreed that it would not manufacture radio sets. See Archer, *Big Business and Radio*, 118.

17. The classic account of the 1924 arbitration and following compromise is Archer, *Big Business and Radio*; see also Sterling and Kittross, *Stay Tuned*.

18. The NBC ran a full-page ad in various publications; it is reprinted online at http:// earlyradiohistory.us/1926nbc.htm, and in Sterling and Kittross, *Stay Tuned*, 118.

19. The commissioner of the FRC in 1931, Henry Lafount, considered the radio to be a "wonderful instrument of commerce." See Robert W. McChesney, *Telecommunications, Mass Media, and Democracy: The Battle for the Control of U.S. Broadcasting, 1928–1935* (Oxford: Oxford University Press, 1993), 34. Generally, the history of these crucial years.

20. On Hoover's broader role in this era, see Richard N. Smith, *An Uncommon Man: The Triumph of Herbert Hoover* (New York: Simon and Schuster, 1984). Zenith company history is quoted in Barnouw, *A Tower in Babel*, 180. The case is *United States v. Zenith Radio Corporation*, 12 F.2d 614 (D.C.Ill. 1926).

21. This idea is emphasized in Philip T. Rosen, *The Modern Stentors: Radio Broadcasters and the Federal Government, 1920–1934* (Westport, CT; Greenwood Press, 1980). Rosen believes that Hoover and the broadcasters agreed on the need for more federal power, but that Congress refused to vest that power in the executive branch and instead created an independent agency.

22. The issues surrounding the Zenith decision and the subsequent formation of the FRC in 1927 are highly contested and subject to numerous interpretations. In contemporary accounts the Radio Act was promoted as a beneficent government response to industry "chaos"; the first to challenge this view, as a normative matter, was the economist Ronald Coase. R. H. Coase, *Journal of Law and Economics* vol. 2 (October 1959), 1–40. As a descriptive matter, the communications historian Robert McChesney's groundbreaking 1993 book *Telecommunications, Mass Media, and Democracy* was among the first to present a highly critical history of the 1927 act, General Order 40, and all that followed— presenting the act as essentially a triumph of large corporate broadcasters. The economist Thomas W. Hazlett, meanwhile, offered one of the first public-choice explanations for the creation of the FRC. Hazlett argued that broadcasters wanted to limit market entry and government wanted to maximize its control, and that it was the prospect of state court recognition of common law rights that spurred Congress to regulate and preempt an emerging property scheme that would have deprived regulators of control. See "The Rationality of U.S. Regulation of the Broadcast Spectrum," *Journal of Law & Economics* 33 (April 1990), 133–75. Hazlett's account has been challenged; see Charlotte Twight, "What Congressmen Knew and When They Knew It: Further Evidence on the Origins of U.S. Broadcasting Regulation," *Public Choice* 95 (June 1998), 247–76. Ultimately, the question of what motivated Congress to pass any law is difficult to answer.

23. The distinction between general public service stations and propaganda stations is described in McChesney, *Telecommunications, Mass Media, and Democracy*, 27.

24. From the FRC Third Annual Report, as quoted in Steven J. Simmons, *The Fairness Doctrine and the Media* (Berkeley: University of California Press, 1978), 32. See also Robert S. McMahon, *Federal Regulation of the Radio and Television Broadcast Industry in the United States 1927–1959* (New York: Arno Press, 1979), 59–60.

25. More information about the General Orders can be found in McChesney, *Telecommunications, Mass Media, and Democracy*, 24–26.

26. Ibid., 29. For Lafount's view, see ibid., 25, 34. See also Steve J. Wurtzler, *Electric Sounds: Technological Change and the Rise of Corporate Mass Media* (New York: Columbia University Press, 2007), 61.

27. See, generally, David Welch, *The Third Reich: Politics and Propaganda* (London/New York: Routledge, 1993), 38–45. The quotes are from Joseph Goebbels, "Der Rundfunk als achte Großmacht," *Signale der neuen Zeit. 25 ausgewählte Reden von Dr. Joseph Goebbels* (Munich: Zentralverlag der NSDAP, 1938), 197–207.

28. The Sarnoff mythology is best understood by reading the published collections of his writings, which lend him the air of a prophet. See Sarnoff, *Looking Ahead.* Sarnoff also helped make himself the first volume of the "Wisdom encyclopedia," See, *The Wisdom of Sarnoff and the World of RCA* (Beverly Hills, CA: Wisdom Society, 1967).

CHAPTER 6: THE PARAMOUNT IDEAL

1. Thomas Tally opened Tally's Broadway Theatre in 1909; it was demolished in 1929. Tally also operated Tally's New Broadway Theatre in Los Angeles. This quote about the theater organ comes from David L. Smith and Orpha Ochse, *Murray M. Harris and Organ Building in Los Angeles* (Richmond, VA: Organ Historical Society, 2005), 87. Tally participated in film's trajectory from "peep show to palace," as David Robinson puts it in *From Peep Show to Palace: The Birth of American Film* (New York: Columbia University Press, 1996). "Thomas L. Tally, Film Pioneer, Dies. Producer First Signed Mary Pickford, Chaplin. A Founder of First National Pictures," *New York Times,* November 25, 1945, Obituaries.

2. For more information about Paramount's block booking scheme and Tally, see Tino Balio, *United Artists: The Company Built by the Stars* (Madison: University of Wisconsin Press, 1976). On the star system, see David Bordwell, Janet Staiger, and Kirstin Thompson, *The Classical Hollywood Cinema: Film Style and Mode of Production to 1960* (New York: Columbia University Press, 1985). See also Eileen Bowser, *The Transformation of Cinema, 1907–1915,* vol. 2 (Berkeley: University of California Press, 1994).

3. Adolph Zukor, *The Public Is Never Wrong: The Autobiography of Adolph Zukor* (New York: G. P. Putnam's Sons, 1953). An entertaining biography of Zukor and figures in the early film industry is Neal Gabler, *An Empire of Their Own: How the Jews Invented Hollywood* (New York: Crown Publishers, 1988). The salesman quote is in Thomas Schatz, *Hollywood: Critical Concepts in Media and Cultural Studies* (London: Routledge, 2004), 81. The Canadian actress Mary Pickford was known as "America's Sweetheart." She wrote an autobiography, *Sunshine and Shadow* (New York: Doubleday, 1955).

4. "The pride and business sense of such men urged them to find means of repressing Zukor before he could acquire dictatorial power." Richard D. MacCann, *The First Tycoons* (Metuchen, NJ: The Scarecrow Press, 1987), 162. See also Benjamin B. Hampton, *A History of the Movies* (Oxford: Oxford University Press, 1932), 176.

5. The portrait of William W. Hodkinson comes from his personal papers, held by the UCLA research library, which include his own journal, letters, transcripts of interviews, and unpublished essays. Also helpful is Bernard F. Dick, *Engulfed: The Death of*

Paramount Pictures and the Birth of Corporate Hollywood (Lexington: University Press of Kentucky, 2001).

6. "Behind-the-scenes Intrigue at Paramount: Testimony of Al Lichtman," *New York Telegraph,* April 25, 1923.

7. This description of Paramount under Hodkinson can be found in Balio, *United Artists,* 9.

8. As quoted in Lewis Jacobs, *The Rise of the American Film: A Critical History* (New York: Harcourt, Brace, 1947), 93.

9. The auteur filmmaking model is contrasted with the central producer model, as described in Joseph Lampel, "The Genius Behind the System: The Emergence of the Central Producer System in the Hollywood Motion Picture Industry," in Joseph Lampel, Jamal Shamsie, and Theresa K. Lant, eds., *The Business of Culture: Strategic Perspectives on Entertainment and Media* (Mahwah, NJ: Lawrence Erlbaum Associates, 2006), 41–56.

10. Jacobs, *Rise of the American Film,* 287.

11. One of Zukor's agents was Benjamin Hampton, who later wrote about this time period. This account is based mostly on Hodkinson's journal and letters, Bernard F. Dick's *Engulfed,* and Benjamin Hampton, *A History of the Movies* (Oxford: Oxford University Press, 1932).

12. From a transcript of a 1966 interview with Hodkinson, in his papers.

13. Hampton, *History of the Movies,* 154–61.

14. Dick, *Engulfed,* 11.

15. Hampton, *History of the Movies,* 153–54. This quote is from Cecil B. DeMille's autobiography. Cecil B. DeMille and Donald Hayne, *The Autobiography of Cecil B. DeMille* (Englewood Cliffs, NJ: Prentice-Hall, 1959), 152.

16. This number comes from Richard Koszarski, *An Evening's Entertainment: The Age of the Silent Picture Feature, 1915–1928* (Berkeley: University of California Press, 1990), 73.

17. Jacobs, *Rise of the American Film,* 166. Details of the agreements with Chaplin and Pickford come from Kozarski, *An Evening's Entertainment,* 74–77.

18. Hampton, *History of the Movies,* 196.

19. To read more about Zukor's bold stock offering, see Tino Balio, *The American Film Industry,* 2nd ed. (Madison: University of Wisconsin Press, 1985), 121. Hampton, *History of the Movies,* 255.

20. Olson's main work on group and organizational behavior is in Mancur Olson, *The Logic of Collective Action: Public Goods and the Theory of Groups,* 2nd ed. (Cambridge, MA: Harvard University Press, 1971).

21. Hampton, *History of the Movies,* 253.

22. Ibid., 255. See also Koszarski, *An Evening's Entertainment,* 75.

23. Koszarski, *An Evening's Entertainment,* 75.

24. On Ford's economics, see Bernard Doray, *From Taylorism to Fordism: A Rational Madness* (London: Free Association Books, 1988).

25. Koszarski, *An Evening's Entertainment,* 75.

26. Hampton, *History of the Movies,* 267.

27. As quoted in Clyde L. King, Frank A. Tichenor, and Gordon S. Watkins, *The Motion Picture in Its Economic and Social Aspects* (Trenton, FL: Ayer Publishing, 1970), 133.

28. Federal Trade Commission v. Famous Players–Lasky Corporation et al., Complaint No. 835, in *The Annual Report of the Federal Trade Commission: For the Fiscal Year Ended June 30, 1922* (Washington, DC: Government Printing Office, 1922).

29. Ibid., 131. See In the Matter of Famous Players–Lasky Corporation et al., 11 FTC 187 (1927).

30. The Frankenstein quote is from an unpublished 1935 essay found in Hodkinson's papers entitled "After Block Booking—What?" The early use of block booking is described in Balio, *American Film Industry,* 117–18.

31. P. S. Harrison, "Give the Movie Exhibitor a Chance!" in Waller, ed., *Moviegoing in America,* 211–13.

32. Ibid., 212.

33. The two U.S. Supreme Court cases are *U.S. v. Paramount Pictures,* 334 U.S. 131 (1948), and *U.S. v. Loew's,* 371 U.S. 38 (1962). These decisions are discussed in Balio, *American Film Industry,* 560–61.

34. *U.S. v. Paramount Pictures,* 157.

35. George J. Stigler, "United States v. Loew's, Inc.: A Note on Block-Booking," *Supreme Court Review* (1963), 152–57.

36. This example is drawn from ibid., 152–53.

37. This idea was developed in Benjamin Klein and Roy Kenney, "The Economics of Block Booking," *Journal of Law and Economics* 26 (1983): 497–540.

38. *U.S. v. Loew's,* 49.

39. Pauline Kael, "Why Are Movies So Bad? or, The Numbers," *New Yorker,* June 23, 1980, 82–93.

40. Balaban himself wrote a history of the early Chicago movie palaces: David Balaban, *The Chicago Movie Palaces of Balaban and Katz* (Chicago: Arcadia Publishing, 2006) 103–6. Other sources on this period of film history are Lee Grieveson and Peter Kramer, *The Silent Cinema Reader* (London: Routledge, 2004), 273; Douglas Gomery, *The Coming of Sound: A History* (New York: Routledge, 2005), 11; and Balio, *American Film Industry,* 223.

41. 2.5 million was the number claimed; see Douglas Gomery, "Fashioning an Exhibition Empire, Promotion, Publicity, and the Rise of Publix Theatres," in Gregory Albert Waller, ed., *Moviegoing in America: A Sourcebook in the History of Film Exhibition* (Malden, MA: Blackwell, 2002). The sale to Warner Bros. is described in Bordwell, Staiger, and Thompson, *Classical Hollywood Cinema,* 399.

42. This account of the film industry lobbying the FTC and the appointment of Myers comes from Louis Pizzitola, *Hearst over Hollywood: Power, Passion, and Propaganda in the Movies* (New York: Columbia University Press, 2002), 248.

43. To read about the success of the Warner brothers and their studio in the words of their descendants, see Cass W. Sperling, Cork Millner, and Jack Warner, Jr., *Hollywood Be Thy Name: The Warner Brothers Story,* 2nd ed. (Lexington: University Press of Kentucky, 1998), 84. For Tally's fate, see "Thomas L. Tally, Film Pioneer, Dies." *New York Times,* November 25, 1945, Obituaries. See also DeMille and Hayne, *Autobiography of Cecil B. DeMille,* 152. "That character" is from Hodkinson's papers.

PART II: BENEATH THE ALL-SEEING EYE

1. These "sustaining" programs and this interesting period in network history, particularly relating to NBC, can be read in Michele Hilmes and Michael Lowell Henry, *NBC: America's Network* (Berkeley and Los Angeles: University of California Press, 2007), 17.

2. This observation would prove prescient as American communications culture would continue to be dominated by mass production for decades. The original may be read in Lawrence P. Lessing, *Man of High Fidelity: Edwin Howard Armstrong, a Biography* (New York: J. B. Lippincott, 1956), 19–20.

CHAPTER 7: THE FOREIGN ATTACHMENT

1. Leo Beranek supplied a copy of Hush-A-Phone's letterhead. The product was the subject of an article in *Popular Mechanics,* February 1941, 230.

2. Much of the Hush-A-Phone story is based on interviews with Leo Beranek and on his autobiography, *Riding the Waves: A Life in Sound, Science, and Industry* (Cambridge, MA: MIT Press, 2008), 91. The hearing and appeal may be found respectively at "In the Matter of Hush-A-Phone Corp. et al., Decision," 20 FCC 391 (1955), and *Hush-A-Phone v. U.S.,* 238 F.2d 266 (D.C. Cir. 1956).

3. The early voice mail machine is described in Mark Clark, "Suppressing Innovation: Bell Laboratories and Magnetic Recording," *Technology and Culture* vol. 34, no. 3 (1993): 516, 529.

4. Compare Richard Posner, "The Social Costs of Monopoly and Regulation," *Journal of Political Economy,* vol. 83, no. 4 (1975).

5. As Clark writes, "the suppression was so effective that historians who have relied on the published record have rendered a highly incomplete picture of Bell Laboratories' activities." Clark, "Suppressing Innovation," 517, 536.

6. Ibid., 534.

7. The descriptions of the Hush-A-Phone hearing with the FCC, and all direct quotes from the hearing, are from *Telecommunication Reports,* January 30, 1950. Additional material comes from Beranek, *Riding the Waves,* 91–92, and from an interview with Beranek.

8. Gregory D. Black, *Hollywood Censored: Morality Codes, Catholics, and the Movies* (Cambridge: Cambridge University Press, 1996), 171. "If you want a Jew to do something," wrote Black, paraphrasing Breen, "you don't ask him politely—you just tell him."

9. This AT&T quality control argument is drawn from the initial 1950 hearing with the FCC, reported in "In the Matter of Hush-A-Phone Corp. et al., Decision," 20 FCC 415 (1955).

10. Nelson and Winter's absorbing "evolutionary" theory—the book's jacket flap calls it "the most sustained and serious attack on mainstream, neoclassical economics in more than forty years"—may be read in full in Richard R. Nelson and Sidney G. Winter, *An Evolutionary Theory of Economic Change* (Cambridge, MA: Belknap Press, 1985).

11. Fred W. Henck and Bernard Strassburg, *A Slippery Slope: The Long Road to the Breakup of AT&T* (New York: Greenwood Press, 1988),38.

12. "In the Matter of Hush-A-Phone Corp. et al., Decision," 20 FCC 419.

13. *Hush-A-Phone,* 238 F.2d 269.

14. Ibid.

CHAPTER 8: THE LEGION OF DECENCY

1. Daniel A. Lord, "George Bernard Shaw," *Catholic World* April–September 1916, 36–37; for a (rather dry) account of his life, see Daniel A. Lord, *Played by Ear: The Autobiography of Daniel A. Lord, S.J.* (Chicago: Loyola University Press, 1956).

2. *Time* magazine, September 13, 1926.

3. On Breen, see Thomas Patrick Doherty, *Hollywood's Censor: Joseph I. Breen and the Production Code Administration* (New York: Columbia University Press, 2007), 21. Excerpts from Breen's letters concerning his views are in Gregory D. Black, *Hollywood Censored: Morality Codes, Catholics, and the Movies* (Cambridge: Cambridge University Press, 1994), 70. This quote comes from a letter Breen wrote to Reverend Wilfrid Parsons. For more on Breen and Jews, see Doherty, 199–225.

4. Doherty, *Hollywood's Censor,* 198.

5. Mark LaSalle, "Pre-Code Hollywood," Green Cine, www.greencine.com/static/primers/precode.jsp.

6. Frank Walsh, *Sin and Censorship: The Catholic Church and the Motion Picture Industry* (New Haven: Yale University Press, 1996), 76.

7. For the entire pledge, and more on the Church's efforts to control the content of films, see Thomas Patrick Doherty, *Pre-Code Hollywood: Sex, Immorality, and Insurrection in American Cinema 1930–1934* (New York: Columbia University Press, 1999), 321.

8. Doherty, *Hollywood's Censor,* 203.

9. For a discussion of the federal government's near-intervention into film censorship, as well as a discussion of the study done on the effects of film on children, see Doherty, *Hollywood's Censor,* 59.

10. Leonard L. Jeff and Jerold L. Simmons, *The Dame in the Kimono: Hollywood, Censorship, and the Production Code* (Lexington: University Press of Kentucky, 2001), 54–55.

11. Ibid., 38.

12. Doherty, *Hollywood's Censor,* 352.

13. Mark LaSalle, "Pre-Code Hollywood," Green Cine, www.greencine.com/static/primers/precode.jsp.

14. Holmes never actually used the phrase, but it derives from his dissenting opinion in *Abrams v. United States,* 250 U.S. 616 (1919).

15. Quoted in Doherty, *Hollywood's Censor,* 7, 75.

16. Ibid., 79.

17. Mark LaSalle, "Pre-Code Hollywood," Green Cine, www.greencine.com/static/primers/precode.jsp.

CHAPTER 9: FM RADIO

1. Engineer Edwin Armstrong, the man behind the research and development of FM radio at Columbia University, was a fascinating, and ultimately tragic, character. As we will see in this chapter, Armstrong spent a great deal of time and money in the latter portion of his life defending FM radio against the FCC and the major broadcasting networks, particularly David Sarnoff's RCA. One excellent biography of Armstrong that is referenced and utilized throughout this chapter is Lawrence P. Lessing, *Man of High Fidelity: Edwin Howard Armstrong, a Biography* (New York: J. B. Lippincott, 1956). Another interesting source, covering more than Armstrong is Tom Lewis, *Empire of the Air: The Men Who Made Radio* (New York: HarperCollins, 1991). Columbia University's Electrical Engineering Department is another useful source: see Yannis Tsividis, "Edwin Armstrong: Pioneer of the Airwaves," Columbia University, www.ee.columbia.edu/ ,isc-pages/armstrong_main.html?mode=interactive (accessed February 2010).

2. Armstrong and David Sarnoff's secretary, Marion MacInnes, were married in 1923. Lessing, *Man of High Fidelity*, 154.

3. This move to the state-of-the-art Empire State Building laboratory and the subsequent experiments conducted there by Armstrong are discussed in Lessing, *Man of High Fidelity*, 219, and Frank Northen Magill, *Great Events from History II: Science and Technology Series* (Pasadena, CA: Salem Press, 1991), 940.

4. This exchange is recounted in Lewis, *Empire of the Air*, 263.

5. For an overview of the Empire State Building experiments and Armstrong's subsequent expulsion to make way for RCA's television broadcast experiments, see Christopher H. Sterling and John M. Kittross, *Stay Tuned: A Concise History of American Broadcasting*, 3rd ed. (London: Routledge, 2002), 156–60.

6. Lawrence Lessing recounts the stunned reaction of listeners to the first public demonstration of the FM radio broadcast conducted by Armstrong in 1935, discussed later in this chapter in more detail. The broadcast, Lessing notes, was revolutionary not only because it was a new technological achievement, but also because the sound was being conveyed "with a life-like clarity never heard on even the best clear-channel stations in the regular broadcast band." Lessing, *Man of High Fidelity*, 209–10.

7. Ibid., 232–38.

8. The radio industry's close relationship to the federal government, and particularly the FCC during the period discussed in this chapter, is explored in Philip T. Rosen, *The Modern Stentors: Radio Broadcasters and the Federal Government,1920–1934*, (Westport, CT: Greenwood Press, 1980).

9. Lessing, *Man of High Fidelity*, 209.

10. The term "radio's second chance" was most famously used in the 1946 book by the same name authored by Charles Siepmann. The driving concept behind Siepmann's book was primarily that radio had not lived up to its early idealistic promise and that FM radio potentially had the power to both increase sound quality and reach beyond the major broadcast networks' stranglehold on content, which was, in his view, negatively over-

commercialized. Charles Arthur Siepmann, *Radio's Second Chance* (New York: Little, Brown, 1946).

11. The shift in band, along with the new rules and their justification, can be found in FCC, *Eleventh Annual Report* (Washington, DC: USGPO, 1945), *Twelfth Annual Report* (1946), and *Thirteenth Annual Report* (1947); see also Lessing, *Man of High Fidelity,* 258–60.

12. See United States Congress, House Committee on Ways and Means, *Hearings* (1950), 197.

13. "In 1979, for the first time, FM passed AM in overall market shares, and, in succeeding years, increased its lead. FM, once the unwanted, ill-treated sibling of AM, had become the desired, admired, and more popular medium." F. Leslie Smith, John W. Wright II, and David H. Ostroff, *Perspectives on Radio and Television: Telecommunication in the United States* (New York: Routledge, 1998), 63.

14. Lessing, *Man of High Fidelity,* 260.

15. Lessing's biography of Armstrong contains an excellent chapter detailing the litigation: "The Last Battle." The chapter details the fairly epic struggle a lone inventor faces when litigating against a corporation with virtually unlimited capital to fund strategically protracted litigation. Lessing, *Man of High Fidelity,* 279–85. Another source is Harold Evans, Gail Buckland, and David Lefer, *They Made America* (New York: Little, Brown, 2004).

CHAPTER 10: NOW WE ADD SIGHT TO SOUND

1. This anecdote about Baird's 1926 demonstration can be found in Antony Kamm and Malcolm Baird, *John Logie Baird: A Life* (Edinburgh: National Museums of Scotland, 2002), 55, 69. For another perspective, see Tom McArthur and Peter Waddell, *The Secret Life of John Logie Baird* (London: Hutchinson, 1986). See also Donald F. McLean, *Restoring Baird's Image* (London: Institution of Electrical Engineers, 2000), 38. To read about Baird in the history of early television, see Albert Abramson, *The History of Television, 1880 to 1941* (Jefferson, NC: McFarland, 1987). A concise biography of Baird can be found in Christopher H. Sterling, "Baird, John Logie (1888–1946)" in Horace Newcomb, ed., *Encyclopedia of Television,* vol. I, 2nd ed. (New York: Fitzroy Dearborn, 2004), 201–2.

2. The Baird undersock was made of unbleached half-hose material sprinkled with borax powder, as described in Russell W. Burns, *John Logie Baird: Television Pioneer* (London: Institution of Electrical Engineers, 2000), 18. Accounts of the Baird undersock and pneumatic shoe can also be found in David E. Fisher and Marshall Fisher, *Tube: The Invention of Television* (Berkeley, CA: Counterpoint, 1996), 24–27.

3. This excerpt from Baird's journal can be found in an interesting book which describes how various household products were born. David Lindsay, "Television," in *House of Invention: The Secret Life of Everyday Objects* (New York: Lyons Press, 2000), 133–42.

4. The *Times* article details the January 1926 demonstration and describes the image that Baird transmitted. "On This Day," *The Times,* January 28, 1985, H13.

5. Jenkins published a description of his discoveries in Charles F. Jenkins, *Vision by Radio, Radio Photographs, Radio Photograms* (Washington, DC: Jenkins Laboratories, 1925). See

also Charles F. Jenkins, *Radiomovies, Radiovision, Television* (Washington, DC: National Capital Press, 1929). Jenkins was named "Father of Television" by *The New York Times,* as quoted by historian Gary R. Edgerton, *The Columbia History of American Television* (New York: Columbia University Press, 2007), 29. A concise biography of Jenkins can be found in Steve Runyon, "Jenkins, Charles Francis (1867–1934)," in Newcomb, ed., *Encyclopedia of Television,* 1218–20.

6. The newspaper article included a picture of Farnsworth holding the parts of his television set. "S.F. Man's Invention to Revolutionize Television," *San Francisco Chronicle,* September 3, 1928, section 2.

7. *The New York Times* credited Baird with leading the international race to achieve a practical television. Clair Price, "A Saga of the Radio Age—and Its Hero," *New York Times,* March 27, 1927, SM6. The Federal Radio Commission issued the first television license to Jenkins in 1928, authorizing him to operate at a power of 250 W. See R. W. Burns, *Television: An International History of the Formative Years* (London: Institution of Electrical Engineers, 1998), 205. The number of viewers was reported in 1929 by the *New York Evening World;* see Edgerton, *Columbia History of American Television,* 30.

8. In the forty-minute broadcast, two actors performed a one-act play in a locked room before three cameras and a microphone. Russell B. Porter, "Play Is Broadcast by Voice and Acting in Radio-Television," *New York Times,* September 12, 1928, 1.

9. As quoted in the biography of Baird by Ronald F. Tiltman, *Baird of Television* (New York: Arno Press, 1974), 170.

10. Jenkins's public offering is described in Edgerton, *Columbia History of American Television,* 31. Baird described this deal in his memoirs, John L. Baird and Malcolm Baird, *Television and Me: The Memoirs of John Logie Baird,* 2nd ed. (Edinburgh: Mercat Press, 2004), 114.

11. A technical description of mechanical television, along with useful diagrams, can be found in A. G. Jensen, "The Evolution of Modern Television," in Raymond Fielding, ed., *A Technological History of Motion Pictures and Television: An Anthology from the Pages of the Journal of the Society of Motion Picture and Television Engineers* (Berkeley: University of California Press, 1979), 235–38.

12. The *Daily News* ran this article on December 30, 1926, as quoted in Burns, *Television: An International History,* 206.

13. This ad can be found at o.tqn.com/d/inventors/1/0/z/2/charles_jenkins.jpg.

14. The advertisement can be found in James N. Miller, "The Latest in Television," *Popular Mechanics,* September 1929, 472–76.

15. In the article, Sarnoff writes that television is still in an experimental stage and will require careful nurturing to grow into a great public service. David Sarnoff, "Forging an Electric Eye to Scan the World," *New York Times,* November 18, 1928.

16. The report, written by Alfred Goldsmith, concluded that only RCA could "be depended upon to broadcast television material with high technical and program quality." As quoted in Michele Hilmes, *Hollywood and Broadcasting: From Radio to Cable* (Champaign: University of Illinois Press, 1999), 28.

17. To read more about Jenkins and his struggles with the FCC, see James A. Von Schilling,

The Magic Window: American Television, 1939–1953 (Binghamton, NY: Haworth Press, 2003), 3, 13.

18. The FCC requirements for a licensed broadcaster can be found reprinted in Robert Stern, *The FCC and Television: The Regulatory Process in an Environment & Rapid Technological Innovation* (Ph.D. diss., Harvard University, 1950).

19. To learn more about the formative years of BBC, see Burton Paulu, *Television and Radio in the United Kingdom* (Minneapolis: University of Minnesota Press, 1981). See also Ronald Simon, *BBC Television: Fifty Years, November 14, 1986–January 31, 1987* (New York: Museum of Broadcasting, 1987). To read about Germany's broadcast of the Olympic games, see Arnd Krüger, "Germany: The Propaganda Machine," in Arnd Krüger and William J. Murray, eds., *The Nazi Olympics: Sport, Politics, and Appeasement in the 1930s* (Champaign: University of Illinois Press, 2003), 27–43. See also David Welch, *The Third Reich: Politics and Propaganda,* 2nd ed. (London: Routledge, 1993).

20. "S. F. Man's Invention to Revolutionize Television," *San Francisco Chronicle,* September 3, 1928. See also Evan I. Schwartz, *The Last Lone Inventor: A Tale of Genius, Deceit, and the Birth of Television* (New York: HarperCollins, 2002), 137.

21. Schwartz, *Last Lone Inventor,* 123.

22. Ibid., 13.

23. This anecdote about the disastrous Crystal Palace fire comes from ibid., 224.

24. This account of Sarnoff at the World's Fair comes from Von Schilling, *Magic Window,* 5. A contemporary description and photograph appear in "Radio Living Room of Tomorrow," *Popular Mechanics,* vol. 72, no. 2, August 1939, 300.

25. For the *New Yorker* quote, David Hillel Gelertner, 1939, "The Lost World of the Fair" (1995), 167. Sarnoff is described in glowing terms in the piece. See Marcy Carsey and Tom Werner, "David Sarnoff," in *People of the Century,* Time / CBS News (New York: Simon and Schuster, 1999), 162–65, 163.

26. Schwartz, 272.

27. See David Sarnoff's testimony before the FCC at Washington, D.C., on November 14, 1938, and May 17, 1939, published in *Principles and Practices of Network Radio Broadcasting: Testimony of David Sarnoff* (New York: RCA Institutes Technical Press, 1939), 16.

28. Walter Lippmann, "The TV Problem," *Today and Tomorrow,* October 27, 1959, in Clinton Rossiter and James Lare, eds., *The Essential Lippmann: A Political Philosophy for Liberal Democracy* (Cambridge, MA: Harvard University Press, 1982), 411–13.

CHAPTER 11: THE RIGHT KIND OF BREAKUP

1. Harold Orlans, *Contracting for Atoms: A Study of Public Policy Issues Posed by the Atomic Energy Commission's Contracting for Research, Development, and Managerial Services* (Washington, DC: Brookings Institution, 1967), 33.

2. For the terms of the 1956 consent decree in the suit against AT&T, see Gerald W.

Brock, *Telecommunication Policy for the Information Age: From Monopoly to Competition* (Cambridge, MA: Harvard University Press, 1998), 71–72.

3. For the statement of the Bell engineers, see Constantine Raymond Kraus and Alfred W. Duerig, *The Rape of Ma Bell: The Criminal Wrecking of the Best Telephone System in the World* (Secaucus, NJ: Lyle Stuart, 1988), 13. For Goldwater's statement, see ibid., 103.

4. The statistics here are drawn from *U.S. v. Paramount Pictures,* 85 F.Supp. 881 (S.D.N.Y. 1949). For a general discussion of statistics on first-run theaters in 1930s–1940s Hollywood, see Andrew Hanssen, "The Block Booking of Films Re-examined," in John Sedgwick and Michael Pokorny, eds., *An Economic History of Film* (New York: Routledge, 2005), 121–51.

5. Thurman W. Arnold, *The Folklore of Capitalism* (New Haven: Yale University Press, 1937), 211. Indeed, in light of these conflicting impulses, Arnold believed that the antitrust laws in this country were systematically underenforced. See ibid., 207–30.

6. Alfred D. Chandler, Jr., *The Visible Hand: The Managerial Revolution in American Business* (Cambridge, MA: Belknap Press, 1977).

7. "Arnold Demands a New Movie Deal," *New York Times,* April 23, 1940, 19. For the case against the AMA, see *United States v. American Medical Association,* 110 F.2d 703 (D.C. Cir. 1940). For the Supreme Court's 1938 remand in the case against the studios, see *Interstate Circuit v. United States,* 304 U.S. 55 (1938).

8. For the final 1948 Supreme Court decision in what has become known as "the Paramount case," see *United States v. Paramount Pictures, Inc.,* 334 U.S. 131 (1948).

9. For Crandall's argument that the Paramount case did not result in downward pressure on ticket prices, see Robert W. Crandall, "Postwar Performance of the Motion-Picture Industry: The Economics," *Antitrust Bulletin* 20 (1975): 61. For Anderson's comment, and a further discussion of the blossoming television industry's impact on the film industry, see Martin Halliwell, *American Culture in the 1950s* (Edinburgh: Edinburgh University Press, 2007), 147. For Anderson's original work, see Christopher Anderson, *Hollywood TV: The Studio System in the Fifties* (Austin: University of Texas Press, 1994), 1.

10. Richard E. Caves, *Creative Industries: Contracts Between Art and Commerce* (Cambridge, MA: Harvard University Press, 2002), 95–96.

11. For a discussion of the key players and events in ushering in the New Hollywood without the constraints of the Production Code, see Peter Biskind, *Easy Riders, Raging Bulls: How the Sex-Drugs-and Rock 'N Roll Generation Saved Hollywood* (New York: Simon and Schuster Paperbacks, 1998), 23–52.

12. Jack Valenti, *The Voluntary Movie Rating System: How It Began, Its Purpose, the Public Reaction* (pamphlet, 1996).

13. The Production Code became progressively less onerous through the 1950s and '60s. In 1968 it was abandoned in its entirety in favor of the MPAA rating system. See Geoff King, *New Hollywood Cinema: An Introduction* (London: IB Tauris, 2002), 31. For a general discussion of the development and identity of the New Hollywood, see ibid., 1–33.

CHAPTER 12: THE RADICALISM OF THE INTERNET REVOLUTION

1. For the full text of the memorandum, see J.C.R. Licklider, "Memorandum for Members and Affiliates of the Intergalactic Computer Network," KurzweilAI.net, www.kurzweilai .net/articles/art0366.html?printable=1.
2. For Licklider's early years and career, see H. Peter Alesso and Craig Forsythe Smith, *Connections: Patterns of Discovery* (Hoboken, NJ: Wiley & Sons, 2008), 60; for his life, M. Mitchell Waldrop, *The Dream Machine: J.C.R. Licklider and the Revolution That Made Computing Personal* (New York: Penguin, 2002).
3. For Rheingold's description of the AN/FSQ-7, see Howard Rheingold, *Tools for Thought: The History and Future of Mind-Expanding Technology* (Cambridge, MA: MIT Press, 2000), 142–44.
4. J.C.R. Licklider, "Man-Computer Symbiosis," *IRE Transactions on Human Factors in Electronics* HFE-1 (1960): 4.
5. John Markoff, *What the Dormouse Said: How the Sixties Counterculture Shaped the Personal Computer Industry* (New York: Penguin, 2005), 9.
6. For an extensive discussion of Baran's career and innovations, see Katie Hafner and Matthew Lyon, *Where Wizards Stay Up Late: The Origins of the Internet* (New York: Touchstone, 1996), 53–67.

CHAPTER 13: NIXON'S CABLE

1. The opening story is based on the author's interview with Ralph Lee Smith, September 14, 2008. His article is "The Wired Nation," *Nation,* May 18, 1970, 582.
2. For a discussion of the project and the controversy it caused, see Richard P. Hunt, "Expressway Vote Delayed by City: Final Decision Is Postponed After 6-Hour Hearing," *New York Times,* December 7, 1962; see also Hilary Ballon and Kenneth T. Jackson, eds., *Robert Moses and the Modern City: The Transformation of New York* (New York: W. W. Norton, 2008).
3. Books on the history of the American cable industry are relatively rare. See Megan Mullen, *The Rise of Cable Programming in the United States: Revolution or Evolution* (Austin: University of Texas Press, 2003), 90–93; Patrick Parsons, *Blue Skies: A History of Cable Television* (Philadelphia: Temple University Press, 2008); and Patrick R. Parsons and Robert M. Frieden, *The Cable and Satellite Television Industries* (Boston: Allyn & Bacon, 1997). See also Brian Lockman and Don Sarvey, *Pioneers of Cable Television: The Pennsylvania Founders of an Industry* (Jefferson, NC: McFarland, 2005).
4. *Hearings Before the Senate Committee on Interstate and Foreign Commerce,* United States Senate, 85th Cong., 2d Sess., 1959 (statement of William C. Grove).
5. House Committee on the Judiciary, 89th Cong., Copyright Law Revision Part 6, Supplementary Report of the Register of Copyrights on the General Revision of the U.S. Copyright Law: 1965 Revision Bill 42 (Comm. Print 1965).

6. U.S. Cong. House Committee on the Judiciary Subcommittee on Courts, Civil Liberties and the Administration of Justice Hearings, 92d Cong. (1972) (statement of Jack Valenti), reprinted in 15 Omnibus Copyright Revision Legislative History 727 (George S. Grossman, ed., 1976).

7. 392 U.S. 390 (1968).

8. Stanley M. Besen and Robert W. Crandall, "The Deregulation of Cable Television," 44 *Law & Contemporary Problems* 77 (1981): 93.

9. On his life, see Ralph Engelman and Morley Safer, *Friendlyvision: Fred Friendly and the Rise and Fall of Television Journalism* (New York: Columbia University Press, 2009); *See It Now's* confrontation with McCarthy is the subject of Thomas Rosteck, *See It Now Confronts McCarthyism: Television Documentary and the Politics of Representation* (Tuscaloosa: University of Alabama Press, 1994), and of the 2005 film *Good Night and Good Luck* directed by George Clooney.

10. Mullen, *Rise of Cable Programming,* 84.

11. Fred Friendly, "Asleep at the Switch of the Wired City," *Saturday Review,* October 10, 1970, 58.

12. Sloan Commission on Cable Communications, *On the Cable: The Television of Abundance: Report* (New York: McGraw-Hill Book Co., 1971), 119. The report examines the history and technology of cable, contemplates its potential, and makes suggestions as to its development.

13. See the Cabinet Committee on Cable Communications (1974), *Cable: Report to the President.* A summary and discussion of the Nixon Cabinet Committee's recommendations is in David Waterman and Andrew A. Weiss, *Vertical Integration in Cable Television* (Washington, DC: AEI Press, 1997), 1–2. For a discussion of the "separations policy," the "open skies" policy, and other developments in the regulation of the cable industry through the end of the Nixon administration, see Parsons, *Blue Skies,* 297–341.

14. The "wringer" quote is in *All the President's Men,* 2nd ed. (New York: Simon and Schuster, 1994), 105.

CHAPTER 14: BROKEN BELL

1. Testimony on S. 1167 before the Subcommittee on Antitrust and Monopoly, July 9, 1974.

2. The picture of John deButts comes mainly from Steve Coll, *The Deal of the Century: The Breakup of AT&T* (New York: Atheneum, 1986); Kristin McMurran, "A.T.& T. Chairman John deButts Puts on Golden Glow and a Big Smile for a Ma Bell TV Pitch," *People,* November 28, 1977; and his obituary in *The New York Times.*

3. For the excerpt from Judge Posner, see Richard Posner, "The Decline and Fall of AT&T: A Personal Recollection," *Federal Communications Law Journal* 61 (2008): 15. For the excerpt on Bell's advocating for absolute control of the system, see "In the Matter of Use of the CarterFone Device in Message Toll Telephone Service," 1968 WL 13208, 4 (FCC, June 26, 1968).

4. For a full discussion of the MCI episode, see Alan Stone, *How America Got Online* (New York: M. E. Sharpe, 1997), 61–81.

5. For the Carterfone case, see *CarterFone Device,* 1968 WL 13208. For a discussion of the FCC's phone jack standardization requirements after *CarterFone,* see Steven M. Besen and Garth Saloner, "The Economics of Telecommunication Standards," in Robert W. Crandall and Kenneth Flamm, eds., *Changing the Rules: Technological Change, International Competition, and Regulation in Communications* (Washington, DC: Brookings Institution, 1989), 210.

6. For a brief discussion of Hayes Corp.'s rise and fall, see Claus E. Heinrich and Bob Betts, *Adapt or Die: Transforming Your Supply Chain into an Adaptive Business Network* (Hoboken, NJ: John Wiley & Sons, 2003), 3–4.

7. For a discussion of the 1971 FCC Computer I decision, see Alan Pearce, "Computer Inquiry I, II, and III—Computers and Communications: Convergence, Conflict, or Policy Chaos," in Fritz E. Froehlich and Allen Kent, eds., *The Froehlich/Kent Encyclopedia of Telecommunications,* vol. 4 (New York: Marcel Dekker, 1992), 219–331.

8. The bill for which AT&T lobbied Congress was often referred to by its critics as the "Bell Bill," or the "Monopoly Protection Act of 1976." See "Communications: A Bill for Ma Bell," *Time,* May 4, 1976.

9. For Faulhaber's views, see, generally, Gerald R. Faulhaber, *Telecommunications in Turmoil: Technology and Public Policy* (Cambridge, MA: Ballinger, 1987).

10. For a full discussion of the MCI episode, see Stone, *How America Got Online,* 61–81.

11. For the full text of Judge Greene's opinion in the case, see *U.S. v. American Tel. & Tel. Co.,* 552 F.Supp. 131 (D.D.C. 1982). For a more in-depth look at the disposition of the antitrust suit against AT&T through the Reagan administration, see Robert Britt Horwitz, *The Irony of Regulatory Reform: The Deregulation of American Telecommunications* (New York: Oxford University Press, 1990), 239–43.

12. Sometime after divestiture, Henry Geller, the distinguished FCC policy insider, interviewed Charlie Brown and Judge Greene on their thoughts as to why AT&T agreed to the breakup; their consensus seems to be that even were AT&T to have prevailed in the instant suit, it would have remained under constant pressure from the Justice Department and would ultimately have been forced to capitulate. See "Questions and Answers with the Three Major Figures of Divestiture," in Barry G. Cole, ed., *After the Break-Up: Assessing the New Post-AT&T Divestiture Era* (New York: Columbia University Press, 1991), 21–50.

13. "9 Years of Litigation Ends: AT&T Clears Way for Bell System Breakup," *Sarasota Herald Tribune,* August 4, 1983, Business section.

CHAPTER 15: ESPERANTO FOR MACHINES

1. Zamenhof first published his idea for an international language in 1889. Ludwik Ł. Zamenhof, *La Lingvo Internacia* (Korn, 1889). Zamenhof introduced the language that

came to be known as Esperanto to Americans in L. Ł. Zamenhof, "What Is Esperanto?" *North American Review* 184: 606 (January 4, 1907), 15–21. For more background information, see Peter G. Forster, *The Esperanto Movement* (The Hague: Mouton Publishers, 1982).

2. Zamenhof's early writing contains a strong sense of idealism and hope. Ludwik Ł. Zamenhof, *An Attempt Towards an International Language,* trans. Henry Phillips (New York: Henry Holt, 1889), 5.

3. Esperanto has been a remarkably strong movement in China; see Gerald Chan, "China and the Esperanto Movement," *Australian Journal of Chinese Affairs* 15 (January 1986): 1–18.

4. Vinton G. Cerf is vice president and chief Internet evangelist at Google Inc. These quotes and opinions were obtained in an interview with the author in June 2008.

5. To read about AT&T at the time of the Internet's inception, see Christopher H. Sterling, Phyllis Bernt, and Martin B. H. Weiss, *Shaping American Telecommunications: A History of Technology, Policy, and Economics* (New York: Routledge, 2006). To read how Vint Cerf, Robert Kahn, and Robert Metcalf interacted with AT&T, see their individual entries in Laura Lambert et al., *The Internet: A Historical Encyclopedia,* vol. 2 (New York: MTM Publishing, 2005).

6. "A Protocol for Packet Network Intercommunication" in Jeremy M. Norman, ed., *From Gutenberg to the Internet: A Sourcebook on the History of Information Technology* (Novato, CA: historyofscience.com, 2005) 871–90.

7. As quoted in Alfred L. Malabre, Jr., *Lost Prophets: An Insider's History of the Modern Economists* (Cambridge, MA: Harvard Business Press, 1994), 220.

8. Friedrich A. Hayek, *The Road to Serfdom* (London: George Routledge & Sons, 1944).

9. This quote comes from Friedrich A. Hayek, *Individualism and Economic Order* (Chicago: University of Chicago Press, 1948), 77.

10. Leopold Kohr, *The Breakdown of Nations* (London: Routledge & Paul, 1957), ix.

11. Schumacher's idea of "enoughness" stemmed from his studies of what he called "Buddhist economics." See Ernst F. Schumacher, *Small Is Beautiful: Economics As If People Mattered* (New York: Harper & Row, 1973).

12. Jane Jacobs, *The Death and Life of Great American Cities* (New York: Vintage Books, 1961).

13. Frederick W. Taylor, *The Principles of Scientific Management* (New York: Harper & Bros., 1911).

14. Jon Postel wrote this into the "Robustness Principle," Section 2.10 of the Transmission Control Protocol (January 1980), available at http://tools.ietf.org/html/rfc761#section-2.10.

15. This paper announced the innovative end-to-end design principle. J. H. Saltzer, D. P. Reed, and D. D. Clark, "End-to-End Arguments in System Design," *ACM Transactions on Computer Systems (TOCS),* vol. 2, issue 4 (November 1984), 277–88.

16. After January 1, 1983, ARPANET users could no longer use NCP, and the shift to TCP/IP was achieved. See Jane Abbate, *Inventing the Internet* (Cambridge, MA: MIT Press, 1999), 141.

CHAPTER 16: TURNER DOES TELEVISION

1. This quote is part of a larger discussion of Turner's burgeoning empire as he moved to found CNN: Harry F. Waters, "Ted Turner Tackles TV News," *Newsweek,* June 16, 1980, 58. Two particularly useful biographies of Turner, alternately known as "Captain Outrageous" and the "Mouth of the South," are Robert Goldberg and Gerald Jay Goldberg, *Citizen Turner: The Wild Rise of an American Tycoon* (New York: Harcourt Brace, 1999), and Ken Auletta, *Media Man: Ted Turner's Improbable Empire* (New York: W. W. Norton, 2004). Others include Porter Bibb, *Ted Turner: It Ain't as Easy as It Looks* (New York: Crown Publishers, 1993), and Christian Williams, *Lead, Follow, or Get Out of the Way: The Story of Ted Turner* (New York: Times Books, 1981).

2. Ted Turner and Bill Burke, *Call Me Ted* (New York: Grand Central, 2008).

3. This statement was the lead-in to Turner's comment to *Newsweek* that he was to take television off the path of destruction. It was followed by this rather melodramatic prophecy: "Someday, somebody will put a bullet in me," he says sadly. "I would like to stay around for a while, but I really do believe that I'll be assassinated." Waters, "Ted Turner Tackles TV News," 58.

4. For further insight into how the FCC drove the division of NBC into the separate NBC and ABC networks, see the Federal Communications Commission's *Report on Chain Broadcasting,* Washington, DC, May 1941.

5. See Auletta, *Media Man,* 32–34, for an explanation of the cable model as pioneered by Turner and his TBS network.

6. Charles Haddad, "Ad Executives Love Turner Tales About Old Times and New," *Atlanta Constitution,* March 5, 1999, H2.

7. Apparently taken from an interview with Bob Hope in the mid-1970s, this ironic quote by the founder of one of the nation's most watched news networks is from Patrick Parsons, *Blue Skies: A History of Cable Television* (Philadelphia: Temple University Press, 2008), 453.

8. This point is drawn from Becker's book on modern cultural identities and sexual politics as examined through the lens of media coverage and representation of gay America: Ron Becker, *Gay TV and Straight America* (New Brunswick, NJ: Rutgers University Press, 2006), 86.

9. Cass R. Sunstein, *Republic.com 2.0* (Princeton, NJ: Princeton University Press, 2007), xi.

10. Ken Auletta, *Three Blind Mice: How the TV Networks Lost Their Way* (New York: Random House, 1991), 5.

CHAPTER 17: MASS PRODUCTION OF THE SPIRIT

1. An account of the rise and fall of United Artists, itself a story of open period filmmaking, may be found in Tino Balio, *United Artists: The Company That Changed the Film Industry* (Madison: University of Wisconsin Press, 1987).

2. Stephen Bach, then a young executive at United Artists, wrote a firsthand account of the failure of *Heaven's Gate* and subsequent fallout in Stephen Bach, *Final Cut: Art, Money and Ego in the Making of* Heaven's Gate, *the Film That Sank United Artists,* rev. ed. (New York: Newmarket Press, 1999). The quotes referenced above are drawn from page 360.

3. Vincent Canby, " 'Heaven's Gate,' a Western by Cimino," *New York Times,* November 19, 1980, Cultural Desk, Late Edition.

4. Yakov Amihud and Baruch Lev, "Managerial Motives for Conglomerate Mergers," *Bell Journal of Economics* 12 (1981): 605–17.

5. Aldous Huxley, "The Outlook for American Culture," *Harper's Magazine,* August 1927.

6. Economic perspectives on the film industry include John Sedgwick and Michael Pokorny, *An Economic History of Film* (New York: Routledge, 2005), and the interesting discussion of uncertainty and the film industry in Arthur S. De Vany, *Hollywood Economics: How Extreme Uncertainty Shapes the Film Industry* (London: Routledge, 2004).

7. Anderson's theory, and his discussion of the interplay of "head" and "tail" consumer demand, may be found in Chris Anderson, *The Long Tail: Why the Future of Business Is Selling Less of More* (New York: Hyperion, 2006).

8. De Vany, *Hollywood Economics,* 4.

9. Steven Ross's biography is gripping reading, and also provides a history of the creation of Time Warner. See Connie Bruck, *Master of the Game: Steve Ross and the Creation of Time Warner* (New York: Simon and Schuster, 1994).

10. General Electric 2008 Annual Report, www.ge.com/ar2008.

11. Roger Cohen, "The Creator of Time Warner, Steven J. Ross, Is Dead at 65," *New York Times,* December 21, 1992.

12. The failure of the Atari *E.T.* game is recounted in Bruck, *Master of the Game,* 180. The *E.T.* game is rated #1 on most lists of the worst games of all time. See, e.g., Emru Townsend, "The 10 Worst Games of All Time," *PC World,* October 23, 2006.

13. Betsy Schiffman, "Michael Eisner: Mouse in a Gilded Mansion," *Forbes,* April 26, 2001.

14. This accusation was delivered by letter after Eisner's resignation from the board. Alex Berenson, "The Wonderful World of (Roy) Disney," *New York Times,* February 15, 2004, Financial Desk, Late Edition.

15. The *Maltese Falcon* case has been effectively overruled; see *Warner Bros. Pictures v. Columbia Broadcasting System,* 216 F.2d 945 (9th Cir. 1954)

16. Edward Jay Epstein's theories on the modern film industry may be found in his *The Big Picture: Money and Power in Hollywood* (New York: Random House, 2005).

17. Richard Roeper, "Throw This God-Awful Sequel a Life Jacket; Even Funnyman Steve Carell Can't Save a Movie That's Drowning in Its Own Low Expectations," *Chicago Sun-Times,* June 22, 2007, Movies.

CHAPTER 18: THE RETURN OF AT&T

1. The secret executive order was first reported in James Risen and Eric Lichtblau, "Bush Lets U.S. Spy on Callers Without Courts," *New York Times,* December 15, 2005. The two

reporters were awarded a Pulitzer Prize for their efforts; Eric Lichtblau later wrote *Bush's Law: The Remaking of American Justice* (New York: Anchor Books, 2008).

2. Whitacre's full statement is available online; see *The AT&T and Bellsouth Merger: What Does It Mean for Consumers?—Hearing Before the Subcommittee on Antitrust, Competition Policy and Consumer Rights of the Senate Committee on the Judiciary*, 109th Cong. (2006) 10–12 (statement of Edward E. Whitacre, Jr., Chairman and CEO, AT&T, Inc.), available at http://ftp.resource.org/gpo.gov/hearings/109s/29938.pdf. Senator Arlen Specter described the hearing in "The Need to Roll Back Presidential Power Grabs," *New York Review of Books* 56:8 (May 14, 2009).

3. AT&T issued this statement of accountability in its annual report of 1911. AT&T, *Annual Report of the American Telephone and Telegraph* (New York: AT&T, 1911), 38. Whitacre's aggressive tactics as CEO of SBC are discussed in Edmund L. Andrews, "Birth of a Giant: A Leader's Vision," *New York Times*, April 2, 1996, and Mark Landler, "Disdaining Regulators, Whitacre Carves Out SBC Empire," *New York Times*, July 21, 1997. Whitacre's statement appeared in *Newsweek* as part of a cover story on his success at SBC. Roger O. Crockett, "Whitacre Steps Up to the Mike," *Newsweek*, April 12, 1999.

4. For Whitacre's explanation of why he doesn't use email, see Roger O. Crockett, "Résumé: Edward E. Whitacre, Jr." *Newsweek*, April 12, 1999.

5. The cover story described how Whitacre built SBC into a "telecom profit machine," but it predicted that the company would soon face fierce competition. Roger O. Crockett, "The Last Monopolist," *Businessweek*, April 12, 1999.

6. As quoted in Albert B. Paine, *In One Man's Life: Being Chapters from the Personal & Business Career of Theodore N. Vail* (New York: Harper & Brothers, 1921), 254.

7. Both Friedman and Stigler won Nobel Prizes in Economics for their work. A small sample includes Milton Friedman, *A Theory of the Consumption Function* (Princeton, NJ: Princeton University Press, 1957); Milton Friedman and Anna J. Schwartz, *A Monetary History of the United States, 1867–1960* (Princeton, NJ: Princeton University Press, 1971); and George Stigler, "The Theory of Economic Regulation," in George Stigler, ed., *Chicago Studies in Political Economy* (Chicago: University of Chicago Press, 1988), 209–33.

8. President Bill Clinton made this statement in his State of the Union address. "Address Before a Joint Session of the Congress on the State of the Union," *Public Papers*, vol. 1 (January 23, 1996), 79–87. For FCC chairman Reed Hundt's remark, see *The State of Competition in the Cable Television Industry: Hearing Before the House Committee on the Judiciary*, 105th Cong. (1997) (statement of Reed E. Hundt, Chairman of FCC), available at www.fcc.gov/Speeches/Hundt/spreh754.html.

9. Telecommunications Act of 1996, Pub. L. No. 104-104, 110 Stat. 56 (codified in scattered sections of 47 U.S.C.). For other sources that discuss the Act, see Patricia Aufderheide, *Communications Policy and the Public Interest: The Telecommunications Act of 1996* (New York: Guilford Press, 1999), and Robert W. Crandall, *Competition and Chaos: U.S. Telecommunications Since the 1996 Telecom Act* (Washington, DC: Brookings Institution, 2005).

10. The article presented SBC as a "case study" to show how the Baby Bells were flagrantly

thwarting competition. Marc Farranti, "Stall Tactics," *Network World,* December 8, 1997, 1, 49–53.

11. A list of lobbyists is maintained by Texas State Ethics Board and is available for 2003 at www.ethics.state.tx.us/tedd/conlob2003c.htm.

12. *Verizon Communications Inc. v. Trinko, LLP,* 540 U.S. 398 (2004).

13. The FCC adopted the Triennial Review Order, which reconsidered the Baby Bells' sharing obligations, on February 20, 2003, and released it on August 21, 2003. FCC, *Triennial Review Order,* 03-36 (2003), available at www.fcc.gov/wcb/cpd/triennial_ review/. *The New York Times* ran a short article discussing the order; see Jennifer Lee, "FCC Discloses New Rules for Telecom Industry," *New York Times,* August 21, 2003.

14. AT&T and SBC made this statement in the initial application of consent to the FCC. *In the Matter of AT&T Corp. and SBC Communications, Inc.,* Docket No. 05-65 (February 22, 2005), available at http://fjallfoss.fcc.gov/ecfs/document/view?id=6517318964. To read about how the deal came about, see Ken Belson and Matt Richtel, "A Telecommunications Architect," *New York Times,* February 2, 2005. Verizon beat out competitor Qwest Communications in a bidding war for MCI; see Ken Belson and Matt Richtel, "Qwest Withdraws Bid After MCI Accepts Verizon Offer," *New York Times,* May 3, 2005.

15. As quoted in Ellen Nakashima, "AT&T Gave Feds Access to All Web, Phone Traffic, Ex-Tech Says," *Seattle Times,* November 8, 2007.

16. Klein's description, along with some of his evidence, can be found at "Whistle-Blower's Evidence, Uncut," Wired.com, May 22, 2005, available at www.wired.com/science/ discoveries/news/2006/05/70944.

17. Ibid.

18. Ryan Singel, "AT&T Sued Over NSA Eavesdropping," Wired.com, January 31, 2006, available at www.wired.com/science/discoveries/news/2006/01/70126. The *Los Angeles Times* also reported on the cooperation between AT&T and the NSA. Josh Meyer and Joseph Menn, "U.S. Spying Is Much Wider, Some Suspect," *Los Angeles Times,* December 26, 2005.

19. The United States moved to intervene on May 13, 2006: http://docs.justia.com/cases/ federal/district-courts/california/candce/3:2006cv00672/175966/123/. The United States also moved to dismiss, invoking the state secrets privilege: http://docs.justia.com/cases/ federal/district-courts/california/candce/3:2006cv00672/175966/124/. To read the final opinion, see *Hepting v. AT&T Corp.,* 539 F.3d 1157 (9th Cir. 2008).

20. The measure amended the Foreign Intelligence Surveillance Act (FISA) of 1978 and granted retroactive immunity to the telecommunications companies that assisted in surveillance. FISA Amendments Act of 2008, Pub. L. No. 110-261, 122 Stat. 2436 (codified in scattered sections of 50 U.S.C.). Obama's remark is quoted in Eric Lichtblau, "Senate Approves Bill to Broaden Wiretap Powers," *New York Times,* July 10, 2008.

21. Whitacre's retirement was reported in Matt Richtel, "AT&T Chief Who Weathered a Sea Change Is Retiring in June," *New York Times,* April 28, 2007. See also Dionne Searcey, "A Pension to Retire For," *Wall Street Journal,* April 27, 2007.

CHAPTER 19: A SURPRISING WRECK

1. As told to *The New York Times* in Tim Arango, "How the AOL–Time Warner Merger Went So Wrong," *New York Times,* January 10, 2010. For other sources on the AOL–Time Warner merger, see Johnnie L. Roberts, "How It All Fell Apart," *Newsweek,* December 9, 2002, and three books: Nina Munk, *Fools Rush In* (New York: HarperCollins, 2004); Alec Klein, *Stealing TIME: Steve Case, Jerry Levin, and the Collapse of AOL Time Warner* (New York: Simon and Schuster, 2003); and Kara Swisher, *There Must Be a Pony In Here Somewhere: The AOL Time Warner Debacle and the Quest for a Digital Future* (New York: Crown Business, 2003).

2. Case and Levin had also served together on the board of the New York Stock Exchange. Munk, *Fools Rush In,* 137.

3. Ibid., 74–76.

4. The succession, of course, was a story of corporate intrigue, and involved the ouster of Steven Ross and Levin's joint enemy, Nick Nickolas, who was technically co-CEO with Ross and ought logically to have been Ross's successor. See Christopher Byron, "As Ross Lay Dying," *New York* magazine, January 4, 1993, 12. On Levin's career, see, e.g., Klein, *Stealing TIME,* 80.

5. Levin was quoted in Roberts, "How It All Fell Apart," cited above. Ted Turner's full quote appears in Saul Hansell, "Media Megadeal: The Overview," *New York Times,* January 11, 2000.

6. Steve Lohr, "AOL Merger Turns Tables on Microsoft," *New York Times,* January 12, 2000.

7. Kramer defends the AOL–Time Warner merger in Larry Kramer, "Why the AOL–Time Warner Merger Was a Good Idea," The Daily Beast, Blogs and Stories, May 4, 2009, available at www.thedailybeast.com/blogs-and-stories/2009-05-04/how-time-warner-blew-it/.

8. You can find the old Pathfinder site on the Internet Archive, http://archive.org.

9. On Disney's total merchandising strategy, see "All the Movies Are Geared to Publicizing . . . and Making Money," *Newsweek,* December 1962, 48–51.

10. This figure was at the time of the merger. Klein, *Stealing TIME,* 259.

11. Ken Auletta, *Media Man: Ted Turner's Improbable Empire,* 96.

12. The FTC and FCC both imposed conditions on the merger, including the "open access" provision referred to in the text, as well as conditions designed to maintain an open market for instant messaging, then thought to be a crucial platform for the future. See "In the Matter of America Online, Inc., and Time Warner Inc., File No. 001 0105, Docket No. C-3989; Applications for Consent to the Transfer of Control of Licenses and Section 214 Authorizations by Time Warner Inc. and America Online, Inc., Transferors, to AOL Time Warner Inc., Transferee," 16 FCC Rcd. 6547 (2001).

13. Jay Greene, "Case vs. Gates: Playing for the Web Jackpot," *BusinessWeek,* June 18, 2001, 42.

14. The power of states to shape the nature of the Internet is the topic of my first book, coauthored with Jack Goldsmith. See Tim Wu and Jack Goldsmith, *Who Controls the Internet* (New York: Oxford, 2006).

CHAPTER 20: FATHER AND SON

1. The quotes in this chapter from Steve Jobs and Eric Schmidt are drawn from the 2007 Macworld conference in San Francisco, or from a February 2010 interview with Eric Schmidt. Jobs's entire keynote address from the 2007 Macworld event may be viewed at www.apple.com/quicktime/qtv/mwsf07/ (last visited March 2010).

2. This title is official; see Google's "Corporate Information" website, www.google.com/corporate/execs.html (last visited March 2010).

3. Tim Bray's comment was made on a personal blog but cleared by Google and widely attributed to it. The blog post is at www.tbray.org/ongoing/When/201x/2010/03/15/Joining-Google.

4. This quote is drawn from the 1927 essay referenced throughout this book: Aldous Huxley, "The Outlook for American Culture," *Harper's Magazine,* August 1927.

5. One particularly interesting history of Wozniak and Jobs's initial meeting and development of what would eventually become Apple, as well as the reinvention of the company in recent years with the development of popular modern Apple technology, may be found in Michael Moritz, *Return to the Little Kingdom: Steve Jobs, the Creation of Apple, and How It Changed the World* (New York: Overlook, 2009). Other descriptions of the early history of Apple include Roy A. Allen, *A History of the Personal Computer: The People and the Technology* (London, Ontario: Allen Publishing, 2001), 36.

6. This quote, as well as much of the Wozniakcentric information in this chapter, is drawn from Steve Wozniak's autobiography, *iWoz—Computer Geek to Cult Icon: How I Invented the Personal Computer, Co-Founded Apple, and Had Fun Doing It* (New York: W. W. Norton, 2006), 103.

7. Wozniak said this at his talk at Columbia University on September 28, 2006.

8. Matthew B. Crawford, *Shop Class as Soulcraft* (New York: Penguin, 2009); Robert Pirsig, *Zen and the Art of Motorcycle Maintenance: An Inquiry into Values* (New York: William Morrow, 1974). Pirsig's book, while generally taken as a meditation on spirituality and technology, actually spends more time on complex epistemological questions that are hard to summarize. Wozniak, *iWoz,* 291.

9. This quote is from Leander Kahney, "How Apple Got Everything Right by Doing Everything Wrong," *Wired,* March 18, 2008. In the article, Kahney also questions Apple and Google's supposedly close relationship: "By Google's definition, Apple is irredeemably evil, behaving more like an old-fashioned industrial titan than a different-thinking business of the future." The book he was promoting with this article: Leander Kahney, *Inside Steve's Brain* (New York: Penguin, 2008).

10. Herbert N. Casson, *The History of the Telephone* (Chicago: A. C. McClurg, 1910), 157.

11. The *New York Times* story is "Psychology of Telephone Girls," *New York Times,* April 4, 1912. The effect of the financial panic on the telephone girls is described in Casson, *History of the Telephone,* 155.

12. The idea of describing Google as a switch comes from my colleague Charles Sabel at Columbia.

13. Siva Vaidhyanathan, *Googlization of Everything: How One Company Is Transforming Culture, Commerce, and Community and Why We Should Worry* (London: Profile Books, 2010).

14. This particular corporate tradition is described in Fred Turner, "Burning Man at Google: A Cultural Infrastructure for New Media Production," *New Media & Society* 11 (2009): 145.

15. As quoted in, among other places, Janet Lowe, *Google Speaks* (Hoboken, NJ: John Wiley & Sons, 2009), 39. Google's origins at Stanford are described in John Battelle, *The Search* (New York: Portfolio, 2005).

16. "At SBC, It's All About 'Scale and Scope,' " *BusinessWeek,* November 7, 2005.

17. SkyNews, interview with Rupert Murdoch, November 9, 2009, available at www .youtube.com/watch?v=M7GkJqRv3BI&feature=player_embedded.

18. These predictions form the thesis of Jonathan Zittrain, *The Future of the Internet and How to Stop It* (New Haven: Yale University Press, 2008).

19. According to Wozniak, in an interview with *Wired* magazine. See Rachel Metz, "iWoz Logs Leap from Geek to Icon," Wired.com, August 24, 2006, available at www.wire .com/gadgets/mac/news/2006/08/7164.

20. The blog post can be found at googleblog.blogspot.com/2007-11-wheres-my-gphone. html.

21. For example, in a 2007 press release, Verizon announced it was committed to allowing any wireless device and any app on its network. See news.vzw.com/news/2007/11/ pr2007-11-27.html.

22. The best account of such a future is a novel by Cory Doctorow, *Down and Out in the Magic Kingdom* (New York: Tor Books, 2003); it is also the evident vision of the Burning Man festival. On the relationship between the tech world and Burning Man, see Fred Turner, "Burning Man at Google," 145.

Index

Numbers in *italics* refer to illustrations.

Abrams, Hiram, 91
acoustics, 19, 21, 103–4, 108, 109, 169
Acoustics (Beranek), 103
Advanced Research Projects Agency (ARPA),
    168, 171, 174–5
Air Force, U.S., 170
airlines deregulation, 142, 161
Alfred P. Sloan Foundation, 183
Allen, Woody, 89, 218
Amazon, 218, 286, 289, 290
American Broadcasting Company (ABC),
    75–6, 177, 209, 210, 214
American Civil Liberties Union (ACLU),
    122
American Medical Association (AMA), 164
American Speaking Telephone Company
    (AST), 26
American Telephone and Telegraph (AT&T),
    3–5, 7, 53–6, 71, 72, 101–14
    breakup of, 160–2, 177, 187–95, *194*, 205,
        239, 240–1, 307
    CEOs of, *see* Vail, Theodore; Whitacre,
        Edward, Jr.
    creation of, 17*n*, 32, 189
    defense installations of, 159–60
    drop-off in quality of service at, 161
    federal relationship with, 102, 105, 147,
        159–60, 187–95, 238–41, 250–2
    foreign attachments and, 108, 111, 113, 188,
        189–90, 192, 240–1
    Internet and, 172–4, 197–9, 249–52, 285–6,
        290
    iPhone and, 271–72
    logo of, 194–5
    long distance service controlled by, 17*n*, 32,
        52, 53–4, 76, 172, 179, 189, 194, 209–10,
        277, 292

    as privately held monopoly, 9, 13, 49, 54,
        59–60, 75–82, 84, 147, 177, 187–9, 192–4,
        199, 304
    radio transmission by, 33–4, 75–82, 84, 129,
        132
    rebuilding of, 205–6, 239–43, 252
    "secret spying rooms" of, 249–52
    suppression of innovative technologies by,
        101–2, 107–10, 112–14, 191–2
    see also Bell System
America Online (AOL), 191, 257–68
    abortive merger of Time Warner and,
        255–68, 272, 284, 286, 318
Anderson, Chris, 220
Android, 293–5, 296
answering machines, 104, 106, 162
antennas, 33, 125, 178, 179
anticommunism, 182
anti-Semitism, 115–16, 117*n*, 119
antitrust laws, 54–5, 72, 96, 160, 163–5, 166*n*,
    167, 187, 192–3, 236, 244, 246, 265, 303,
    312
Apocalypse Now, 217, 220
Apple, 51, 258, 269–79, 289, 291–7, 305,
    314
Apple 1, 274, 276
Apple II, 274–5, 276–7, 278, 292–3
Armstrong, Edwin, 125–9, 131, 133–4, 150, 339
Army, U.S., 39, 131
Arnold, Thurman, 163–5, 194
ARPANET, 197–9, 202
Artificial Intelligence, 170
Associated Press (AP), 22–3, 31, 324
Atlanta Braves, 211
Auletta, Ken, 214, 264
automobiles, 9, 12–13, 26, 162
Avatar, 220, 229

Baby Bells, 194, 240–8, *248*
Bach, Stephen, 218, 349
Baird, John Logie, 136–41, 145–6, 148, 151, 152, 340
Baird Television Limited, 140, 141, 145–6, 148, 151
Balaban and Katz, 97
Balio, Tino, 69, 330
Baran, Paul, 172–4
Bardèche, Maurice, 69
Barr, William, 247
Battelle, John, 282
Bazelon, David, 113
Becker, Ron, 213
Bell, Alexander Graham, 7, 17–19, 20–2, 26–7, 29–32, 63, 112, 134, 136, 140, 153, 199, 273
    depression and illness of, 27, 30
    first functional phone constructed by, 17–18, 20, 46, 107, 141, 274
    patent infringement lawsuit brought by, 30–2
    teaching of the deaf by, 17
    telephone patents of, 17, 18, 25, 30–1, 46, 107
Bell, Smith, 244, 248
Bell Atlantic, 241, 244
Bell Company, 17–18, 24, 45
    advertising by, 25
    business model of, 46
    investment in, 17, 20, 21, 50–3
    as monopoly, 9, 13, 17, 31, 32, 51–3, 57
    reorganization of, 29–30
    Vail as general manager of, 29–30, 46
    Western Union's efforts against, 22, 25–7, 29–32, 52
Bell Laboratories, 104–7, 108, 180, 193, 194, 306
Bell System, 3–5, 7, 9, 17, 26, 32, 78, 99, 173–4, 178
    brain trust of, 241–2
    break up of, 17*n*, 51, 113, 160–2, 187–95, 240–1, 248–9
    as common carrier, 57–9
    decline of, 50
    dominance of wire communications by, 11, 51, 80, 189
    local service of, 17*n*, 194, 244
    predatory pricing of, 49, 192
    as public trust, 55, 244
    rebuilding of, 205–6, 239–43, 252–3, 255, 285
    resistance to Independents by, 49–53, 55, 71, 142, 189
    shareholders of, 46, 52, 60, 105
    see also American Telephone and Telegraph (AT&T); Baby Bells
Benkler, Yochai, 36, 255

Beranek, Leo, *102*, 103–4, 108–11, 169, 337
Bernhardt, Sarah, 62
Bernstein, Carl, 185
Besen, Stanley, 181
Bible, 38, 139, 236
Bill of Rights, 300
Black, Alfred, 93, 94
Black, Gregory, 120–1
blogosphere, 36, 286, 293
Book of Wireless, The (Ross), 35
Bork, Robert, 56
Borneman, Ernest, 163
boxing matches, 33–5, 79, 85, 210, 325
Brandeis, Louis, 301
Brave New World (Huxley), 12, 99
Bray, Tim, 272, 353
Brazil, 114
Breen, Joseph, 116–17, 119–20, 122, 123–4, 165–6, 338
Briggs, Asa, 42
British Broadcasting Company (BBC), 40–4, 79, 83, 115, 296
    television broadcasts of, 136–42, 147, 150–2, 181
British Empire, 8, 82, 140, 142
Broadcast over Britain (Reith), 41
Brookings Institution, 165
Brown, Charlie, 194–5
Bruck, Connie, 225, 226
Burch, Edmund, 45–6, 47, 48, 53, 327
Burning Man festival, 281, 354
Bush, George W., 238, 247, 251, 349
Bush v. Gore, 24
Business Week, 240, 265, 285

Cabinet Committee on Cable Communications, 177, 184–5, 345
Cable News Network (CNN), 208, 212, 264, 284
cable television, 6, 139, 157, 194, 205, 207–16, 244, 258, 263–4, 285, 303, 316
    advertising on, 178, 179, 210, 211
    deregulation of, 180–6, 207, 209
    film and broadcast networks recombined with, 205, 210, 211, 255
    network development and, 208–12
    Nixon and, 177, 184–6, 207, 209, 242
    origins of, 18, 26, 176–86
    potentials of, 176–7, 180, 182, 183–4, 207, 208
    subscriptions to, 179, 180, 210–11, 216
    suppression of, 179–81, 308
    technology of, 178–9, 180, 189, 209–10
    video-on-demand, 232
Canby, Vincent, 218

Cannes Film Festival, 232, 234
capitalism, 8–9, 27–8, 160, 163, 300
Capitalism, Socialism, and Democracy
(Schumpeter), 28*n*
Carnegie, Andrew, 271, 301
Carpentier, Georges, 33, 35, 40
cartels, 6, 11, 62–71, 164, 222–3
Carter, Jimmy, 185–6, 242
Carterfone, 190, 191
Case, Steve, 257–9, 262–5, 267–8, 284
Casper, Mark, 38
Casson, Herbert N., 26, 279
Caves, Richard, 165
CBS News, 13, 155, 213
censorship, 70, 100, 116–24, 161, 165–7, 314
Cerf, Vint, 145, 197–9, 268, 270, 347
Chandler, Alfred, 163
Chaplin, Charlie, 92
Chasing Amy, 234
Chicago, University of, 242
China, People's Republic of, 257, 267, 314
Christensen, Clayton, 20
Christian Broadcasting Network, 209–10
Chrysler Corporation, 164
Churchill, Winston, 42
Cimino, Michael, 217–18, 232, 237
circuit switching, 173, *173,* 174
civil rights movement, 12
Civil War, U.S., 22–3, 301
Clark, David, 201–2
Clark, Mark, 106
Cleopatra, 229
Clerks, 234
Clinton, Bill, 55, 243, 309, 350
CNBC, 235
Coase, Ronald, 284*n*
Cold War, 159, 199, 250
Coll, Steve, 188
Columbia Broadcasting System (CBS), 11,
75–6, 83, 99, 128, 132, 139, 143, 153–5, 177,
178, 181–2
Columbia University, 126, 131, 276, 282, 339
Comcast, 259, 264, 311
common carriage, 23, 57–60, 100, 184, 311
Communication Act of 1934, 145, 243–4
communications empires, 5, 10–12
break up of, 11, 17*n*, 51, 113, 160–2, 187–95
genesis of, 10–11
ideas and forms repressed by, 11, 100–156,
180–1
political influence of, 22–4
reconstitution of, 11–12, 205–53
revolution in, 5, 11, 161–2, 168–75, 255
state support and overview of, 11, 159–60
see also specific communication companies
Communism, 214, 257

Community Antenna Television (CATV), 178,
179, 183
competition, 8–9, 28
barriers to entry and, 48
promotion of, 167, 187–95, 241–7
suppression of, 22, 25–7, 29–32, 49–53, 55,
71, 101–4, 107–10, 112–14, 179–81, 189,
191–2, 244–9
"Computer as a Communication Device, The"
(Licklider and Taylor), 174
Computer Inquiries, 191
computers, 168–75, 195, 198
handheld, 162
laptop, 319
mainframe, 197, 202, 270, 275, 298
music industry threatened by, 31
networking of, 190–1, 195, 197–9, 201–3
see also personal computers
Congress, U.S., 59, 82–3, 122, 124, 129*n*, 191,
241, 251, 252
see also Senate, U.S.
Conlon, Tom, 292
Constitution, U.S., 121–2, 299–301, 308–9
Commerce clause in, 300
Control Video Corporation, 258
Coolidge, Calvin, 76, 78, 98
Coons, Horace, 31–2
Coppola, Francis Ford, 89, 217–18, 233
copyright law, 180, 186, 231–2
infringement of, 284*n*, 287–8
Copyright Office, U.S., 179
corporations, 6, 271
merger and acquisition of, 53
public duties of, 52, 205–6
reconstitution of, 205–53
scientifically organized, 9
see also cartels; media conglomerates;
monopolies
Crandall, Robert, 165, 181, 343
creative destruction, 28, 30, 49, 135, 164, 181,
195
Crockett, Roger, 285
Crown Corporation, 43
C SPAN, 215

Daniels, Joseph, 7
Darwin, Charles, 19, 28, 122, 139
Darwinism, 8, 9, 297
data processing, 170, 189, 190–1
Davisson, Clinton, 105
Death and Life of Great American Cities, The
(Jacobs), 201
deButts, John, 188, 191, 193, 195
Deep Throat, 166
Deer Hunter, The, 217, 218

Defense Department, U.S., 159–60, 168
"Defining Moguls," 29, 85
De Forest, Lee, 35, 37, 39, 85, 326
DeMille, Cecil B., 91, 98
democracy, 184, 213, 304
Democratic Party, 23–4, 140, 177
Dempsey, Jack, 33–5, 40, 85, 125, 325
De Niro, Robert, 114
Depression, Great, 127, 146, 147, 154, 242,
    310
De Vany, Arthur S., 221
digital subscriber lines (DSL), 107, 263
Diller, Barry, 222
Disney, Roy, 227, 235
Disney Company, 222, 227, 234, 235, 261, 264,
    311
Doherty, Thomas, 116
domain names, 266, 280
Douglas, William O., 164
Drawbaugh, Daniel, 18–19, 20
Drinkwater, John, 65
Duering, Al, 161
DVDs, 232, 237

Eastman Kodak, 64, 68, 73
eBay, 145, 259, 290
economies, 7, 8–9, 27
    controlled, 148, 199, 306
    free-market, 144, 147–8
    innovation and, 27–8, 147–8
    rise and fall of, 300–1
Edison, Thomas, 4, 64, 70–1, 140, 223*n*, 235
    improvements of telephone by, 26, 30, 141
Edison Motion Picture Patents Company of
    New Jersey, 55, 61–2, 71, 73, 222
    see also Film Trust
Eisner, Michael, 222, 226–7, 235, 271
elections, U.S., 22–24, 35, 140
electricity, 5, 19, 58
Electronic Frontier Foundation (EFF), 251
electronic privacy, 24
email, 21*n*, 34
Empire and Communications (Innis), 302*n*
Empire of Their Own, An (Gabler), 71
"End-to-End Arguments in System Design"
    (Reed, Clark and Saltzer), 202
Engelbart, Douglas, 171–2, 175
English language, 42, 198
Entertainment and Sports Programming
    Network (ESPN), 211, 212, 213
Epstein, Edward Jay, 232, 349
Esperanto, 196, 197, 346–7
E.T.: The Extra Terrestrial, 226
Evan Almighty, 236–7
Eveready Hour, 77, 78

Facebook, 202, 256, 290
facsimile machines, 107, 129, 190, 260
Famous Players, 87, 228*n*
Famous Players–Lasky Corporation, 91*n*, 335
Farnsworth, Philo, 138, 148–51, 152–4
fascism, 59*n*, 199, 214
Faulhaber, Gerald R., 192, 346
Federal Communications Commission (FCC),
    11, 54, 102, 108–9, 112–13, 128–33, 154,
    177, 184, 213, 240–1, 252, 265, 272, 286,
    309, 312–13
    cable television and, 177, 180–1, 184, 207
    competition promoted by, 187–94, 240, 243,
        246–7
    suppression of FM by, 130–3, 144
    suppression of mechanical television by,
        143–8
Federal Radio Commission (FRC), 82–4, 143*n*
Federal Trade Commission (FTC), 80, 95, 98,
    265, 312, 336, 352
fiber optics, 107, 178, 180
Fight with an Octopus, A (Latzke), 50
film festivals, 224, 232–4
film franchises, 228
film stock, 64, 68
Film Trust, 62–72, 88, 89, 100, 223*n*, 306, 307,
    313
    Independents challenge of, 65, 71, 86–7,
        90, 94
financial panic of 1907, 279
First Amendment, 13, 57, 83, 121–2, 144, 310
First Bell Monopoly, 32, 46
First National Exhibitors' Circuit, 87, 92–4,
    97, 98
Fisher, William, 18
Folklore of Capitalism, The (Arnold), 163
Fonda, Jane, 208
Ford, Gerald R., 185–6
Ford, Henry, 9, 162, 297, 336
    business theory of, 12–13, 94, 147, 199
    mass production techniques of, 8, 12–13
Ford Motors, 164, 280
Foreign Intelligence Surveillance Act (FISA),
    251, 351
Forsher, James, 63
Fortnightly Corp. v. United Artists, 180
Fox, William, 67, 68, 70, 72, 73, 87, 92
Fox Broadcasting, 67
Fox Features, 68
Fox News, 67, 213
free speech, 13, 85, 121–2, 124
frequency modulation (FM), 125–35, *127*, 339
    development of, 125–9, 131, 133–4, 150
    high fidelity of, 129, 131
    suppression of, 126–35, 150, 153, 308
Freud, Sigmund, 169

Friedman, Milton, 242, 350
Friendly, Fred, v, 13, 155, 181–4, 207, 280, 345

Gabler, Neal, 71
galvanic current, 18
Gates, Bill, 225, 258
General Electric (GE), 79, 128, 140, 154, 264
  Universal Studios merger with, 223–4, 235, 237
General Film Exchange, 67
General Motors (GM), 26, 164, 183
Germany, Imperial, 18, 65
Germany, Nazi, 27, 77, 83, 84–5, 147–8, 164, 188
Gifford, Walter, 78, 81
Gladwell, Malcolm, 19
Godfather, The, 162, 233
Goebbels, Joseph, 84–5, 302
Goldsmith, Alfred N., 38, 326, 341
Goldwater, Barry, 161, 242
Gone with the Wind, 99, 116
Google, 36, 55, 104, 197, 261, 266, 267, 268, 270–3, 275, 279–90, *285,* 292–8, 318, 353
  complaints about, 287–9
  corporate structure of, 281–3
  as most popular Internet switch, 279–80, 282
  origins of, 18, 282, 287
  specialization of, 283–4, *283*
Googlization of Everything (Vaidhyanathan), 281
Gore, Al, 243
Gould, Jay, 31
Grable, Betty, 160
Graham Act of 1921, 59
Gray, Elisha, 18, 21, 30
Green Acres, 48, 211
Greene, Harold H., 193–4, 240, 244
Griffith, D. W., 6, 98
Grimmelmann, James, 72
Griswold, A. H., 78
Gulf & Western, 224, 235

Hafner, Katie, 173
Hampton, Benjamin, 90–1, 92, 93
Hand, Learned, 231*n*
Hanselman, John, 108
Harrison, P. S., 95–6
Harry Potter books, 220, 228
Harvard University, 26, 36, 103, 109, 290, 309
Hayek, Friedrich, 144–5, 199–201, 202, 297
Hayes, Dennis, 190
Hayes, Rutherford B., 22–4

Hayes Modem, 190
Hays, William, 116
Heaven's Gate, 217–18, 219, 220, 223, 232, 237
Hickman, Clarence, 104, 106
Hitler, Adolf, 84, 123, 164, 199, 230
hobbyists, 6, 48
  computer, 190, 276
  radio, 34–5, 37, 39, 43–4
Hodkinson, William W., 67, 73, 86, 88–91, 95, 98, 217, 330
Hollywood studios, 11, 61, 63, 64, 66–73, 86–100, 116–24, 160–7, 178, 214, 217–37
  antitrust action against, 160–1, 164–6, 166*n,* 167
  block sales practice of, 95–7, 98, 163, 164, 334
  media conglomerates' acquisition of, 218–19, 222–37
  vertical integration of, 162, 164, 223
  see also motion pictures
Holmes, Oliver Wendell, 13, 122
Home Box Office Network (HBO), 210, 211, 212, 258
Homebrew Computer Club, 276
Hoover, Herbert, 74–5, 82–3, 85, 145, 331, 332
Hopp, Julius, 34
Hubbard, Gardner Green, 17, 20, 21, 25, 29–30, 154
Huber, Peter, 309
Hundt, Reed, 243, 309
Hush-A-Phone, 101–4, *102,* 107–10, 112–14, 146, 169, 190, 191, 337
Huxley, Aldous, 12–13, 99, 155, 219, 273, 316, 336–7, 353

IBM, 170
IBM AN/FSQ-7, 170
ICBMs, 159
I Love Lucy, 12, 182, 211, 213
Immelt, Jeff, 223
I'm No Angel, 118
Independent Motion Picture Company (IMP), 68
Independent telephone movement, 46–50, 55–6, 327–8
  Bell campaign against, 49–50, 51–3, 55, 71, 142, 189
industry:
  centralized factories in, 12
  disruption of, 19–20
  evolution of, 160
  monopolistic vs. competitive, 8–9, 32
  vertically integrated, 147

information technology, 5–13
  battle for territory in, 289–8
  central control of, 5, 6, 7–9, 11–12, 13
  Cycle from open to closed systems in, 6–7,
    9–12, 18, 20, 25, 28, 30, 68, 84, 98, 156,
    160, 177, 218, 252–3, 256, 260, 289, 297,
    303, 313, 316
  entrepreneurial motives in, 36, 39, 208
  evolution of, 10–12, 14
  free use of, 6, 7, 11, 12
  humanitarian motives in, 36–7, 205–6
  industrial combat in, 25–7
  rise of, 3–14, 15–98
  social transformation and, 5–7, 12–13, 139
innovation, 21, 27–8, 167
  centralized, 107, 110–11
  conceptual vs. technical, 46–7
  disruptive, 19–20, 25, 107, 128, 138, 195, 207
  economic growth and, 27–8, 147–8
  entrepreneurial, 48
  evolutionary model of, 111–12
  radical, 168–75
  suppression of, 26–7, 101–3, 107–14, 126–35,
    166*n*
  sustaining, 19–20
Institute of Radio Engineers, 131
intellectual property, 179, 228–32, 236
  brand names and, 230–1, 235
  copyright and, 231–2
International Projecting and Producing
  Company, 66
Internet, 32, 35, 73, 145, 151*n*, 154, 162, 236
  AT&T and, 172–4, 197–9, 249–52, 285–6,
    290
  broadband access to, 262, *262*, 263, 267,
    285, 303
  collaborative projects on, 36
  decentralized design of, 5, 12, 170, 197–9,
    201–3, 256, 317
  dialup access to, *262*, 263
  diversity of content and services on, 5, 256,
    272–3, 317–18
  early development of, 110, 168–75, 197–9,
    201–3, 256
  encapsulation concept of, 198
  explosive growth of, 5, 12, 256
  federal spying on, 249–52
  funding of, 198, 215
  idealistic promise of, 36, 290
  mass consumption of, 190, 191
  monitoring of transactions on, 238
  net neutrality principles at core of, 23, 202*n*,
    260, 267, 286, 311
  openness of, 5, 7, 11, 12, 35, 36, 174, 256, 303,
    317–18
  paid downloads on, 37
  technology of, 169, 173–4, *173*, 190–1, 197–9,
    201–3, 256
  transmission speed of, 5
Internet Protocol (TCPab/IP), 196, 202
Internet Revolution, 5, 11, 168–75, 255
Internet Service Providers (ISPs), 262–3, 266,
  284, 309
invention, 47
  amateur, 6, 17–18, 34
  craftsmanship vs. miracle-working of, 19
  evolution of information industries from,
    10–11, 14
  freedom of thought in, 20
  importance of loners and outsiders to
    process of, 19–20, 22, 30, 66, 112
  patents for, 17, 18, 25, 30–1
  secret research and, 21
  simultaneous, 18–19, 20–1, 137–8
Invention of Love, The (Stoppard), v, 6
iPad, 272, 277, 291, 292, 293
iPhone, 269–72, 277, 291, 292, 293, 294, 295,
  314
iPod, 145, 269, 277, 291, 293
It's a Wonderful Life, 123
iTunes, 272
iWoz (Wozniak), 291, 353

Jackson, Andrew, 301
Jacobs, Jane, 200–1, 202, 297
Jacobs, Lewis, 68
James Bond films, 228
Japan, 159, 195*n*
Jazz Singer, The, 98
Jefferson, Thomas, 162, 300
Jenkins, Charles Francis, 64, 137–8, 139–41,
  142, 144, 145, 148, 152, 341
Jenkins Radiovisor, 142, 146
Jobs, Steve, 51, 258, 269–72, 273–4, 276–9,
  291–2, 294, 297, 353
  see also Apple
Justice Department, U.S., 55, 161, 240, 241, 307
  Antitrust Division of, 164, 191–4, 236, 244,
    312

Kael, Pauline, 97
Kaempffert, Waldemar, 37
Kahn, Robert, 197–9
Kennedy, Anthony, 300
Kennedy, Jeremiah, 61, 63, 68
Kennedy, John F., 174
Keynes, John Maynard, 199, 301
Kingsbury Commitment, 55–6, 59, 244
Kinsley, Michael, 260, 264
Klein, Benjamin, 96

Klein, Mark, 249–51
Kline, Ronald, 47
Kohr, Leopold, 200, 202
Koszarski, Richard, 94
Kramer, Larry, 259
Kraus, Raymond, 161
Kronos effect, 25, 28, 64, 107, 112, 128, 139, 156, 179

Laemmle, Carl, 65–8, 70, 72, 73, 87, 92, 223*n*
Lafount, Henry, 82, 84
LaSalle, Mick, 118, 121, 123
Last Lone Inventor, The (Schwartz), 149
Latzke, Paul, 49, 50
Legion of Decency, 118–19, 121, 122, 123, 338
Lessig, Lawrence, 36, 309
Lessing, Lawrence, 100, 125, 133, 337, 339, 340
Levin, Gerald, 257–8, 259, 261–3, 264–8
Lewinsky, Monica, 106
libertarians, 28*n*, 59, 185, 199, 239
Licklider, J. C. R., 108, 109–10, 168–75, 197, 268, 344
Lincoln Center, 103
Lindsay, John V., 182
Lippmann, Walter, 41, 155–6
Lohr, Steve, 259
Long Tail, The (Anderson), 220
Lord, Daniel, 115–21, 122, 165
Lost in Space, 170
Low, Archibald M., 142
Lucasfilm, 230
Luddite effect, 240
Lumière, Louis, 64
Lynch, S. A., 93, 94
Lyon, Matthew, 173

Macintosh, 171*n*, 269, 274*n*, 276–8, 291–2, 295
magnetic recording, 104, 106, 107, 111, 168, 307
"Man-Computer Symbiosis" (Licklider), 170
Manhattan Institute, 309
Manhattan Project, 159
Mann Elkins Act of 1910, 57*n*
Mao Zedong, 257
Marconi, Guglielmo, 140
markets, 13, 122
    efficiency of, 8, 28
    failure of, 200*n*
    free, 28*n*, 122, 144, 147–8, 310
Markoff, John, 171
Martin, W. H., 108
Massachusetts General Hospital, 27, 30
Massachusetts Institute of Technology (MIT), 103, 180
    Lincoln Laboratories at, 169

mass production, 8, 12–13
McCarthy, Joseph, 182
McChesney, Robert, 84
McDonald, Eugene, 82
McQuiston, J. C., 74
media conglomerates, 11, 205–6, 218–37, 255–6, 259–68, 288
    accounting practices of, 219, 222, 232
    capitalization of, 219
    empire building and material enrichment of, 226–7, 258–68
    Hollywood studios acquired by, 218–19, 222–37
    loss of content value in, 227
    managing risk in, 219, 221–2, 223–4, 226, 227–8, 232
    profit and shareholder value basis of, 206, 224–5
    vertical integration of, 259
Méliès, George, 73
Mesa Telephone Company, 46, 47, 53, 329
Metro-Goldwin-Mayer (MGM), 92, 98, 218
Microsoft Corporation, 104, 263–4, 271, 278–9
    Windows operating systems of, 54, 55, 279, 291–2, 294, 295
Microwave Communications Inc. (MCI), 34, 114, 183, 241, 244–7
    AT&T vs., 188, 189, 191–2, 245–7
microwave towers, 178–9, 189, 209
Miramax, 232–4, 235
mobile telephones, 4, 48, 107, 114, 269, 316, 319
modems, 190, 309
modular telephone jack (RJ-45), 190, 240–1
Moglen, Eben, 36
monopolies, 11–13
    break up of, 11, 51, 55, 160–7, 177, 187–95
    commitment to the public good and, 8, 9, 10, 41–4, 206, 239
    common carriage and, 23, 57–60
    competition vs., 8–9
    deregulation of, 160–7, 180–95, 205, 207, 209, 240–7
    efficiency of, 161, 195
    government support of, 55, 105, 159–60, 177, 238–42
    return on capital in, 10
Moonbeam sanctuary, 267
Morgan, J. P., 50–4, 72, 189, 225, 328
Moses, Robert, 177, 201
Motion Picture Association of America (MPAA), 166
motion picture cameras, 64
Motion Picture Producers and Distributors of America, 116
motion picture projectors, 4–5, 64, 137–8

motion pictures, 6, 9, 61–73, 86–98, 217–20, 222–4
 blockbuster, 220
 censorship of, 70, 100, 116–24, 161, 165–7
 concert, 232
 customer demand for, 220, 221–2, *221*
 director-centered, 217–18, 224, 234, 334
 distribution of, 65, 86–7, 88–9, 91, 93–7, 98, 162, 163, 232–4
 evolution of, 10–11, 13, 61–73
 failure of, 217–18, 219, 220, 236–7
 foreign, 62–3, 64, 66, 71, 73, 166
 independently produced, 220, 233–5
 introduction of sound in, 117–18
 invention of, 6, 63–4
 patents on technology of, 62–3, 64, 71
 pornographic, 166
 pre-Code moral standards of, 117–19, 121, 123
 production and cast of, 13, 73, 88–91, 95, 162, 165–7, 228–9, 234–5, 236
 rating of, 166
 star-centered, 62–3, 66, 71, 228
 transformation of, 165–7
 West Coast–East Coast feud and, 68–73
 see also Hollywood studios; *specific films*
motion picture theaters, 61–2, 63, 64, 65, 86, 87, 93–8, 162
 block booking of, 95–7, 98, 163, 164, 334
 separation of studios and, 164–6
Motley Fool, 262
movable type, 6
Moving Picture World, 62
MSNBC, 264
Mueller, Milton, 51, 59
multimedia presentations, 4–5
Murdoch, Rupert, 288–9
Murrow, Edward R., 181–2
Music Television (MTV), 212, 213
Myers, Abram, 98
My Life and Work (Ford), 9, 12

National Association of Broadcasters, 82
National Broadcasting Company (NBC), 11, 75–6, 81–4, 85, 99, 125, 128, 130, 132, 138, 139, 143, 146–7, 177
National Broadcasting System (NBS), 75, 76, 77, 81
National Geographic, 3–5
National Geographic Society, 3, 9, 323
National Security Agency (NSA), 238, 250, 251, 308
Navy, U.S., 39, 82
Nelson, Richard, 111–12

New Deal, 242, 304
New York Telephone Company, 279
New York Times, 18*n*, 22, 23, 86, 101, 129, 137, 139–40, 143, 153, 193, 218, 226, 238, 259, 279, 284, 294
NFSNET, 203
Nichols v. Universal Pictures Corp., 231*n*
nickelodeons, 62
Niptow disk, 136*n*
Nixon, Richard M., 106, 182, 184–7, 192, 243, 247, 309
 cable television promoted by, 177, 184–6, 207, 209, 242
 resignation of, 185, 187
nuclear bombs, 148, 159, 172

Obama, Barack, 251
Office of Technology Policy (OTP), 192
Office of Telecommunications Policy (OTP), 184
oil industry, 50, 51, 53, 55, 304
Olson, Mancur, 93, 335
online computer services, 190–1, 194
online networking companies (ISPs), 191
Open Handset Alliance, 293
Open Skies policy, 185
Orton, William, 25
"Outlook for American Culture, The" (Huxley), 13

Pacific Telesis Group, 244, 248
packet networking, 169, 173–4, *173,* 197, 198
Paine, Albert, 4
Palo Alto Research Corporation, 277*n*
Paramount Pictures, 63, 67, 68, 73, 86–7, 88–93, 95, 97, 118, 119, 222, 235
 breakup of, 161, 164, 195
 Zukor at, 91–3, 146, 163, 164
Pathé-Frères, 61, 73
Pearlstein, Norm, 260
Pedrick, Gale, 41
Pershing, John, 4
personal computers (PCs), 11, 190, 203, 274–93
 challenges to, 291–3
 desktop components of, 270–1
 origins of, 18, 171–2
Phantom Public, The (Lippmann), 41
Phantoscope, 64
Philadelphia Story, 123
phonographs, 4–5, 47, 64
photography, 141
Physical Society of Frankfurt, 18
Pickford, Mary, 86–7, 92, 138, 228*n*

Pixar Animation Studios, 294
PlayStation, 141
pornography, 166, 180
Posner, Richard, 188, 345
Postel, John, 201
Post Office, U.S., 23, 58
printing press, 176
Production Code, 116, 119–24, 155, 162, 318, 343
    collapse of, 161, 165–7
prostitution, 163
public broadcasting, 182–3, 210, 215
Public Broadcasting System (PBS), 210, 215
public interest programming, 41–4, 184
Public Utility Regulatory Act of 1995 (PURA 95), 246
Publix Theater Corporation, 97

Queen Elizabeth, 62–3, 66, 71, 228n
Queen's Work, The, 115–16
Quigley, Martin, 116–17, 119–20

radio, 4–6, 9–11, 32, 33–44, 125–35, 138–9, 302
    advertising on, 43, 74–5, 75, 76–7, 82, 99, 139, 331–2
    amateur, 34–5, 37, 39, 43–4, 82, 85
    amplitude modulation (AM), 126, 127, 127, 128–30, 132–3, 139, 308
    British, 40–4, 79, 83
    centralized control of, 13, 84–5, 128
    demonstration of, 4–5
    education on, 38, 39, 74
    evolution of, 10–11, 13, 33–8
    federal oversight of, 74–5, 82–4, 128–31
    idealistic hopes for, 36–44, 74
    influence of, 13
    localism in, 40, 43–4
    mass audiences reached with, 33–5, 36
    music and talk on, 35, 36, 37–8, 40, 76
    national conferences on, 74
    openness of, 35–6, 39, 41, 46, 77–8
    origins of, 18, 34–5, 178
    propaganda on, 40, 77, 83
    public service role of, 41–4
    ship transmission of, 34
    sportscasting on, 33–5, 40, 79, 85
    telephone transmission of, 33–4, 75–82, 84, 129, 132
    two-way communication one, 34–5, 39
    uniting disconnected communities by, 37, 38
    see also frequency modulation (FM)
Radio Broadcasting Preservation Act of 2000, 129n

Radio Corporation of America (RCA), 33–4, 38, 39, 132–5, 138, 143–4, 146, 235
    Sarnoff and, 40, 79–81, 85, 125–8, 130, 133–5
    as television company, 135, 149–50, 151–4
radios:
    FM reception on, 130
    manufacture and sales of, 35, 38, 39, 43, 74, 76–7, 78n, 80, 82, 125, 146
    mobile, 190
radio stations, 34–5, 75–7
    AM, 129, 132
    FM, 129, 130–1, 132
    frequencies assigned to, 39, 78n, 82, 84, 132
    licensing of, 39, 132
    networks of, 11, 75–82, 84
    programs on, 39–40, 77, 78
radio transmitters, 33
Radio Trust, 82, 128, 131, 306
Radulovich, Milo, 182
RAND Institute, 172, 174
Rape of Ma Bell, The (Kraus and Duerig), 161
Reagan, Ronald, 192, 193, 242, 247
Redford, Robert, 233
Reed, David, 201–2
Reid, John, 22–3
Reis, Johann Philip, 18
Reith, John, 40–3, 115, 296, 326–7
Republican Party, 22–4
Revolution LLC, 267–8
Rheingold, Howard, 170
Road Runner, 265
Robertson, Pat, 209–10
Rockefeller, John D., 9, 53, 225, 271, 301
Rocky, 220, 233
Roeper, Richard, 236, 249
Roman Catholic Church, 115–19, 123–4
Roosevelt, Eleanor, 123
Roosevelt, Franklin Delano, 76, 119, 153, 166
Roosevelt, Theodore, 4, 54–5
Ross, Steven, 73, 218–19, 222–3, 224–7, 235, 258, 259, 260, 271, 349
Rothafel, Samuel "Roxy," 87
Rubin, Andy, 293

Saenger Amusement Company, 94
Saltzer, Jerome, 201–2
Sandia National Laboratories, 159
San Francisco Chronicle, 138, 148
Sarnoff, David, 34, 138–9, 314, 325–6, 332, 342
    FM suppressed by, 125–8, 130, 133–5, 150, 153
    at RCA, 40, 79–81, 85, 125–8, 130, 133–5
    television and, 138–9, 143–50, 151–5
satellite transmission, 180, 185, 210
SBC, 245–7, 249

Scalia, Antonin, 247
Schlesinger, Arthur, Jr., 300
Schmidt, Eric, 270–2, 283, 295, 297, 298
Schumacher, E. F., 200, 202, 347
Schumpeter, Joseph, 21, 27–8, 31, 73, 80–1, 135,
    138, 154, 160, 208, 266, 301, 308, 325
Schwartz, Evan, 149
science fiction, 169, 170
Scientific American, 26, 37, 45
Search, The (Battelle), 282
search engines, 55, 131, 266
Second Bank of the United States, 301
Second Bell Monopoly, 51–3
See It Now, 181–2
Senate, U.S., 164
    Judiciary Committee of, 238–9
Separations Principle, 304–16
Sex, Lies, and Videotape, 232–3
Sex and the City, 118
Shaw, George Bernard, 42, 115, 327
She Done Him Wrong, 118
Sherman Act, 54, 163–4, 236, 246, 307, 312
Silicon Valley, 14, 266, 281, 294
Sinclair, Upton, 66, 182
Skype, 292, 314
Slate, 264, 294
Slate.com, 260, 264
Sloan Foundation, 6, 207
Small Is Beautiful (Schumacher), 200
Smith, Adam, 8, 28, 80, 301
Smith, Alfred, 140
Smith, Kevin, 234
Smith, Ralph Lee, 176–7, 182, 183–4, 207, 344
Smith, William Henry, 22
socialism, 8, 28n, 67, 199
social networking, 35, 162
software, 277, 294
Sound Waves, 49–50
Southwestern Bell, 240, 241, 245, 248
Soviet Union, 27, 144, 164, 172, 199, 306
Specter, Arlen, 238–9
Spielberg, Steven, 226
Sprint, 54, 241
Stalin, Joseph, 123, 164, 188, 199
Stallone, Sylvester, 220, 233
Standard Oil, 50, 53, 55
Stanford Research Institute Augmentation
    Research Center, 171
Stanford University, 131, 282
Starr, Paul, 70–1
Star Wars, 170
Stewart, Potter, 180
Stigler, George, 96, 242, 350
Stone, Alan, 3
Stoppard, Tom, v, 6
Strand Theater, 87

Strassburg, Bernard, 112
Sundance Film Festival, 224, 232, 233, 234
Sunstein, Cass, 213
Supreme Court, U.S., 19, 56, 73, 96–7, 164,
    180, 247, 309, 316

Taft, William Howard, 54–5, 311
Tally, Thomas, 86, 87, 92, 98, 334
Tally's Broadway, 86, 334
tape recorders, 104, 106
Taylor, Frederick, 8, 199, 201, 323
Taylor, Robert, 174
Telecommunications Act of 1996, 243–4, 245,
    247, 248, 350
telegraph, 17, 20, 26, 129, 214
    decline of, 56
    efforts at improvement of, 21–2
    invention of, 19
    telephone as rival of, 22–8, 30–2
telephone, 3–6, 22, 195
    controversy over invention of, 18–21, 30, 324
    demonstrations of, 3–5
    "dual service" era in, 48
    Edison's improvement of, 26, 30, 141
    evolution of, 9–10, 17–19, 25–7, 30, 32
    first transmission of speech by, 22
    local service on, 17n, 26, 32
    party line service on, 47
    primitive prototypes of, 18–19, 20, 25, 29
    as social technology, 47
    telegraph as rival of, 22–8, 30–2
    universal connections of, 9, 51, 56
    wiring of, 45–6, 48, 170
    see also specific telephone companies
Telephone, Telegraph, and Cable Company of
    America, 53–4
telephone jacks, 190, 240–1
telephone networks, 45–6
telephone operators, 47, 49, 279
Telephony, 49
television, 6, 9, 32, 131, 135–56
    advertising on, 155–6
    black-and-white, 207–8
    British, 136–8, 139–41, 142, 147, 150–1, 152,
        181
    capitalization of, 138, 149
    color, 151
    criticism of, 155–6
    electronic, 128, 138, 141, 147, 148–51, 154,
        307
    federal regulation of, 145
    FM as sound for, 133
    free reception of, 179–80
    Golden Age of, 184, 211, 304
    high definition, 141, 151

as mass medium, 140
mechanical, 128, 138, 140–9, 341
network control of, 11
origins of, 18, 26, 128, 136–8, *137*, 149–56
patents for, 148, 149, 153–4
picture resolution on, 141, 147, 148–9, 151
popular and social culture shaped by, 139
programming on, 12, 140, 147, 155, 182, 211, 214
public access channels on, 186
radio affiliation with, 133, 141, 145, 146–7, 153
radio frequency spectrum and, 145
Sarnoff and, 138–9, 143–50, 151–5
simultaneous broadcast of radio and, 140
struggle for control of, 138–9, 143–55, 179–81
technology of, 141, 147, 149–51, 157, 181, 207–8
television cameras, 148–9, 150
television sets, 150–1, 152
manufacture and sales of, 141, 142, 154, 195*n*
television stations, 140, 145–6, 147, 154
Terminator 2, 11
Tesla, Nikola, 5–6
Texaco Star Theater, 99
text messages, 21*n*
Thin Man films, 228
Tilden, Samuel J., 23
Time, 152, 260
Time Inc., 258
Time Warner Inc., 205, 218–19, 224, 235, 257–68
abortive merger of AOL and, 258–68, 272, 284, 286, 318
Titanic, 8, 220
TMZ.com, 267
Traction Kings, 53–4
Transformers, 229
Transmission Control Protocol (TCP), 198 9, 201
Truman, Harry, 159
trusts, 54, 157
Turner, Ted, 157, 207–12, 258, 268, 348
Tuttle, Henry, 101–4, 108–10, 112–14
Twelfth Night (Shakespeare), 231*n*
Twentieth Century–Fox, 67, 73, 92, 119, 222
Twitter, 143
2XG radio station, 35

ultra high frequency (UHF), 157, 181, 207–8, 209
United Artists, 98, 217–18, 233, 348–9
United Media, 230

United Nations, 103
United Press (UP), 23
Universal Studios, 68, 69, 73, 92, 98, 119, 236–7
GE merger with, 223–4, 235, 237
utopianism, 6, 8, 14, 77

vacuum tubes, 80, 151
Vaidhyanathan, Siva, 281, 354
Vail, Theodore, 3–5, 7–10, 78, 105, 125, 128, 154, 208, 225, 271
AT&T headed by, 3–4, 32, 50–60, 105, 142, 187, 188, 195, 239, 240, 241, 252, 280
centralized monopolies believed in by, 7–9, 60, 187, 188, 193, 199–200
as general manager of Bell Company, 29–30, 46
Valenti, Jack, 166, 179, 343
vaudeville, 62
Verizon, 240, 241, 244, 246–9, 251, 296, 354
Verizon Communications v. Trinko, 246–7
vertical integration, 147, 162, 164, 223, 259, 284*n*, 305
Viacom, 288
Victorian morality, 41, 42, 116
Vietnam War, 182, 185, 217
virtual communities, 39
Vitascope, 64
Vonage, 285

Wallace, Alfred Russel, 19
Walson, John, 178
Warner, Harry, 66
Warner, Jack, 66
Warner, Sam, 66
Warner Bros., 66, 73, 92, 97–8, 119, 154*n*, 258, 284, 330
Warner Bros.–Seven Brothers, 222
Warner Communication, Inc., 218, 222–3, 227, 235
Warren, Earl, 180
Watergate scandal, 185
Watson, Thomas, 17–18, 22, 273
WEAF, 75, 77
Weinstein, Bob, 232–4
Weinstein, Harvey, 232–4
West, Mae, 118
Western Electric, 159, 194
Western Union, 81, 128, 324
Bell divestiture of, 55–6
Bell takeover of, 52, 153
development and sales of telephones by, 26–7, 30–2
diminishment of, 24, 107
1876 presidential election role of, 22–4

Western Union *(continued)*
  efforts to destroy Bell Company by, 22,
    25–7, 30–2, 52
  monopoly of, 17, 20, 22, 24, 25, 32
  wealth and prestige of, 26, 30–1
Westinghouse, 39, 74, 79, 128
Where Wizards Stay Up Late (Hafner and
  Lyon), 173
Whitacre, Edward, Jr., 238–41, 245, 247, 248,
  249, 252, 272, 285, 350, 351
White, J. Andrew, 33–5
White, Samuel S., 21
Whitehead, Clay, 177, 184–5, 187, 192
Whitehead Report, 185
Who Controls the Internet? (Wu and
  Goldsmith), 201*n*
Wikipedia, 36, 267, 284, 286, 287, 289, 290
Wilbur, Zenas F., 18*n*
Wilson, Woodrow, 3, 56
Winter, Sidney, 111–12
Wired Nation, The (Smith), 177
wires, 19, 32–3
  "squirrel line," 48
  telegraph, 26, 32
  telephone, 45–6, 48, 170
Wizard of Oz, The, 99
WJRL, 207–8, 209
Woodward, Robert, 185

word processors, 20
World's Fair of 1939, 151–2, 153
World War I, 4, 33, 61, 73
World War II, 12, 27, 103, 131–2, 154, 159, 160
World Wide Web, 202, 260–1, 280, 287, 288,
  291
Wozniak, Steve, 258, 273–8, 291, 292, 353
Wright, Robert, 260, 264
WTCG, 210, 211
W3XK, 140

Xerox, 277*n*

Yahoo!, 202, 259, 261, 266, 267, 280
Yale Law School, 163
YouTube, 141, 281, 287, 288

Zamenhof, Ludwik Lazarz, 196, 346–7
Zenith Corporation, 82, 195*n*, 333
Zittrain, Jonathan, 26, 290–1
Zukor, Adolph, 61–3, 64–8, 71, 73, 86–7,
  89–95, 97–8, 125, 128, 163, 205, 228*n*
  at Paramount, 91–3, 146, 163, 164
  at Warner Bros., 118–19, 123
Zworykin, Vladimir, 149, 150, 153

ABOUT THE AUTHOR

Tim Wu is an author, a policy advocate, and a professor at Columbia University. In 2006 he was recognized as one of fifty leaders in science and technology by *Scientific American* magazine, and in 2007, *01238* magazine listed him as one of Harvard's one hundred most influential graduates. He writes for *Slate,* where he won the Lowell Thomas gold medal for travel journalism, and he has contributed to *The New Yorker, TIME, The New York Times, The Washington Post,* and *Forbes.* He is a fellow of the New America Foundation and the chairman of the media reform organization Free Press.

A Note on the Type

This book was set in Adobe Garamond. Designed for the Adobe Corporation by Robert Slimbach, the fonts are based on types first cut by Claude Garamond (c. 1480–1561). Garamond was a pupil of Geoffroy Tory and is believed to have followed the Venetian models, although he introduced a number of important differences, and it is to him that we owe the letter we now know as "old style." He gave to his letters to a certain elegance and feeling of movement that won their creator an immediate reputation and the patronage of Francis I of France.

Composed by North Market Street Graphics, Lancaster, Pennsylvania

Printed and bound by Berryville Graphics, Berryville, Virginia

Designed by Maggie Hinders